MODEL ENGINEERING FOR SIMULATION

MODEL ENGINEERING FOR SIMULATION

Edited by

LIN ZHANG

BERNARD P. ZEIGLER

YUANJUN LAILI

ACADEMIC PRESS

An imprint of Elsevier

Academic Press is an imprint of Elsevier
125 London Wall, London EC2Y 5AS, United Kingdom
525 B Street, Suite 1650, San Diego, CA 92101, United States
50 Hampshire Street, 5th Floor, Cambridge, MA 02139, United States
The Boulevard, Langford Lane, Kidlington, Oxford OX5 1GB, United Kingdom

Notices
Knowledge and best practice in this field are constantly changing. As new research and experience broaden our understanding, changes in research methods, professional practices, or medical treatment may become necessary.

Practitioners and researchers must always rely on their own experience and knowledge in evaluating and using any information, methods, compounds, or experiments described herein. In using such information or methods they should be mindful of their own safety and the safety of others, including parties for whom they have a professional responsibility.

To the fullest extent of the law, neither the Publisher nor the authors, contributors, or editors, assume any liability for any injury and/or damage to persons or property as a matter of products liability, negligence or otherwise, or from any use or operation of any methods, products, instructions, or ideas contained in the material herein.

Library of Congress Cataloging-in-Publication Data
A catalog record for this book is available from the Library of Congress

British Library Cataloguing-in-Publication Data
A catalogue record for this book is available from the British Library

ISBN 978-0-12-813543-3

For information on all Academic Press publications visit our
website at https://www.elsevier.com/books-and-journals

Working together
to grow libraries in
developing countries

www.elsevier.com • www.bookaid.org

Publisher: Katey Birtcher
Acquisition Editor: Katey Birtcher
Editorial Project Manager: Karen R. Miller
Production Project Manager: Denny Mansingh
Cover Designer: Miles Hitchen

Typeset by SPi Global, India

Contents

7. A Practical Approach to Model Validation

KE FANG, MING YANG

8. Quantitative Measurements of Model Credibility

MEGAN OLSEN, MOHAMMAD RAUNAK

9. A Comprehensive Method for Model Credibility Measurement

YUANJUN LAILI, LIN ZHANG, GENGJIAO YANG

10. Quality Assessment and Quality Improvement in Model Engineering

U. DURAK, I. STÜRMER, T. PAWLETTA, S. MAHMOODI

11. Validation of DEVS Models Using AGILE-Based Methods

L. CAPOCCHI, J.F. SANTUCCI

12. DEVS Activity Tracking Based on Model Engineering and Simulation

L. CAPOCCHI, A. MUZY, J.F. SANTUCCI

13. Generic Concept and Architecture for Efficient Model Management

GÜNTER HERRMANN, AXEL LEHMANN, ROBERT SIEGFRIED

Contributors

L. Capocchi   SPE Laboratory (UMR CNRS 6134), University of Corsica, Corte, France

Changbeom Choi   Handong University, Pohang, South Korea

U. Durak   German Aerospace Center (DLR), Brunswick, Germany

Ke Fang   Control & Simulation Center, School of Astronautics, Harbin Institute of Technology, Harbin, China

Günter Herrmann   Institut für Technik Intelligenter Systeme, Neubiberg, Germany

Tag Gon Kim   KAIST, Daejeon, South Korea

Yuanjun Laili   School of Automation Science and Electrical Engineering, Beihang University, Beijing, China

Axel Lehmann   Institut für Technische Informatik, Universität der Bundeswehr München, Neubiberg, Germany

Feng Li   School of Automation Science and Electrical Engineering, Beihang University, Beijing, China

Ying Liu   School of Automation Science & Electrical Engineering, Beihang University, Beijing, China

S. Mahmoodi   Clausthal University of Technology, Clausthal-Zellerfeld, Germany

Saurabh Mittal   The MITRE Corporation, United States

A. Muzy   I3S Laboratory (UMR CNRS 6070), Team Bio-Info, Batiment Algorithme, Sophia Antipolis, France

James Nutaro   Computational Sciences and Engineering Division, Oak Ridge National Laboratory, Oak Ridge, TN, United States

Megan Olsen   Department of Computer Science, Loyola University Maryland, Baltimore, MD, United States

T. Pawletta   Wismar University of Applied Sciences, Wismar, Germany

Mikel D. Petty   University of Alabama in Huntsville, Huntsville, AL, United States

Mohammad Raunak   Department of Computer Science, Loyola University Maryland, Baltimore, MD, United States

José L. Risco-Martín   Universidad Complutense de Madrid, Madrid, Spain

J.F. Santucci   SPE Laboratory (UMR CNRS 6134), University of Corsica, Corte, France

A. Schmidt   Wismar University of Applied Sciences, Wismar, Germany

Robert Siegfried   aditerna GmbH, Riemerling, Germany

Xiao Song   School of Automation Science, Beihang University (BUAA), Beijing, China

I. Stürmer   Model Engineering Solutions UK Ltd, London, UK

Gary Tan   Singapore-MIT Alliance Research & Technology (SMART), Singapore, Singapore

Mamadou K. Traoré   University of Clermont Auvergne, LIMOS CNRS UMR 6158, Aubière, France

Zhongshi Wang   ITIS GmbH, Neubiberg, Germany

Eric W. Weisel   Old Dominion University, Norfolk, VA, United States

Yan Xu   Singapore-MIT Alliance Research & Technology (SMART), Singapore, Singapore

Ming Yang   Control & Simulation Center, School of Astronautics, Harbin Institute of Technology, Harbin, China

Gengjiao Yang   School of Automation Science and Electrical Engineering, Beihang University, Beijing, China

Laili Yuanjun School of Automation Science and Electrical Engineering, Beihang University; Engineering Research Center for Complex Product Advanced Manufacturing Systems, Ministry of Education, Beijing, China

Bernard P. Zeigler RTSync Corp., Chandler; Arizona Center for Integrative Modeling and Simulation, Tucson, AZ, United States

Lin Zhang School of Automation Science and Electrical Engineering, Beihang University; Engineering Research Center for Complex Product Advanced Manufacturing Systems, Ministry of Education, Beijing, China

Weicun Zhang School of Automation and Electrical Engineering, University of Science and Technology Beijing, Beijing, China

Chun Zhao School of Automation Science and Electrical Engineering, Beihang University, Beijing, China

Fuwang Zhao School of Automation Science, Beihang University (BUAA), Beijing, China

Bios of Editors

Lin Zhang is a professor at Beihang University. He received his BS degree in 1986 from the Department of Computer and System Science at Nankai University, China. He received his MS degree and the PhD degree in 1989 and 1992 from the Department of Automation at Tsinghua University, China. From 2002 to 2005 he worked at the US Naval Postgraduate School as a senior research associate of the US National Research Council. He has served as President of the Society for Modeling and Simulation International (SCS) (2015–2016). He is a Fellow of SCS and the Federation of Asian Simulation Societies (ASIASIM), the Executive Vice President of China Simulation Federation (CSF), an IEEE senior member, a chief scientist of the national 863 key projects, and associate editor-in-chief and associate editor of six peer-reviewed international journals. He authored and coauthored 200 papers, and 18 books and chapters. His research interests include service-oriented modeling and simulation, cloud manufacturing and simulation, model engineering, etc.

Bernard P. Zeigler, RTSync Corp. and Arizona Center for Integrative Modeling and Simulation, Tucson. AZ, is best known for his theoretical work concerning modeling and simulation based on systems theory. Zeigler has received much recognition for his various scholarly publications, achievements, and professional service. His 1984 book, Multifaceted Modeling and Discrete Event Simulation, was published by the Academic Press and received the outstanding Publication Award in 1988 from The Institute of Management Sciences (TIMS) College on Simulation. Zeigler is a Fellow of the Institute of Electrical and Electronics Engineers (IEEE) as well as The Society for Modeling and Simulation International (SCS), which he served as President (2002–2004). He is a member of the SCS Hall of Fame. In 2013, he received the Institute for Operations Research and Management Sciences (INFORMS) Simulation Society Distinguished Service Award. In 2015, he received the INFORMS Lifetime Professional Achievement Award, which is the highest honor given by the Institute for Operations Research and the Management Sciences' Simulation Society.

Yuanjun Laili is currently a lecturer (Assistant Professor) at the School of Automation Science and Electrical Engineering, Beihang University (Beijing University of Aeronautics and Astronautics), China, since September 2015. She obtained her BS, MS, and PhD degrees in 2009, 2012, and 2015, respectively, from the School of Automation Science and Electrical Engineering in Beihang University, China. From 2017 to 2018 she worked as a research scholar at the Department of Mechanical Engineering, University of Birmingham, UK. She was elected in the "Young Talent Lift" project of China Association for Science and Technology, and was awarded as the "Young Simulation Scientist" of the Society for Modeling and Simulation International (SCS). She is an associate editor of "International Journal of Modeling, Simulation and Scientific Computing". Her research interests include system modeling and simulation, evolutionary computation, and optimization in manufacturing systems. She is the author of one monograph and 28 journals and conference articles.

Preface

A model is an abstract expression of objects to study and embodies the high intelligence of human beings in perception of the real world. Simulation is an important way of understanding and even changing the real world. Modeling is taken as the foundation for simulation, especially for the simulation of complex systems (e.g., systems of systems). With the continuous growth of complexity and diversity of systems to be studied, models are becoming more complicated and diversified. How to build a right model is the core issue in simulation. Although the importance of the engineering idea is gradually recognized in applications involving the full model lifecycle, currently no complete theory, philosophy, or technology system is available.

To meet the challenges in the development and management of complex system models, this book gives a systematic introduction to the concept of model engineering (ME). We aim at setting up a generic, normalized, and quantifiable engineering methodology for ME. Our approach is to explore the basic principles in model construction, management, and maintenance to best deal with the data, processes, and organizations/people involved in the full life cycle of a model. We believe this is the key to guarantee the credibility of the model life cycle.

This book shows state-of-the-art research of the authors that relates to ME for simulation, including model construction, model lifecycle process management, model library management, model description, management and execution, model composition and reuse, quantitative measurement of model credibility, model validation and verification, applications of model engineering, etc.

This book appears in conjunction with the "Theory of Modeling and Simulation, 3rd Edition," in which the system theoretic and DEVS-based concepts developed for modeling and simulation (M&S) will contribute to a solution of problems raised by the holistic approach of ME for simulation.

As the first book that systematically introduces the concept of model engineering, this book tries to draw the attention of researchers to establish a generic methodology on ME independent of specific application fields. ME should be a subdiscipline of M&S.

Moreover, ME can be used not only in the domain of M&S, but also in other fields that need modeling and model management.

Lin Zhang
Bernard P. Zeigler
Yuanjun Laili

1

Introduction to Model Engineering for Simulation

Lin Zhang*,†, Bernard P. Zeigler‡, Laili Yuanjun*,†

*School of Automation Science and Electrical Engineering, Beihang University, Beijing, China
†Engineering Research Center for Complex Product Advanced Manufacturing Systems, Ministry of Education, Beijing, China ‡RTSync Corp. and Arizona Center for Integrative Modeling and Simulation, Tucson, AZ, United States

1 BACKGROUND

Simulation has been widely accepted as an important computational approach for human beings to understand and change the objective world (Mittal et al., 2017).

Modeling and simulation (M&S) play a vital role in many critical fields, such as economy, aerospace, information, biology, material, energy, advanced manufacturing (Fowler and Rose, 2004), agriculture, education, military, transportation (Mahmassani, 2001), ecosystems (Holling, 2001), pharmaceuticals, and health.

The systems to which M&S are being applied are increasingly complex, thus requiring significant advances in its science and technology. Various sorts of mathematical theories have been applied to approximate, analyze, predict, and optimize complex systems. However, our understanding of the features that make systems complex, for example, nonlinearity, uncertainties, emergence, and the dynamic interactions among components, continues to remain in an initial stage. As an alternative approach to understanding of complex systems, simulation has special advantages in the design, analysis, development, optimization, control, maintenance, and training. In the past decades, many simulation technologies have been developed, which include discrete event simulation (Fishman, 2013), agent-based simulation (Drogoul et al., 2003), runtime infrastructure (RTI)-based federation simulation (Perumalla et al., 2003), and so on. Simulation can facilitate quick understanding of a system's behavior and can enable analysis, training, and decision-making on the system without actually testing it in the real world (Zeigler et al., 2018; De Jong, 2002; Karnopp et al., 2012; Ören and Zeigler, 1979; Ören et al., 2012; Ouyang, 2014; Stevens et al., 2015; Zeigler et al., 2016).

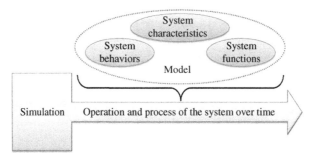

FIG. 1   A model within a simulation.

Simulation is a model-based activity; it uses models as a basis to imitate the operation of a real-world process or system over time for managerial or technical decision-making (NSF, 2006; DoD, 2007). A model that is adequate for the purposes of simulation adequately represents the key characteristics, behaviors, and functions of the simulated object for these purposes. If the object is a complex system, a set of models is required for representing its components, which are then composed to create a model of the whole system. The composed model can be seen as a surrogate for the targeted system, whereas simulation represents the operation and process of the system over time, as shown in Fig. 1.

It is critical to formulate the objectives (or requirements) underlying a simulation activity. Constructing a model that adequately achieves these objectives is critical for the success of a simulation application. Take the development of missiles as an example. The early developed "Bristol Bloodhound" was successfully launched with 79 model-based events and required only 92 calibrated launches. In contrast, not using M&S, the development of "NIKE-I missile" had to be physically tested 1000 times. Of course, if the simulation model is established with mistakes or unstable factors, the decision and analysis results applied to the real system may lead to catastrophic consequences. The "Three Mile Island" accident is a typical example of using an unreliable system model to train people. This model flaw led to misoperation on the real system and caused more than a billion dollar economic loss.

Later we will present a detailed definition of "credibility" and how constructing the right model is critical to credible simulation results.

As the number of system components continues to grow and their interactions change dynamically, the performance of a complex system varies constantly and evolves gradually to a System of Systems (SoS) (Keating et al., 2003). Such system refers to a collection of dedicated subsystems that pool their components and capabilities together to offer more functionality and performance than simply the sum of these constituent systems. An SoS is featured by:

- a large number of components and complex relationships,
- decentralized dynamics with strong uncertainty,
- huge amounts of data to be processed.

To analyze it using simulation, basic models designed from bottom up must be established to support testing the mechanisms needed to coordinate the component behaviors so as to enable the SoS to achieve its global requirements (Zeigler et al., 2016). That means that the engineers (who typically come from different domains) need to take the time to establish collaboratively new models for every component of every subsystem in an SoS. If the simulation requirement changes slightly, specific models may need to be rebuilt. To avoid the cumbersome remodeling process, model reuse and composition become critical model-based activities.

Many studies have been carried out on the reuse of existing models for different kinds of SoS from the perspective of system engineering. The most typical research includes model-driven simulation (MDS) (McGinnis and Ustun, 2009), dynamic data-driven application system (DDDAS) (Darema, 2004), and model-based system engineering (MBSE) (Wymore, 1993). The research on MDS has fully shown the importance of models and established a framework for the design of tool-independent metamodel. This metamodel contains the simplest description of the state, action, and process of the corresponding executable simulation model. By using a unified modeling language, it will be easily reused to perform the simple workflow of different systems and then guide the refinement of the existing simulation model for new system analysis. As shown in Fig. 2, such metamodel is taken as a middleware to bridge the gap between a real-world system (or object) and a similar or matched simulation model. On the contrary, the research on DDDAS tries to directly reuse the entire system model by a feedback control loop, as shown in Fig. 3. On one hand, the simulation data is collected to reconfigure the running model. On the other hand, the model itself will control the adaptation of the simulation process for generating the required result. Such a two-way control mechanism enables the existing model to be executed in a dynamic way and adapt to a wide range of systems.

Both of these paradigms put their main focus on how to establish a reusable model for a wide range of objects and how to adapt it to different simulation requirements. However, only a single model or a small group of models are considered for reuse. For different classes of models, the metamodel or the closed-loop control mechanism must be specifically redesigned.

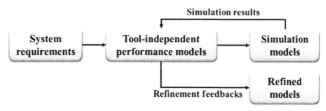

FIG. 2   The framework of a model-driven simulation.

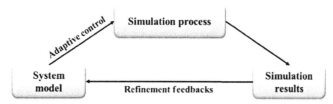

FIG. 3   The framework of a dynamic data-driven application system.

Different from these two paradigms, MBSE was proposed to manage the models and their related knowledge established in the whole life cycle of system engineering instead of just the documentation (Estefan, 2007). It refers to the use of a model as a basic element to describe customer requirements and to design, analyze, and verify a required system. Most research on MBSE considers the design, storage, collaboration, and test of reusable models with respect to different stages of system engineering on a conceptual level and passes the technical implementation mechanisms to different domain engineers. Consequently, MBSE research has not addressed how to manage existing multidisciplinary simulation models and compose them together to form a valid and credible simulation system in a rapid manner.

This chapter analyzes in depth the current challenges in the modeling of complex systems (especially SoSs). It also introduces the concept of model engineering (ME), its key technologies, and provides an overview of this book.

2 MAIN CHALLENGES ON MODEL LIFE CYCLE

A system, especially a complex system, can be generally divided into three layers, the system layer, the subsystem layer, and the component layer, as shown in Fig. 4. It is the dynamic collaboration between different components that enables the whole SoS perform much more functionalities than the simple sum of multiple subsystems (Nielsen et al., 2015). To form a complete and credible SoS simulation, the model established for each component should possess the following properties:

- the model must be credible,
- the model should be fully adapted to multidisciplines,
- the model should be able to connect to each other and respond to different inputs dynamically, and
- the model may process large amounts of data.

These properties have brought many obstacles for engineers to establish an efficient (or even a right) SoS. Although the concept and framework of SoS have been addressed by researchers many times, there have been very few attempts to establish a complete SoS simulation system or a group of extendable models for SoS simulation. The main challenges are summarized as follows.

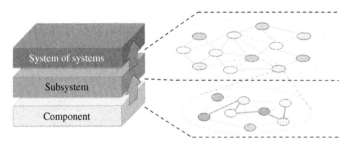

FIG. 4   Three layers of system of systems.

2.1 The Credibility of a Composed SoS System Model Is Hard to Verify

Most of the existing verification-and-validation (V&V) methods are designed for a single model under a specific environment (Law, 2008; Sargent, 2009; Tremblay and Dessaint, 2009). They focus mainly on the consistency, uncertainty, and sensitivity validation of simulation results compared with the desired results (or real-system results) (Moriasi et al., 2007; Moss, 2008; Park et al., 2010). The design knowledge, the development environment, the refinement process, and the maintaining way which also directly influence the credibility of the model are ignored. When the model is rebuilt, refined, or reused in a new environment, its credibility is hard to be reassessed by the previous V&V method.

In addition, all of the component models are right and independently verified does not mean that the composition of them is also right and credible. On one hand, it is still lack of complete and unified criteria to verify if a composed system model is fully reliable under a specific circumstance. On the other hand, the methods for verifying and validating the composed system model with dynamic changes and inside interactions are still very rare.

2.2 Existing Models Are Hard to Extend and Interoperate

Without a uniform specification for the modeling, simulation, and model maintenance process, the model can be built into any shape. By hundreds of simulation software and tools, millions of simulation models have been well established for different domains. However, as most of them cannot be extended into a new environment, the increased number of models actually makes no sense. Model reuse can only be carried out under a standard M&S architecture. Till now, there is still lack of a basic scalable model, standard process, and interdisciplinary rules to support fast model reuse and system construction. It is even unable to determine if a model is reusable before testing it in the targeted environment.

Even if a model is reusable, it is hard to interoperate with others in a new environment. First of all, the inputs and outputs (I/O) of the model are usually fixed. Without full matching with the I/O, it cannot be set up in a right way. Second, due to the runtime environment barrier and nonuniform simulation process, these models cannot even be executed under cooperative circumstances. That is why most of the existing collaborative simulation is carried out in a distributive manner with independent execution environment. Moreover, each component of such a collaborative simulation requires a member engineer to maintain it in real time. How to make a model more adaptable for a new execution environment and how does it respond to both the environment and other models autonomously are still two crucial problems in the domain of M&S.

2.3 Crucial Data and Knowledge Produced and Processed by the Model Is Lost

Traditionally, engineers or researchers primarily concern about the simulation results produced by a system model. Nonetheless, the data produced throughout the process that the model is defined, implemented, refined, and maintained is ignored. To refine a model in a changing situation, some real-time model calibration methods have been proposed, which enables a series of real-time simulation data to be analyzed and stored. Nowadays, engineers start to put emphasis on the knowledge extraction and process normalization of M&S. History information becomes more and more important in M&S and system construction. However, large amounts of interaction data, disciplinary knowledge, and management data, which is a fundamental basis to determine if a model will be credibly reused, if a group of models are composable and suitable for new requirement, and if the whole composed system model is reliable has yet to be well stored, analyzed, and applied.

On the whole, very few studies have focused on the life cycle of the model, the management of interdisciplinary models, and make use of them for SoS construction. How can a model or a group of models efficiently evolve, adapt, and cooperate for complex systems is still an open question to be solved.

3 THE CONCEPT OF ME

To focus on the life-cycle management of simulation models and guarantee the credibility of SoS modeling and simulation, the concept of ME is introduced in this section. We will formally define the life cycle of a model and connect each stage in the life cycle, so as to make the existing model better to reuse, combine, and evolve in a credible and standard way.

The life cycle of a model defined in ME is shown in Fig. 5. It contains six steps, that is, problem definition, model design, model construction, model configuration, VV&A (verification, validation, and accreditation), model application, and model maintenance.

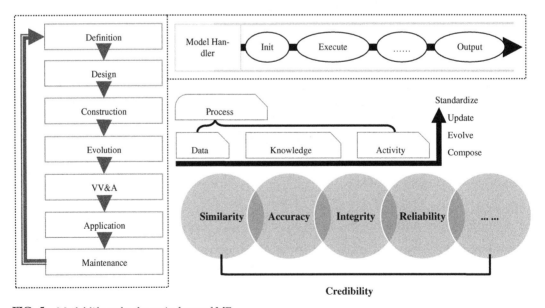

FIG. 5   Model life cycle: the main focus of ME.

Based on this cycle, ME is defined as a general term for theories, methods, technologies, standards, and tools relevant to a systematic, standardized, quantifiable engineering methodology that guarantees the credibility of the full life cycle of a model with the minimum cost (Zhang, 2011, Zhang et al., 2014a):

(1) ME regards the full life cycle of a model as its object of study, which studies and establishes a complete technology system at the methodology level based in order to guide and support the full model life-cycle process such as model construction, model management, and model use for complex systems.

(2) ME aims to ensure credibility of the full model life cycle, integrate different theories and methods of models, study and find the basic rules independent of specific fields in the model life cycle, establish systematic theories, methods and technical systems, and to develop the corresponding standards and tools.

(3) ME manages the data, knowledge, activities, processes, and organizations/people involved in the full life cycle of a model, and takes into account the time period, cost, and other metrics of development and maintenance of a model.

(4) Here the credibility of a model includes functional and nonfunctional components. Functional components are a measurement of the correctness of functions of the model compared to the object being modeled. Nonfunctional components include features related to the quality of a model, such as availability, usability, reliability, accuracy, integrity, maturity, ability of modelers as well as management of the modeling process. Credibility is a relative index with respect to the purpose of modeling and simulation. Evaluation of credibility includes objective and subjective evaluation. Objective evaluation is mainly based on data and documents, while subjective evaluation is mainly based on expertise. Quantitative definition and measurement of credibility will be one of the most important research topics of ME.

4 KEY TECHNOLOGIES OF ME

According to the framework of the Body of Knowledge of Model Engineering given in Zeigler and Zhang (2015) and Zhang et al. (2014a,b), technologies involved in ME can be divided into the following categories (Fig. 6) including general technologies, model construction technologies, model management technologies, model analysis and evaluation technologies, and supporting technologies. Some key technologies in the categories will be discussed in this section.

4.1 General Technologies

4.1.1 *Modeling of the Model Life-Cycle Process*

In accordance with the standards of ME, modeling of the model life-cycle process means to build a structural framework of activities that usually happen in the life cycle. As demonstrated in Fig. 7, the framework is a visible pipeline to show the state of a model related to the key stages, key elements, and key data of its life-cycle management. It is also designed

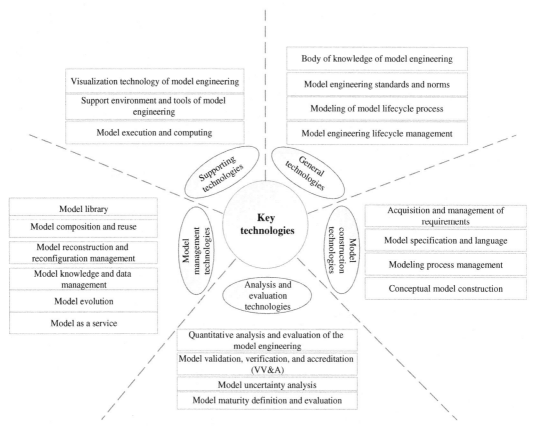

FIG. 6   Key technologies of model engineering.

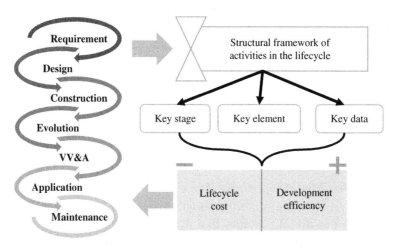

FIG. 7   A model of the model life-cycle process.

as a reference to evaluate the life-cycle cost and comprehensive efficiency and to improve both the model and the management strategy.

4.1.2 Model Engineering Life-Cycle Management

The life-cycle management of ME is carried out for managing data (development data and runtime data), knowledge (common knowledge shared by different models and the domain knowledge), activities, tools (especially M&S tools and model evaluation tools), and person (the modeler, the tester, and the user). Data/knowledge management technology focuses mainly on the data and knowledge in model, runtime environment, and the whole model life cycle. It includes the methods for key data extraction during the ME life cycle, knowledge classification from multidisciplines, information learning throughout modeling and simulation, and data/knowledge storage for further improvement, as shown in Fig. 8.

In the near future, we expect that the number of multidisciplinary models will grow and the assembly and disassembly of systems, data, and models will continue to become more complex.

Accordingly, data mining strategies and knowledge extraction algorithms used in ME must become much more: (1) *scalable*, to adapt to a wider arrangement of domain information, (2) *efficient*, to implement intelligent system construction, and (3) *stable*, to ensure credible simulation and model management.

ME life-cycle management also consists of monitoring the processes of model reconfiguration, evolution, and maintenance, and the multilayer optimization of modeling practices, operational workflows, and maintenance schemes to realize an efficient risk/cost control and speedup throughout the whole life cycle of a model.

4.2 Model Construction Technologies

A large amount of research on model construction (modeling) has accumulated over the years in the M&S domain. From the point of view of ME, some issues for modeling methods are of most concern. Such issues include: (1) acquisition and management of model requirements, (2) model specifications and modeling languages, (3) modeling process management, and (4) conceptual model construction.

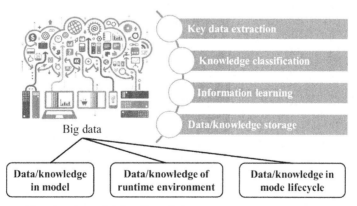

FIG. 8   The main considerations in model data/knowledge management.

4.2.1 Acquisition and Management of Model Requirements

The model life cycle starts with requirements. Accurate requirement acquisition is the key to credible M&S. However, requirement acquisition and management is very challenging due to the uncertainty and ambiguity in the systems being modeled. Research on requirement acquisition is needed to improve the means to extract, describe, parse, and validate requirements via automated or semiautomated means. Similarly, research is needed on the management of requirements to formulate how to reflect changing requirements to influence model construction and maintenance in an accurate and timely manner.

To acquire accurate model requirements for system simulation, we need to get as much information as we can to understand (1) the underlying modeling objectives of the simulation, (2) the nature of the targeted system, and (3) the kind of environmental conditions that are required. Therefore, an analytical strategy is particularly important in which we hierarchically decompose the structural requirements for further design and model matching, as illustrated in Fig. 9.

Additionally, system features (user demand, system structure, and environmental conditions) extracted from the above strategy must be stored and managed by category of the facet. When a new requirement comes in, these facets will be used to match similar models and existing domain knowledge to support rapid system construction.

4.2.2 Model Specification and Language

Model specification is informed by the detailed description of system simulation requirements, model input/output, model functionalities, model activities/states, and the related domain rules. As shown in Fig. 10, the model requirement is not only a documental or structural description about what kind of model we need, but also a simple and uniform representation of the general features possessed by a targeted system. These features should specify some common rules corresponding to system components and their interconnections. Similarly, at the component level, the specification for the metamodel (which performs some domain-independent functionalities/states/activities) and the domain rule (which represents some domain-dependent state transformation mechanisms) should be prebuilt. By

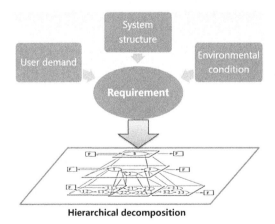

FIG. 9   Acquisition and decomposition of model requirement.

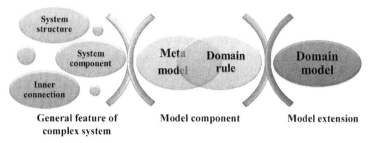

FIG. 10 Different levels of model specification.

dividing a general domain model into a metamodel and domain rule, a domain model is easily assembled and disassembled for efficient reuse. In addition, this division enables the modeling data/knowledge to be clearer and easier to manage.

Although a model can be built by different sorts of lower-layer model languages, a unified upper-layer language is still necessary for the engineers to construct a new system model efficiently or to transform an existing one to meet the simulation demand. Following the vision of the above specifications, an upper-layer model language should also clearly describe the outer structure and inner behavior of a system and be able to transform into multiple lower-layer model languages, as represented in Fig. 11.

4.2.3 *Modeling of Process Management*

Two kinds of efforts are necessary to guarantee the credibility of a model. One is to do VV&A after the model is built. The other is to manage and optimize the modeling process. VV&A has important implications to discover model problems and defects, but it clearly does not and cannot solve the problem of how to acquire a correct model. Especially for complex systems, due to the complexity and uncertainty of the system, the modeling process can be very complicated, which makes VV&A of a model also extremely difficult. Even if the defects are found via VV&A, the modification of the model will be very difficult and expensive.

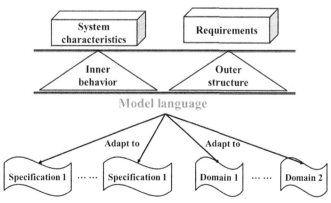

FIG. 11 Upper-layer model language.

Therefore, it is very important to structure and optimize the modeling process. Consequently, methods are needed to measure the degree of formality and optimization (maturity) of modeling and simulation processes. A highly structured level of organizational capabilities and the use of proven processes applied to modeling can guarantee, to a large extent, the credibility of a model (Fujimoto et al., 2017).

Capability maturity model (CMM) and CMM integration (CMMI), originating in software engineering, can be introduced to establish a capability maturity model for the modeling and simulation process (MS-CMMI) (Zhang and Mckenzie, 2017). Such an approach can enhance the management efficiency and personnel capabilities in high-quality model construction and management. Related research opportunities include MS-CMMI evaluation, optimization, risk analysis, and control of modeling processes, notional mappings with CMMI, etc.

4.3 Model Management Technologies

Model management consists of core methodologies and technologies that guarantee highly efficient and credible composition, sharing, reuse, evolution, and maintenance.

4.3.1 Model Library

A scalable model library is key to implement efficient ME. It should be able to handle heterogeneous models by using a formal description language, recognize multidisciplinary model features, and enable a fast indexing and location of similar/suitable model for a task specification, as demonstrated in Fig. 12. The main techniques with respect to model library are model classification criteria, model storage mode, model indexing schemes, and ways of searching for models. In contrast to a database, the model library stores and processes not only model descriptions, but also their instances and interconnection relationships. Currently, there is lack of techniques for establishing a model library.

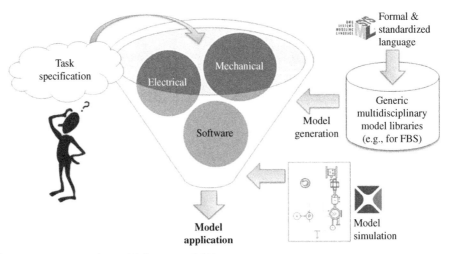

FIG. 12   A simple scheme for establishing a model library.

4.3.2 Model Evolution

Model evolution is one of the most important innovations proposed in ME. From the separated metamodel and domain knowledge point of view, ME aims at using models as a service and enabling them to evolve autonomously to a new improved version instead of through manual refinement. Borrowing the idea of incremental learning, model evolution refers to an incremental adaptation process to make the specific model more scalable and credible to the current system requirements.

The factors including parameters, states, behaviors, and functions in metamodel and different sorts of domain knowledge such as additional domain parameters, action rules, constraints, and domain-related functions are all able to be updated as modules of a simulation system, as shown in Fig. 13. To enable the metamodel and domain knowledge to update autonomously over time, we need to establish dynamic connections between system requirements and these lower-layer factors in line with the historical data from the model life cycle. These connections, which can be trained by the existing incremental learning algorithms and intelligent multiagent system-based strategies, will then guide the factors of a model to change toward the ones connected to the most similar requirements.

4.3.3 Model Reconfiguration

Different from model refinement in the stages of application and maintenance, model reconfiguration means to change part of the model during its runtime. It is an important basis to implement model reuse with a minimum roll-back design cost. According to the above-mentioned model specification, a model should be designed to hold multiple functionalities and flexible interfaces. How to dynamically choose suitable functionalities and domain

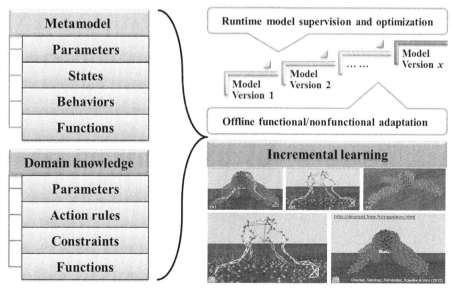

FIG. 13 Time-dimensional model evolution scheme.

knowledge to ensure accurate response in simulation is the main concern of model reconfiguration methods. Specifically, it can be divided into lower-layer reconfiguration and upper-layer reconfiguration.

LOWER-LAYER RECONFIGURATION

The lower-layer model reconfiguration method is only for a metamodel which is independent of the domain knowledge. It includes functional reconfiguration, structural reconfiguration, and parameter reconfiguration, as shown in Fig. 14. Specifically, structural reconfiguration refers to a combined multiple metamodel to form a larger one with more functionalities.

UPPER-LAYER RECONFIGURATION

On the contrary, upper-layer reconfiguration is directly related to domain knowledge and the practical simulation environment, as shown in Fig. 15. Thus, it can be divided into domain-related reconfiguration and simulation-related reconfiguration. The domain-related part is responsible for selecting an add-on domain knowledge (i.e., domain functionalities, domain parameters, and constraints) to a model, while the simulation-related reconfiguration is set up to determine the environmental parameters and simulation engine-related settings to assure a correct and fluent simulation process. Obviously, model reconfiguration is a complex dynamic optimization problem, in which the two-level variables can be either determined in two steps or in one time.

4.3.4 Model as a Service

With the development of cloud-based technologies, a heterogeneous model with its execution engine can be integrally encapsulated as a service. That is to say, not only a metamodel

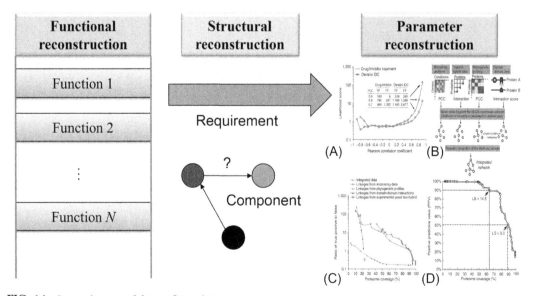

FIG. 14   Lower-layer model reconfiguration.

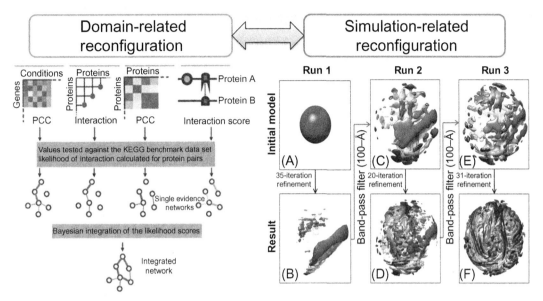

FIG. 15 Lower-layer model reconfiguration.

and a domain component model can be encapsulated, a composed system model is also capable of being encapsulated. This makes the execution of different sorts of models easier in any cloud-based environment. To implement flexible model sharing, the scheme of cloud-based servitization[1] of simulation model can be illustrated in Fig. 16. First of all, a uniform servitization template (i.e., a service class) should be designed previously for different levels of model. When a system construction requirement arrives, the suitable models will be deployed or replicated into different virtualized cloud resources to support a distributed simulation.

4.3.5 Model Composition

Model composition is a technology established on the flexible model reuse scheme. It is designed to realize more intelligent model collaboration and system construction when the number of models is too large to be implemented with manual selection. In the research of model composition, the two critical problems are how to match suitable models to form a valid candidate set and how to select the best models for a collaborative system simulation. The former matching problem can be solved by some feature-based or domain-based model

---

[1]Servitization is a new concept that has two meanings from IT and business perspectives. From the IT perspective, it means the service encapsulation of objects with service-oriented technology. From the business perspective, it is defined as "the innovation of organization's capabilities and processes to better create mutual value through a shift from selling product to selling Product-Service Systems," where a product-service system is "an integrated product and service offering that delivers value in use" and a "servitized organization is one which designs, builds and delivers an integrated product and service offering that delivers value in use" (http://andyneely.blogspot.com/2013/11/what-is-servitization.html).

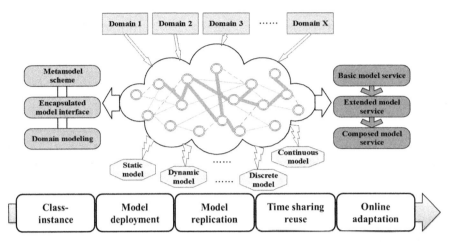

FIG. 16   Cloud-based servitization of simulation models.

classification and model clustering methods, while the latter model selection has to be considered under different conditions, that is, offline condition and online condition.

OFFLINE MODEL COMPOSITION

Different from the general service composition, the collaboration between models are not usually performed in a strict sequential or directed acyclic manner. Feedback cycles and concurrencies may exist simultaneously in their collaborative topology. Therefore, traditional algorithms designed for service composition may not applicable. As listed in Fig. 17, an offline model composition method should be able to recognize each kind of connection in a specific collaborative topology (extracted from a standardized system simulation requirement) and generate feasible solutions with the consideration of sequencing rules, cycling rules, and concurrency rules between the candidate models.

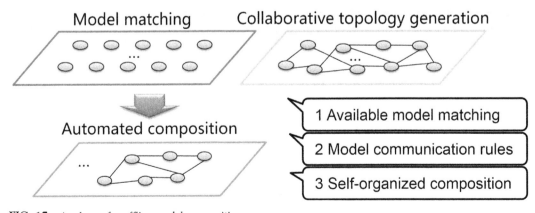

FIG. 17   A scheme for offline model composition.

ONLINE MODEL COMPOSITION

Online model composition is executed during the real-time simulation process based on a given offline composition scheme. The main workflow of an online model composition method is drawn is Fig. 18. Specifically, it is driven by online evaluation of the current system state compared with a desired state. If the online adjustment threshold is reached, the method will perform the adjustment in four steps, that is, candidate adjustment, domain rule adjustment, connection adjustment, and parameter adjustment from the top down. After the online refinement, the evaluation model will continue to monitor the system state and determine whether a further modification is required. In other words, an online model composition method should be performed at a high speed and provide a feasible solution at different levels and thus is more difficult to design.

4.4 Analysis and Evaluation Technologies

Model evaluation is a very traditional topic in the domain of M&S. In ME, it means not only the VV&A of a model, but also the evaluation of the whole process of ME.

4.4.1 The VV&A of a Model

In the past decades, different authoritative organizations and researchers have established a few standards for the VV&A of a nonseparable model. However, little research has been done on the evaluation of models in a composed situation and in its further maintenance process. As demonstrated in Fig. 19, the bridge between the existing model evaluation indicators and the models in different layers of system construction is a key to implement the efficient evaluation of the model life cycle.

In addition, most current research focuses on qualitative analysis, and quantitative and formalized analysis methods are lacking, so VV&A quantitative analysis and formalized analysis technology are still the main research content in the ME.

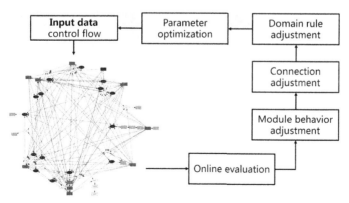

FIG. 18   A general workflow of an online model composition method.

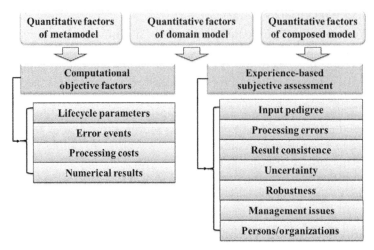

FIG. 19   Gap existing between the evaluation indicators and the models in different layers.

4.4.2 The Evaluation of the Whole Process of ME

The quality of the ME process will directly determine the quality of a model in its design, implementation, application, and maintenance. To ensure the control and management quality of ME, key issues from both the control stages and the management process should be extracted first for the construction of evaluation indicators, as illustrated in Fig. 20. With a suitable set of indicators to cover the whole process of ME fully, the existing expert scoring mechanisms such as Fuzzy-AHP and TOPSIS can be directly applied to assess it. Because the process of ME is not a one-time execution workflow but a long-accumulated management framework, the evaluation must be carried out based on historical information. Thereby, a case library is also a fundamental element in evaluation to quantify the quality of the ME process and so as to form a hybrid evaluation mechanism for fast ME evaluation.

Research topics related to evaluation also include quantitative analysis of the complexity and uncertainties in risk analysis and control of ME processes, quantitative measurement of model life-cycle quality and cost, etc.

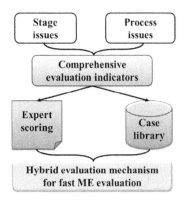

FIG. 20   A scheme for the evaluation of the ME process.

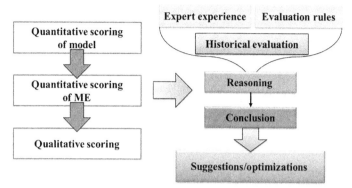

FIG. 21   A comprehensive evaluation scheme.

With the combination of model life-cycle evaluation and ME process evaluation, a comprehensive evaluation scheme can be drawn as in Fig. 21 to guide further optimization and calibration targeted to the whole ME framework.

4.4.3 *Model Maturity Definition and Evaluation*

The maturity of a model is a very important index for model composition, sharing, and reuse. Maturity definition of a model is not an easy job since different models have different features, different application requirements, and different execution environments. Model maturity will be a comprehensive index related to multidimensional features as illustrated in Fig. 22. Research effort need to be made on the definition and evaluation of model maturity.

4.5 Supporting Technologies

The supporting technologies for the implementation of ME primarily consist of the transparent visualization ways for model life cycle and operational platform to enable fundamental execution of activities involved in the whole process of ME, as illustrated in Fig. 23.

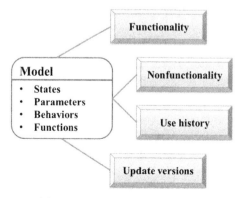

FIG. 22   Factors for determining model maturity.

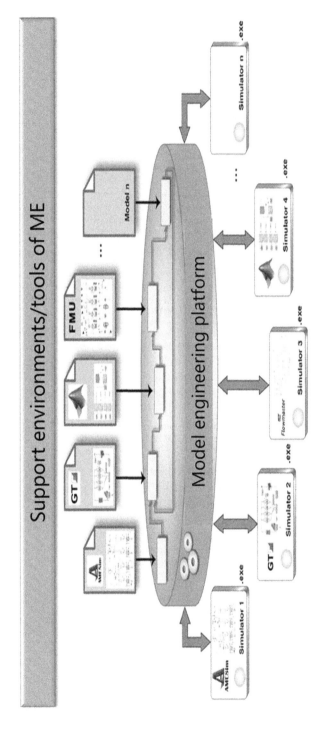

FIG. 23   A vision on the support environments for the implementation of ME.

5 THE LAYOUT OF THE BOOK

This book contains 19 chapters that cover some of the technologies listed above. Chapter 2 discusses a computational approach to representing and evaluating the parameters and approximate morphisms between pairs of models within ME for simulation, which lays a basis for later chapters that discuss the support for model repositories. Chapter 3 discusses unified approaches to modeling. Chapters 4 and 5 deal with challenges and methods of model composition and reuse. Chapters 6 and 7 discuss the requirements, resolutions, and trends of verification, validation, and accreditation (VV&A) for complex simulation models. Chapters 8–10 address the quantitative analysis and evaluation of different characteristics of models, which include quantitative methods for model credibility evaluation, quality assessment and improvement approaches, and the methodologies for graphical models. Chapters 11 and 12 discuss validation and implementation of DEVS models. Chapter 13 describes a generic concept for the development of a model management system. Chapters 14–16 deals with model execution and simulation algorithms. Chapter 17 introduces a modeling and simulation environment with DEVSim++. Chapters 18 and 19 provide two application examples of ME.

6 ROLE OF THEORY OF MODELING AND SIMULATION

The theoretical concepts developed specifically for M&S can contribute to a solution of problems raised by the holistic approach of ME for simulation. This book is appearing in conjunction with a companion volume, "Theory of Modeling and Simulation, third Edition" (Zeigler et al., 2018). This theory presents a framework for M&S that formalizes the basic entities (Models, Simulators, Experimental Frames) and relationships between them (Modeling Validity, Simulation Correctness) that are fundamental to productive formulation of research on key ME technologies. For example, the theory of M&S is an important element in the Body of Knowledge supporting ME research. Acquisition and management of requirements can be structured around the essential role of experimental frames in formalizing modeling objectives. The theoretical concepts of iterative system specification and model formalisms underpin theory-based conceptual model construction and model specification languages. Research on ME analysis and evaluation technologies can profit from the use of the extensive theory of system morphisms that underlie the validity and correctness relationships expressed in the theory. Finally, model management technologies can fruitfully formulate the necessary domain-independent generic model construction and composition processes. These can be based on the structure/behavior and hierarchical closure under coupling properties established within the theory of M&S.

7 CONCLUSION

ME deals with the model life-cycle credibility and scalability problem for complex systems, especially Systems of Systems. ME is supposed to be taken as a subdiscipline of M&S, which aims at providing standardized, systematic, and quantifiable management and control to guarantee the credibility of the model life cycle. In this book, we systematically introduce the concept of ME, present some state-of-the-art research on ME-related theories and technologies, and provide some solutions to ME problems.

Moreover, ME can be used not only in the domain of M&S, but also in other fields that need modeling and model management.

References

Darema, F., 2004. In: Dynamic data driven applications systems: a new paradigm for application simulations and measurements. International Conference on Computational Science. Springer.

De Jong, H., 2002. Modeling and simulation of genetic regulatory systems: a literature review. J. Comput. Biol. 9 (1), 67–103.

DoD, 2007. DoD Modeling and Simulation (M&S) Management. NUMBER 5000.59.

Drogoul, A., Vanbergue, D., Meurisse, T., 2003. Multi-Agent Based Simulation: Where Are the Agents? Springer Berlin Heidelberg, Berlin, Heidelberg.

Estefan, J.A., 2007. Survey of model-based systems engineering (MBSE) methodologies. Incose MBSE Focus Group 25 (8), 1–12.

Fishman, G.S., 2013. Discrete-Event Simulation: Modeling, Programming, and Analysis. Springer Science & Business Media, New York, NY.

Fowler, J.W., Rose, O., 2004. Grand challenges in modeling and simulation of complex manufacturing systems. Simulation 80 (9), 469–476.

Fujimoto, R., Bock, C., Chen, W., Page, E., Panchal, J.H., 2017. Research Challenges in Modeling and Simulation for Engineering Complex Systems. Springer-Verlag, London, UK.

Holling, C.S., 2001. Understanding the complexity of economic, ecological, and social systems. Ecosystems 4 (5), 390–405.

Karnopp, D.C., Margolis, D.L., Rosenberg, R.C., 2012. System Dynamics: Modeling, Simulation, and Control of Mechatronic Systems. John Wiley & Sons, Hoboken, NJ.

Keating, C., Rogers, R., Unal, R., Dryer, D., Sousa-Poza, A., Safford, R., Peterson, W., Rabadi, G., 2003. System of systems engineering. Eng. Manag. J. 15 (3), 36–45.

Law, A.M., 2008. How to build valid and credible simulation models. Winter Simulation Conference, WSC 2008. IEEE.

Mahmassani, H.S., 2001. Dynamic network traffic assignment and simulation methodology for advanced system management applications. Netw. Spat. Econ. 1 (3), 267–292.

McGinnis, L., Ustun, V., 2009. A simple example of SysML-driven simulation. Winter Simulation Conference.

Mittal, S., Durak, U., Ören, T. (Eds.), 2017. Guide to simulationased disciplines. In: Advancing Our Computational Future, first ed. Simulation Foundations, Methods and Applications, Springer International Publishing, Cham, Switzerland.

Moriasi, D.N., Arnold, J.G., Van Liew, M.W., Bingner, R.L., Harmel, R.D., Veith, T.L., 2007. Model evaluation guidelines for systematic quantification of accuracy in watershed simulations. Trans. ASABE 50 (3), 885–900.

Moss, S., 2008. Alternative approaches to the empirical validation of agent-based models. J. Artif. Soc. Soc. Simul. 11 (1), 5.

Nielsen, C.B., Larsen, P.G., Fitzgerald, J., Woodcock, J., Peleska, J., 2015. Systems of systems engineering: basic concepts, model-based techniques, and research directions. ACM Comput. Surv. 48 (2), 18.

NSF, 2006. Report on Simulation-Based Engineering Science. National Science Foundation (NSF) Blue Ribbon Panel.

Ören, T.I., Zeigler, B.P., 1979. Concepts for advanced simulation methodologies. Simulation 32 (3), 69–82.

Ören, T.I., Zeigler, B.P., Elzas, M.S., 2012. Simulation and Model-Based Methodologies: An Integrative View. vol. 10 Springer Science & Business Media, Berlin, Heidelberg.

Ouyang, M., 2014. Review on modeling and simulation of interdependent critical infrastructure systems. Reliab. Eng. Syst. Saf. 121, 43–60.

Park, I., Amarchinta, H.K., Grandhi, R.V., 2010. A Bayesian approach for quantification of model uncertainty. Reliab. Eng. Syst. Saf. 95 (7), 777–785.

Perumalla, K.S., Park, A., Fujimoto, R.M., Riley, G.F., 2003. Scalable RTI-based parallel simulation of networks. Proceedings of Seventeenth Workshop on Parallel and Distributed Simulation (PADS 2003), pp. 97–104.

Sargent, R.G., 2009. In: Verification and validation of simulation models. Simulation Conference (WSC), Proceedings of the 2009 Winter. IEEE.

Stevens, B.L., Lewis, F.L., Johnson, E.N., 2015. Aircraft Control and Simulation: Dynamics, Controls Design, and Autonomous Systems. John Wiley & Sons, Hoboken, NJ.

Tremblay, O., Dessaint, L.-A., 2009. Experimental validation of a battery dynamic model for EV applications. World Electr. Veh. J. 3 (1), 1–10.

Wymore, A.W., 1993. Model-Based Systems Engineering. vol. 3. CRC Press, Boca Raton, FL.

Zeigler, B.P., Zhang, L., 2015. Service-oriented model engineering and simulation for system of systems engineering. In: Concepts and Methodologies for Modeling and Simulation. Springer, Cham, Switzerland, pp. 19–44.

Zeigler, B.P., Sarjoughian, H.S., Duboz, R., Souli, J.-C., 2016. Guide to Modeling and Simulation of Systems of Systems. Springer International Publishing, Switzerland.

Zeigler, B.P., Muzy, A., Kofman, E., 2018. Theory of Modeling and Simulation: Discrete Event and Iterative System Computational Foundations, third ed. Academic Press, USA.

Zhang, L., 2011. In: Model engineering for complex system simulation. The 58th CAST Forum on New Viewpoints and New Doctrines, Li Jiag, Yunnan, China, October 14–16.

Zhang, L., Mckenzie, R., 2017. Maturity Models. In: Fujimoto, R., Bock, C., Chen, W., Page, E., Panchal, J.H. (Eds.), Research Challenges in Modeling & Simulation for Engineering Complex Systems. Springer International Publishing, Switzerland. September 1, 2017.

Zhang, L., Shen, Y., Zhang, X., Song, X., Tao, F., Liu, Y., 2014a. The model engineering for complex system simulation. The 26th European Modeling & Simulation Symposium (Simulation in Industry), Bordeaux, September 10–12.

Zhang, L., Li, F., Song, X., Liu, Y.K., 2014b. In: A supporting environment of model engineering for complex systems. The 11th Chinese Intelligent System Conference (CISC'14), Beijing, China, October 18–19.

Simulation-Based Evaluation of Morphisms for Model Library Organization

Bernard P. Zeigler

RTSync Corp., Chandler, AZ, United States Arizona Center for Integrative Modeling and
Simulation, Tucson, AZ, United States

1 INTRODUCTION

This chapter is written in conjunction with the companion book to this volume, Theory of
Modeling and Simulation, third edition. Readers are encouraged to refer to that book for theoretical background that is not explicitly covered here. In particular, that forum presents a
framework, called the modeling and simulation framework (MSF) which includes an approach to test the relation between detailed models and their abstractions (Zeigler et al.,
2018; Zeigler and Nutaro, 2016; Zeigler, 2016).

The framework recognizes that the complexity of a model can be measured objectively by
its resource usage in time and space relative to a particular simulator, or class of simulators.
Furthermore, properties intrinsic to the model are often strongly correlated with complexity
independent of the underlying simulator. Successful modeling can then be seen as *valid
simplification*, that is, reduction of complexity to enable a model to be executed on
resource-limited simulators and at the same time, creating *morphisms that preserve behavior
and/or structural properties*, at some level of resolution, and within some experimental frame
of interest. Indeed, according to the framework, there is always a pair of models involved, call
them the *base* and *lumped* models.

In this chapter we lay the basis for later chapters that discuss support for model repositories. We discuss a computational approach to represent and evaluate parameter and to approximate morphisms between the base and lumped models within model engineering for
simulation. Here, the base model is typically "more capable" and requires more resources

for interpretation and simulation than the lumped model. By the term "more capable," we mean that the base model is valid within a larger set of experimental frames (with respect to a real system) than the lumped model. Here we note that the terms "base" and "lumped" are terms employed with the framework to denote the full range of possible pairs of models in which the first is more capable (e.g., more detailed, disaggregated, high resolution, fine-grained) than the second (less detailed, aggregated, low resolution, coarse-grained.)

Some typical distinctions often drawn between base and lumped models with respect to agent modeling are presented in Table 1 (Zeigler, 2016; NPS Faculty, n.d.; Bathe et al., 1988).

However, the important point is that within a *particular experimental frame of interest* the lumped model might be just as valid as the base model. Furthermore, the trade-off between the performance and the accuracy (Tekinay et al., 2012) is a fundamental consideration where performance refers to the computational resources used in a simulation run and accuracy refers to the validity of a model with respect to a referent system within an experimental frame (Zeigler et al., 2018). Use of computational resources, tied to a simulator's time and space demands, in generating the model's behavior are correlated with its scope/resolution product (Zeigler et al., 2018).

TABLE 1   Some Fundamental Distinctions Between Base and Lumped Models

	Base Model	Lumped Model
Objectives	1. Results traceable to specific performance data and assumptions 2. Evaluate subtle differences in weapons, sensors, or tactics 3. Understand how different inputs affect combat performance	1. Predict overall results 2. Include small numbers of parameters 3. Parameter values amenable to identification from feasibly obtainable data
Representation	Individual agents as separate entities	Aggregate entities into groups typically respecting command hierarchy
Entity attributes and variables	Location in space and time, position in social or other hierarchies, perception of the situation: threats and opportunities, capabilities, etc. updated at event occurrences or time steps	Averaged entity values attributed to groups Discrete events compounded into rates for groups Global state sets, cross products of individual state sets
Interaction processes	Decomposed into sequences of events and activities Tracking of individual behaviors	Processes aggregated into group level formulae abstracting individual behavior
Timing mechanisms	Coordinate the event sequences for the numerous participants so that subtle interaction patterns can be modeled	Micro stochastic sequences can be aggregated into macro behaviors using law of large numbers expressed more simply in stochastic or deterministic form
Computational complexity (Scope/resolution/ interaction product)	Lean toward large scope, high resolution, and unconstrained interaction	Lean toward smaller scope, low resolution, and constrained interaction

In this chapter we discuss a computational approach to represent and evaluate parameter and to approximate morphisms between pairs of models within model engineering for simulation. We set the foundation for simulation-based computation of metrics for departure from strict structure preservation, tying approximate satisfaction of structural conditions to the resulting behavioral error. In the following, we relate the approach to multiresolution modeling (MRM) and discuss the advantages of maintaining base/lumped model pairs as well as the global organization of repositories around such pairs.

2 MULTIRESOLUTION MODELING

We now place this overview of base model/lumped model concepts within the context of MRM, the construction of a family of models at multiple levels of resolution. Nearly two decades have passed since Davis and Bigelow (1998, 2002) recommended that MRM should become the underlying paradigm behind the development of major defense simulation projects. They stated that MRM is essential for exploratory analysis of military design spaces because it is neither cognitively nor computationally possible to keep track of all relevant variables and causal relationships and the associated resource-based metrics of model complexity grow faster than available computational power. Today, the importance of having models with multiple levels of abstractions is better appreciated and research in the last decades has formed a foundation toward a better understanding of the roadblocks to such endeavors (see Hofmann (2004) for an array of references). However, despite significant advances (Davis and Tolk, 2007; Davis, 1993, n.d.; Davis and Hillestad, 1993, 1992), much research is still needed, for example, in how to use high-resolution models to better understand when deterministic and stochastic aggregations are valid, a paradigm shift away from using such models to solely drive high-fidelity simulations.

A typical multiresolution scenario applicable to defense investigates the operational differences between low-level military entities such as individual tanks and the aggregated high-level units, for example, battalions or platoons when moving in a battlefield. Attributes of an aggregated entity like a tank battalion are often determined by applying an aggregation mapping to the attributes of its individual entities. The mapping can group a set of tanks to a single tank battalion together with a function to derive holistic attribute values, for example, an average speed of a tank battalion, from the constituent individual tank speeds. (Disaggregation is the inverse mapping.) Tekinay et al. (2012) confirmed earlier results that showed significant performance gains within acceptable accuracy when using such spatiotemporal aggregations for battlefield scenarios using different spatial resolutions and scales (Moon, 1996).

2.1 Distinguishing Abstraction, Resolution, and Fidelity for Systems of Systems

However much of such research needs to be better framed to form a stronger foundation for the future. As pointed out recently, simulationists often conflate the terms abstraction, resolution, and fidelity, taking them to be closely or inversely related when in fact closer examination reveals they are largely independent. Using preliminary formalization—*abstraction level is inversely related to the size of a model's structural content, resolution is an ordering* relation based on inclusion of structural content, and *fidelity* refers to the degree of intersection between a model's structural content and that of a "complete" model of the real world—Moon and Hong (2013) show that only two out of six pairwise implications hold. A deeper analysis can be made by employing the MSF to cast the issue into one or more problems concerning the morphisms (structure/behavior preservation relations) between pairs of models viewed as dynamic systems at different levels of specification.

Systems of systems (SoS) may be modeled at different levels of abstraction, resolution, or fidelity but both low and high levels are perhaps equally lacking in their ability to truthfully reflect the system properties. Models may be constructed at lower abstraction levels that support faster simulation and logic-based verification, for example, model checking of properties. Models may also be constructed at abstraction levels that are purported to be closer to reality. In the first case, what is left out may invalidate the proof of a property and in the second case, doubt may be raised concerning the extent and credibility of the available knowledge to support the details demanded by the model (Zeigler and Nutaro, 2016; Zeigler, 2016). In general, the question becomes, how can we ensure that errors introduced in the aggregation/disaggregation processes lie within acceptable limits? Furthermore, interoperation of models at different levels of abstraction presupposes effective ways to develop and correlate the underlying abstractions (Zeigler, 2016; Davis and Tolk, 2007).

In the sequel, we discuss computational evaluation of morphisms within model library organization including execution control of experimentation and simulation for base/lumped model behavior comparison. We then discuss an example of such evaluation involving combat attrition modeling including characterization of approximate morphisms, lumpability zone evaluation, and lumpability zone dependence on parameters. Finally, conclusions are drawn and several implications for model engineering are discussed including the advantages of maintaining base/lumped model pairs in addition to isolated model developments, and the global organization of repositories around such pairs.

3 MORPHISM EVALUATION WITHIN MODEL LIBRARY ORGANIZATION

To set the backdrop for the focus on parameter and approximate morphisms for model library organization we overview the process of model construction and execution as illustrated in Fig. 1. The process begins with the model building objective underlying the simulation study to be performed. This leads to the extraction of particular model and experimental frame components from a repository of such components and the construction of model and experimental frame pair under consideration. The means for such objective-based extraction may be supplied by the System Entity Structure/Model Base (SES/MB) methodology described in the literature (Rozenblit and Zeigler, 1986) and the companion volume (Zeigler et al., 2018). Chapter 18 of this volume (Pawletta and Durak, n.d.) reviews the SES/MB framework and presents an infrastructure based on the SES/MB and some extensions that can be used for modeling and simulation of versatile dynamic systems.

The model and frame pair is an input to the execution process involving simulation and observation as shown in Fig. 1. We breakdown the execution control process in the next

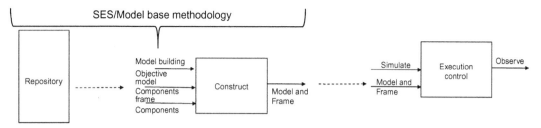

FIG. 1   Sketch of overall process of model building and simulation-based observation.

section in order to lay the basis for the operationalization of the morphism concept as a comparison of base and lumped models within a common (or interconvertible) experimental frame.

However, the subsequent exploration of the model within the frame is not dependent on the genesis process.

3.1 Execution Control of Experimentation

The internals of the execution control of simulation are sketched in Fig. 2. As indicated in Fig. 1, the ModelAndFrame component is derived from the model construction phase and the simulator is attached to it so that simulation functions can be applied to it. Such functions include the ability to start and stop the execution as well moving forward for one iteration (which is equivalent to resuming and pausing). The ability to employ the same standardized simulator with all types of models pulled from the repository is a feature of the Discrete Event

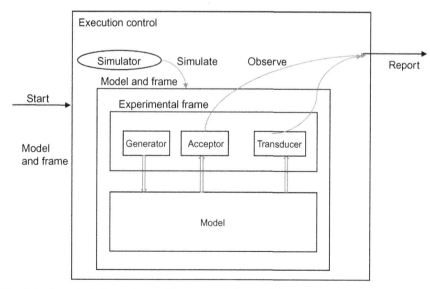

FIG. 2   Simulator in control of experimental frame coupled to model.

System Specification (DEVS) formalism and greatly eases the process but is not an essential requirement. The functions just enumerated, however, are indispensable for the following formulation and may not be easily achieved with legacy software components. The three types of components in an experimental frame, EF are as follows:

Generators—generate the input trajectories to the model.
Acceptors—check conditions for termination.
Transducers—collect data from the model while it is being executed.

More background on the functionality of these components and their modular representation is available in Zeigler et al. (2018). The overall functionality of the experimental frame is not necessarily decomposable into the explicit modular elements depicted in Fig. 2. The explicit form is convenient for portraying the execution control process presented next.

Execution control process (see Fig. 2)

1. Create a Simulator, *sim* and attach it to the ModelAndFrame, *mf*
2. Access the Acceptor, and Transducer (*acc*, and *trans* resp.) components of *mf*
3. Tell *sim* to start (which starts the Generator and thereby the rest of the *mf*)
4. Until all iterations done {
5. For each iteration
6. If simulator is not running return to Step 1
7. Tell *sim* to do one iteration,
8. Query *acc* for termination condition
9. If *acc* returns true {
10. Tell *sim* to stop
11. Query *trans* for results and report them
12. }}

We illustrate the basic computational approach to testing of model abstraction by starting with a simple example. A *homomorphism* illustrated in Fig. 3A is centered on a correspondence between the base and lumped model states. The correspondence is defined by a mapping from the base model states to the lumped model states. The mapping becomes a homomorphism if it is preserved under transitions, that is, for every base state and its corresponding lumped state the respective states to which they transit also correspond to the mapping.

Calling upon the functionalities of the executive control process we show how to computationally test the existence of a prospective homomorphism in the following pseudocode:

LISTING 1 COMPUTATIONAL TEST OF HOMOMORPHISM

1. Create a Simulator, *bsim* and attach it to the BaseModelAndFrame, *bmf*
2. Create a Simulator, *lsim* and attach it to the LumpedModelAndFrame, *lmf*
3. Access the Transducer, *btrans* component of *bmf*

LISTING 1 COMPUTATIONAL TEST OF HOMOMORPHISM   31

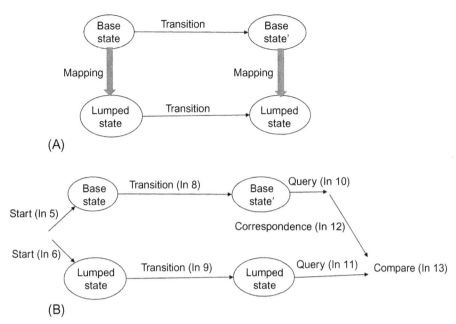

FIG. 3   Homomorphism concept (A) and a computational test pattern (B).

4. Access the Transducer, *ltrans* component of *lmf*
5. Tell *bsim* to start (which starts *bmf* in its initial state)
6. Tell *lsim* to start (which starts *lmf* in its initial state, assumed to be corresponding to that of *bmf*)
7. Until all iterations done {
8. Tell *bsim* to do one state transition of *bm*
9. Tell *lsim* to do one state transition of *lm*
10. Query *btrans* for the current state of *bm*, *bmState*
11. Query *ltrans* for the current state of *lm*, *lmState*
12. *bmState' = Correspondence(bmState)*
13. *Compare: If bmState' ~= lmState, break and report not homomorphic*
14. }
15. Tell *bsim* to stop
16. Tell *lsim* to stop
17. *Report homomorphism confirmed*

Referring to Fig. 3B and Listing 1, lines 1 and 2 create simulators for each of the base and lumped model pairs. In lines 3 and 4, access (e.g., a pointer is defined) to the transducers of each model pair is obtained. Each transducer receives the name of the state of the model it is linked after a state transition. In lines 5 and 6, the simulators are each model pair are started which in turn start their attached models that are set to be in corresponding states. Line 7 starts a cycle in which a number of iterations is performed in which each model is made to take one state transition at each iteration. Successive states are compared for correspondence. If all corresponding state pairs satisfy such preservation of correspondence and of output, a successful confirmation of homomorphism is declared otherwise a failure of homomorphism is declared.

3.2 Control of Simulation for Morphism Evaluation

An elaboration of the overall methodology depicted in Fig. 1 that shows how to implement the test of homomorphism in the pseudocode of Listing 1 is shown in Fig. 4.

The process assumes that parameter values and initial states are inputs that are used to direct it toward meeting the objectives of interest. These parameter values inform the construction of the base and lumped model-frame pairs under the SES/MB methodology. The lumped pair construction may also be informed by output parameters from the construction of the base pair and only known from its construction. After construction of the lumped and base model pairs, simulators are attached and started which starts the models in default states. Once started, the desired initial states conforming to the correspondence to be checked are established and the simulators are executed in iterative fashion according to the pseudocode for homomorphism testing given above.

The homomorphism depicted in Fig. 3 can be stated more generally to be applicable to a wider set of aggregation processes as illustrated in Fig. 5. For such a morphism to hold, the models are set in corresponding states and simulated for possibly different numbers of transitions until corresponding states are encountered. As noted before, all pairs of corresponding states must be congruent in this manner for the morphism to hold. A specific case is where one (macro) transition in the lumped model is realized by several (micro) transitions in the base model.

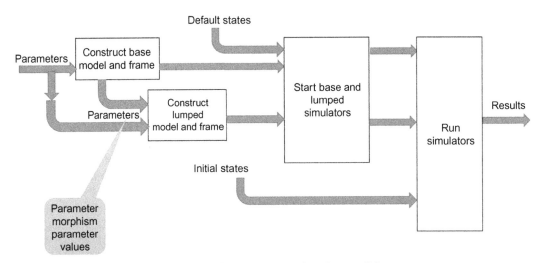

FIG. 4   Control of model and experimental frame simulation for relative validity.

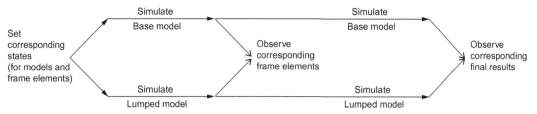

FIG. 5   Generalized morphism between the base and lumped models.

3.3 Implementing Parameter Morphism Evaluation

In the following, we refer the reader to Chapter x of this volume (Pawletta and Durak, n.d.) for an introduction to the SES/MB methodology.

Referring to the execution control process of Listing 1, pseudocode for the comparison of results obtained from the base and lumped models is outlined as follows:

LISTING 2 SES-BASED CONTROL OF BASE AND LUMPED MODEL EXECUTION

1. Specify SESs for the families of base model/frame pair and lumped model/frame pairs, respectively

2. For desired number of samples {

    a. Set the external parameter values for both base and lumped models (including random number seeds)

    b. Prune the SES to obtain a Pruned Entity Structure (PES)

    c. Use the PES to automatically construct base model and frame pair

    d. Extract the parameter values emerging from the base model and set them in the lumped model

    e. Similar to 2(b) and (c), invoke the SES and PES to automatically construct lumped model and frame pair

    f. Call the executive control processes for the base model/frame and lumped model/frame pairs to obtain the results at the termination state

3. }

4. Compare the obtained results (perform statistical computations)

In the next section, we provide an example of lumpability evaluation in the domain of combat modeling. This as an instance of empirically testing the quality of an approximate parameter morphism.

4 EXAMPLE: COMBAT ATTRITION MODELING

Davis (1993) set forth an example of combat simplified model construction that could be understood and worked through in detail. Zeigler (2016) showed that such a pair-of-models approach leads to better results overall than the development of low- and high-resolution models separately and independently. Benefits include the ability to perform mutual cross-calibration avoiding difficulties in harmonization of the underlying ontologies and ability to better correlate system outcome predictions. We revisit a pair of models similar to those developed in Zeigler (2016) where continuous time Markov (CTM) modeling in DEVS (Zeigler et al., 2018) was used to express base and lumped models with an explicit morphism relation between them. Two opposing forces of equal size and unit fire power are arrayed against each other (Fig. 3).

We start with the lumped model which is simpler to work with than the more detailed base model. The state of the lumped model is a pair (n, m) where n and m are the number of units (e.g., tanks) contained within blue and red forces. (We use the blue/red terminology instead of attacker/defender in line with the symmetry of the equal capabilities.) Each force has the same firing rate of FR (e.g., in rounds/min) and assuming units fire independently the combined rate of x units is $x*FR$. Assuming each receiving unit has the same vulnerability—probability of being killed, pk—the probability of y units being killed is $y*pk$. So in state (x,y) the rate of killing the other side is $(x*FR)(y*pk) = x*y*pk*FR$. In CTM modeling terms, the state (m,n) transitions to the states $(m-1,n)$ or $(m,n-1)$ with the same rate $m*n*pk*FR$, respectively (Zeigler, 2017). (Note that, as we will return to later, the kill rates will be different if firing capabilities and vulnerabilities are different.) Here in the usual way, a CTM model assumes that at most one event occurs to cause a transition, namely, one of the units is killed, in any state. The base model treats the units of the opposing forces individually so that the pair (m,n) is disaggregated to a pair of vectors of dimensions m and n populated by 0s (dead) and 1s (alive) which sum to m and n, respectively. The starting state (n,n) is disaggregated to a pair of all 1 vectors of dimension n, for example, $(1,1, …1)$ representing all tanks in the group having the same firing rate. When a unit is killed its corresponding coordinate is set to 0. Assuming uniformity of interaction (from which we back-off soon) the fire is uniformly distributed over the opposing side so that each opposing unit has the same probability of receiving fire and being killed, pk. Thus when there are x blues and y reds, the probability of killing an individual red unit is $x*FR*pk$. Strictly this holds for large numbers of units and then the average number of y red units killed will be $y*FR*pk*x$. Mapping the disaggregated state (xi) (yi) back to lumped model state (x,y) we see that the attrition rate agrees with the lumped model formulation.

This argument is given more rigor by using the concepts of morphism of systems in this case applied to Markov models. Let us consider the situation in a base model state shown in the upper left corner of Fig. 6. The aggregation mapping makes a correspondence to the lumped model state in which the individual units are summed to get the totals of each side still alive. Now a transition in the lumped model will be an event in which one of the units is killed thereby reducing the total by 1 as shown (shown as a blue casualty, but the same holds for a red one). If there is a transition in the base model such as shown which kills a particular blue unit in the same average elapsed time then the aggregation mapping will continue to relate base and lumped states. By induction on transitions, the average time to reach a terminating state (in which one of the side's numbers reach zero) for each model is the same.

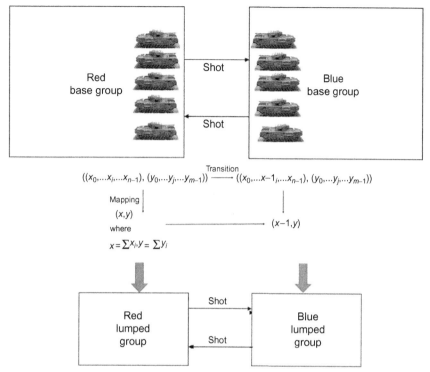

FIG. 6 Opposing tank forces with attrition morphism.

4.1 Approximate Morphisms

Let us examine the condition that justifies the homomorphism, called the *lumpability* condition. Considering fire from the blue to the red side, uniformity can be formulated as the following requirements: (a) emitted fire uniformity: every blue unit distributes the same total fire volley to the red group; (b) received fire uniformity: every red unit receives the same fire volley from the blue group. While the effect of departure from uniformity due to couplings between pairs of units, here we assume such coupling uniformity pertains but that the base model vulnerability parameters, pk_i may not be the same for all units i within a force group.

To consider such heterogeneous distributions of values, we set up a mapping from base model parameters to lumped model parameters that can be defined as follows:

For each block of components (blue and red force groups) the mapping of vulnerability parameters is an aggregation which sends the individual values to their average (mean) value:

$$F_{\text{Block}} : \text{Base Parameters} \rightarrow \text{Lumped Parameters}$$
$$F_{\text{Block}}\left(\{pk_i|\ i=1,..,N_{\text{Block}}\}\right) = pk_{\text{Block}} = \text{Mean}_{\text{Block}}\left(\{pk_i|\ i=0,..,N_{\text{Block}}\}\right)$$

where pk_i are the individual vulnerability values in a block and pk_{Block} is their mean value assigned to the lumped block.

This is a particular instance of a *parameter morphism*, where the morphism not only sets up a correspondence between states but also maps base model parameter values to lumped values for which the correspondence is expected to hold.

Since the morphism is expected to hold exactly only when exact uniformity within bocks holds, we measure the departure from uniformity by employing

$$\text{LumpedSTD}_{\text{Lumped}} = \text{STD}\left(\{pk_i | \ i = 0, .., N_{\text{Block}}\}\right)$$

namely, the standard deviation relative to the mean. Note that strict uniformity holds (all block vulnerability parameter values are equal) if, and only if, the standard deviation vanishes as required for such a measure.

The relaxed morphism concept, enabled by allowing departure from uniformity in parameter mappings, is called an *approximate morphism* (Zeigler et al., 2018). To operationalize this concept, we need to characterize the error we can expect from a lumped model which assumes exact uniformity. In other words, an approximate morphism must also be accompanied by information that indicates the departure from exact homomorphism in relation to the departure from uniformity. This information can be compiled into a *lumpability zone* statistic that characterizes the neighborhood of the LumpedSTD near zero in which acceptable predictions of the lumped model may be obtained.

4.2 Lumpability Zone Evaluation

To illustrate the lumpability zone information about an approximate morphism that might be carried along with its definition, we performed an empirical investigation of the attrition model in Fig. 6 using the process outlined in the SES-based comparison of results obtained from the base and lumped models. Referring to Fig. 4, in this case, the external parameter values passed to both base and lumped models are the mean vulnerability and firing rates. The parameter values emerging from the construction of the base model and passed to the lumped model are the initial sizes of the opposing forces.

All runs started with 100 blue units while the number of red units ranged from 40 to 60. The fire rate was uniform at 0.1 shots/unit time and the same for both sides and the output to input couplings were uniform all to all. However the vulnerability was variable, sampled from a beta distribution with mean 0.5 and standard deviation (std) ranging from 0.05 to 0.25 in steps of 0.05. The beta distribution std. was verified to give a good prediction of the LumpedSTD for the vulnerability actually observed. Note: The largest variance possible for the beta distribution with mean 0.5 is 0.25. Also, since the beta distribution does not accept 0 as a std., the starting point for the base model bypassed the distribution and set the vulnerability to a constant 0.5. Overall, 30 pseudorandom samples were taken for each setting of the base model. The lumped model was constructed with the same fire rate and mean vulnerability as the base model and 30 samples were taken for each red/blue configuration as determined by the base model. The mean and standard deviation of the termination times obtained for the base model are presented in Table 2 and drawn in Fig. 7. We see that the termination

TABLE 2   Termination Time Comparison for Base and Lumped Models

LumpSTD	Base Mean Termination Time	Base STD Termination Time	Lumped Mean Termination Time	Error	%Error
0	184	126	182	2	1.09
0.05	149	90		−33	−22.15
0.1	154	100		−28	−18.18
0.15	159	117		−23	−14.47
0.2	354	195		172	48.59
0.25	568	488		386	67.96

Remark: *If base and lumped disagree it could be that the base is at fault. In the present case, at first we failed to note the restriction of the beta function to a nonzero standard deviation. Setting this parameter to zero led to very small termination times for the base model relative to the lumped model. This led us to search for the cause of the disparity and eventually led to detecting the source of the error as violation of the beta function parameter in the base model.*

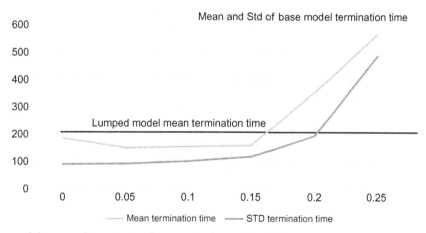

FIG. 7   Lumpability zone determination for opposing force ratio of 2:1.

time sharply rises after the LumpSTD of 0.15 suggesting that the lumpability zone is in the region [0,0.15]. The lumped model mean termination time was 182 which agrees well with the base model with fixed vulnerability. We note that it is higher than the values within the zone but lower than those outside it. We note that the standard deviation of the lumped model termination time is high compared to the mean, as are those of the base model. This indicates that the processes are highly variable (cf. a log normal distribution where the standard deviation can be much higher than the mean).

4.3 Lumpability Zone Dependence on Parameters

The formulation of morphisms in terms of shared (or interconvertible) parameters, and the computational implementation just discussed, promotes exploration of the quality of

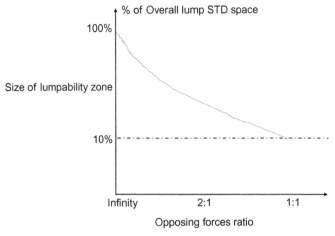

FIG. 8   Dependence of relative size of lumpability zone on model parameters.

approximation over parts (or all) of the parameter and state spaces of the models. For example, as illustrated in Fig. 8 the size of the lumpability zone significantly depends on parameter values. The relative size of the zone for the attrition example with 2:1 force ratio in Fig. 8 is approximately 60% (=0.15/0.25). This size shrinks when the forces are equal in strength because of the sensitivity to nonuniformity when firing rates and vulnerabilities are equally balanced. On the other hand, the size expands to 100% when one side completely dominates the other and the sensitivity to small variations becomes only a second-order effect.

Another look at the dependence of approximate morphism quality is illustrated in Fig. 9. Here the mean termination times of the base and lumped models are plotted for increasing opposing force sizes. Agreement is very close for sizes larger than 50, while the approximation is not so good for smaller numbers. Note that termination times increase to a maximum for small numbers and then decrease with increasing sizes. This suggests that the increase in aggregate fire intensity with increasing shooters overcomes the low vulnerabilities to result in high-casualty rates. Interestingly, the simulation run times for the base model grow much

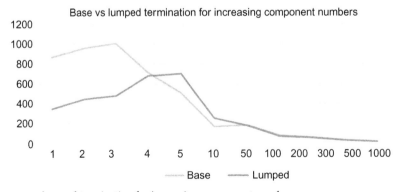

FIG. 9   Base versus lumped termination for increasing component numbers.

more rapidly for larger sizes than for the lumped model. Taking the lumpability zone as [100,1000] we see that in this zone we have high accuracy with short executions of the lumped model without trade-offs.

The computational approach for the evaluation of approximate morphism error makes it possible to quantify, test, and characterize the extent to which, and conditions under which, aggregations and simplifications of simulations may be usefully employed.

5 CONCLUSIONS AND FURTHER RESEARCH

5.1 Summary

This chapter presented a computational approach to evaluate the approximate morphisms (structure/behavior preservation relations) between pairs of models within model engineering for simulation. Based on the SES/MB methodology and multifaceted modeling we set the foundation for simulation-based computation of metrics for departure from strict structure preservation, tying approximate satisfaction of structural conditions to the resulting behavioral error.

5.2 Implications for Research in Model Engineering

There are several implications for model engineering.

5.2.1 Creation of Base/Lumped Pairs

As suggested base/lumped model pairs offer significant advantages over isolated model developments. We first review such advantages before proceeding to discuss the global organization of repositories around such pairs. The methodology offers an approach seeking and focusing on lumping processes with high effectiveness. This should inform the development and performance measurement of automated aggregation algorithms to construct good lumped models. The dependence of error on departure from strict structure preservation, once quantified, can then drive the choice of level of resolution (given a level of tolerable error, the maximum allowable reduction in resolution can be determined).

The implications of this characterization of approximate system morphisms for multiresolution modeling and simulation can be enumerated as follows:

- From the examples studied, approximate lumpability as measured by lumpability zone size may be quite restricted but nevertheless much less restrictive than strict lumpability.
- In view of this, the proposed process can support seeking to develop, and focusing on, processes with high effectiveness (low error with respect to uncontrolled construction).
- The relative volume of the lumpability zone can be taken as the probability of getting a useful approximation using the lumping process. The smaller this probability, the greater the challenge for automated aggregation algorithms to construct good lumped models. The performance of such a method should be measured against the expected number of random trials to hit upon a lumped model within the lumpability zone which is the inverse of the lumpability ratio.

- The size of the lumpability zone significantly depends on the parameter values. Although potentially involved in computation performance/accuracy trade-offs, it may happen that in a lumpability zone we have high accuracy with short executions of the lumped model without trade-offs.

5.2.2 Organization of Repository Around Base/Lumped Model Pairs

Once created, base/lumped model pairs need to be stored in the repository for retrieval and use. Morphisms between pairs automatically organize models into lattice-like structures that allow inference of induced pairings by transitivity. Focusing on experimental frames, it is critical to have an ability to ask whether there are any frames that meet given objectives and whether there are models that can work within such frames. The relation that determines if a frame can logically be applied to a model is called *applicability* and its converse is called *accommodation.* Notice that validity of a model in a particular experimental frame requires, as a precondition, that the model accommodates the frame. The degree to which one experimental frame is more restrictive in its conditions than another is formulated in the *derivability* relation. A more restrictive frame leaves less room for experimentation or observation than one from which it is derivable. So it is easier to find a model that is valid in a restrictive frame for a given system. In fact, applicability may be reduced to derivability. Let the *scope* frame of the model represent the most relaxed conditions under which it can be experimented with (this is clearly a characteristic of the model). Then a frame is applicable to a model, if it is derivable from the scope frame of the model. Thus parameter morphisms interconnect within an overarching structure of derivability relations.

The morphism-based organization just discussed can extend a recently proposed MSF to support a holistic analysis of health-care systems (Traoré et al., 2017). This provides a stratification of the levels of abstraction into multiple perspectives and their integration in a common simulation framework. In each of the perspectives, models of different components of health-care system can be developed and coupled together to accommodate experimental frames that are characteristic of these perspectives. Influences of other perspectives on the parameter values of any one of them are reflected through explicit assumptions and simplifications. Components of the various perspectives are integrated to provide a holistic view of the health-care problem and system under study. The resulting global model can be coupled with a holistic experimental frame to derive results that cannot be accurately addressed in any of the perspective taken alone.

5.2.3 Tool Sets for Parameter Morphism Evaluation and Model Base Organization

Creation of base/lumped pairs and their imposition of an organization of models in a repository need to be supported by appropriate data structures and tool sets. Support for such tools could consist of implementations of operations and workflow orchestrator implications such as those roughly suggested in Table 3.

5.2.4 Time Scale Relations and System Entity Structure

Santucci et al. (2016) propose an extension of the SES that introduces new elements to enable it to support the concepts of abstraction and time granularity in addition to its existing expression of composition and component specialization. In relation to abstraction, we note that the approach of this chapter is to consider a *pair* of models, base and lumped, with a

TABLE 3   Example Operations and Workflow Orchestrators

	Simulation Control	Evaluation
n-array operations e.g., Operation(base,lumped, ef)	makeSimulation(sim,base,lumped,ef) StartSimulation(sim) RunSimulation(sim,iteration) getResults(sim)	Compare(base,lumped,ef) ComputeLumpabilityZone(base, lumped,ef) ComputeBackgroundLevel(base, ef)
Workflow implications	StartSimulation(base,lumped) => StartSimulation(base) and StartSimulation(lumped) etc.	CompareWith(base,lumped) => StartSimulation(base,lumped)

morphism relation between them. To extend the SES to represent such pairs, and indeed hierarchies of such pairs, amounts to compressing the separate SESs for such models into a family of models represented by a single SES. To do this Santucci et al. introduced two types of specialization relations for abstraction and time granularity and associated specification of mappings. Further research is required to investigate whether there are better ways to manage abstraction and time granularity with the SES, possible implementation approaches, formalizations, and computational support along the lines of this chapter.

We note that Chapter x of this volume (Pawletta and Durak, n.d.) offers a computational infrastructure that can support the operations needed in Table 3.

References

Bathe, M.R., Maxwell, J.G., McNaught, K.R., 1988. Modelling Combat as a Series of Mini-battles. Royal Military Coll of Science Shrivenham (United Kingdom).

Davis, n.d. Stochastic Multi-Resolution Model (PEM) Calibrated to a High Resolution Simulation. MR-1138-OSD, The RAND Corporation, Santa Monica, CA.

Davis, P.K., 1993. An Introduction to Variable-Resolution Modeling and Cross-Resolution Model Connection. RAND, Santa Monica, CA (R-4252-DARPA).

Davis, P.K., Bigelow, J.H., 1998. Experiments in MultiResolution Modeling (MRM). RAND, Santa Monica, CA. MR-100-DARPA.

Davis, P.K., Bigelow, J.H., 2002. Motivated Metamodels: Synthesis of Cause-Effect Reasoning and Statistical Modeling. The RAND Corporation, Santa Monica, CA.

Davis, P.K., Hillestad, R., 1993. Families of models that cross levels of resolution: issues for design, calibration and management.Proceedings of the Winter Simulation Conference. pp. 1003–1012.

Davis, P.K., Hillestad, R.J. (Eds.), 1992. Proceedings of Conference on Variable Resolution Modeling, CF-103-DARPA. RAND's National Defense Research Institute.

Davis, P.K., Tolk, A., 2007. Observations on new developments in composability and multi-resolution modeling.Proceedings of the 2007 Winter Simulation Conference, pp. 859–870.

Hofmann, M.A., 2004. Criteria for decomposing systems into components in modeling and simulation: lessons learned with military simulations. Simulation 80 (7–8), 357–365.

Moon, I.-C., Hong, J.H., 2013. Theoretic interplay between abstraction, resolution, and fidelity in model information. Winter Simulations Conference (WSC), pp. 1283–1291. https://dx.doi.org/10.1109/WSC.2013.6721515.

Moon, Y.K., 1996. High Performance Simulation-Based Optimization Environment for Large Scale Systems (Doctoral Dissertaion). University of Arizona.

NPS Faculty, Aggregated Combat Models. http://faculty.nps.edu/awashburn/Washburnpu/aggregated.pdf. [(Accessed 3 April 2017)].

T. Pawletta, U. Durak, n.d. Modeling and Simulation of Versatile and Adaptable Systems with an Application in Engineering, Chapter 18 of this volume.

Rozenblit, J.W., Zeigler, B.P., 1986. Entity-based structures for model and experimental frame construction. In: Modeling and Simulation Methodology in the Artificial Intelligence Era. North-Holland, Amsterdam.

Santucci, J.F., Capocchi, L., Zeigler, B.P., 2016. System entity structure extension to integrate abstraction hierarchy and time granularity into DEVS modeling and simulation. Simulation 92 (8), 747–769.

Tekinay, C., Seck, M.D., Verbraeck, A., 2012. Exploring multi-level model dynamics: performance and accuracy (WIP).Proceedings of the 2012 Symposium on Theory of Modeling and Simulation, San Diego, CA, pp. 20:1–20:6.

Traoré, M.K., Zacharewicz, G., Duboz, R., Zeigler, B., 2017. Modeling and simulation framework for value-based healthcare systems. Simulation. https://dx.doi.org/10.1177/0037549718776765.

Zeigler, B.P., 2016. The role of approximate morphisms in multi-resolution modeling: can we relax the strict requirements? J. Def. Model. Simul. https://dx.doi.org/10.1177/1548512916659826.

Zeigler, B.P., 2017. Constructing and evaluating multi-resolution model pairs: an attrition modeling example. J. Def. Model. Simul. 14(1).

Zeigler, B.P., Muzy, A., Kofman, E., 2018. Theory of Modeling and Simulation, third ed. Academic Press, San Diego, CA.

Zeigler, B.P., Nutaro, J.J., 2016. Towards a framework for more robust validation and verification of simulation models for systems of systems. J. Def. Model. Simul. 13 (1), 3–16. https://dx.doi.org/10.1177/1548512914568657.

Unified Approaches to Modeling

Mamadou K. Traoré

University of Clermont Auvergne, LIMOS CNRS UMR 6158, Aubière, France

1 INTRODUCTION

During model development of complex systems, two questions arise at the early stage of the process:

- A model for what?
- How to specify it?

1.1 Model for What?

Models are built to support system requirements, design, analysis, verification, and validation activities beginning in the conceptual design phase and continuing throughout development and later life cycle (Estefan, 2007). The core underlying principle is that the efficient design and development of a complex system requires an iterative process of modeling, performance evaluation, logical analysis for requirement verifications, and implementation (prototyping) for runtime testing (Hong and Kim, 2006). This iterative process is necessary to reveal subtle knowledge about the systems, which are, in most cases, beyond intuition. Moreover, a violation of requirement(s) or undesired behavior at this stage can be a signal of a fundamental flaw in the design that must be resolved early to forestall costly errors in the final system (Zeigler and Nutaro, 2016). Therefore, the system under study (which may be a physical system or a nonexistent conceptualized system) is represented as a model, possibly consisting of interacting components, using an appropriate formalism. The information represented in the model, and the choice of an appropriate formalism to write the model are determined by a number of factors such as the questions to be answered about the system, the properties of the system to be analyzed, the capabilities of the available model solver, the analyst's experience, etc. A set of algorithms and protocols, referred as the model solver, manipulate this model to generate some results. These results provide feedback as more knowledge about the system under study to the analyst. Hence, the knowledge gained from each iteration may serve as a guide for a deeper understanding of the system's behavior and the influences it may have on its environment or which the environment may have on it. Alternatively, it may serve as a feedback to revise the designed model and/or requirements until an acceptable level of satisfaction of critical requirements is guaranteed before committing time and resources to the implementation of the system. Efforts to automate such as process are referred to as Model-Driven Engineering (Mittal and Martín, 2013; Bocciarelli and D'Ambrogio, 2014).

As depicted in Fig. 1, depending on the questions to be answered about the system under study, models employed in the iteration loops to mine the desired knowledge of the system are often built for one of the three major analysis methodologies: (1) simulation, (2) formal analysis, and (3) enactment. These methodologies promote reasoning about systems from somewhat divergent viewpoints. A viewpoint can be defined, in the general context of software and systems engineering, as a description of appropriate machinery consisting of domain, languages, specifications, and methodologies to capture and process one or more related engineering or technical concerns about a system and the information associated with such concerns (Finkelstein et al., 1992; Kurpjuweit and Winter, 2007). However, often they are necessarily required in combinations for sound system designs.

Simulation allows compressing time (i.e., use logical time approximation) and evaluating or analyzing a model over a specified period and under scenarios or environment defined by the experimental frame (EF). In modeling and simulation (M&S), an EF defines the objective(s), assumptions, and constraints of a simulation study and the context within which a system is observed or the validity of the model is evaluated (Zeigler et al., 2000; Traoré and Muzy, 2006). The results obtained from a simulation study may be used to infer or forecast a system's behavior and performance, identify problems and their causes, etc. Comprehensive lists of problems that are suitable for simulation and the possible uses of simulation results are provided in Maria (1997) and Carson (2004).

Formal analysis methods (FM) are mathematically based languages, techniques, and tools that permit the specification, verification, and development of software and hardware systems in a systematic manner (Wing, 1990; Clarke and Wing, 1996). The goal of FM is to unveil and correct subtle and very expensive errors that may result in system failures (Clarke and Wing, 1996). The likelihood of having such errors in a system design increases as the system grows in scale and functionality; hence, the need for a systematic use of FM at some strategic

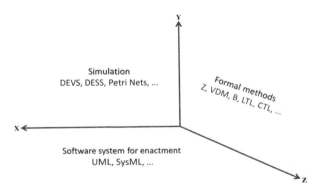

FIG. 1   Isolated analysis practices and some related modeling formalisms.

phases of a systems development process to enhance the developers' understanding of the system through early revelations of ambiguities, inconsistencies, and violations of specification requirements. The overall motivation for this will be the eventual construction of complex systems that operate reliably (Clarke and Wing, 1996). Similar to the way an EF defines the objectives and context of a simulation study of a system, the requirement proposes the premise on which the logical and symbolic reasoning about a system is based using FM. In essence, the requirement is an abstract specification of a collection of properties (or evaluation criteria) that need to be satisfied by a system under study (Lamsweerde, 2000). It serves as a contract between the client and system developers (Wing, 1990). The logical analysis of the combined operational model of the system and requirement helps to verify that the former satisfies the latter. In some cases, counterexamples are generated to illustrate the violations where requirements are not met. An empirical survey of the whys and wherefores of using FM is presented in Hall (2007).

Enactment is, in a general software engineering context, the mechanism for the execution or interpretation of software process definitions, which may involve live interactions with the environment and external actors like human-in-the-loop, to provide supports that are consistent with the process definitions (Dowson and Fernström, 1994). It is most known in business process management (BPM) (Van Der Aalst et al., 2003; Jeston and Nelis, 2014) as the execution of a business workflow (Kouvas et al., 2010), that is, the (semi) automation of business processes during which information and work lists are passed from one participant to another for necessary actions (Ottensooser and Fekete, 2007). In service engineering and human-computer interaction, enactment refers to the manifestation of the functionalities represented by a system's prototype (Holmlid and Evenson, 2007). A system prototype is described as a representation of the functionality, but not the appearance, of a finished artifact that can be used as a proof that a certain theory, concept, or technology works, or otherwise (Holmquist, 2005). Here, we describe enactment as executing a system operational model, that is, one that can be executed in a suitable software environment (Bruno and Agarwal, 1995), to act out its behavior by using the physical clock time (as opposed to the simulation logical time), as the reference for the scheduling and execution of events (Aliyu et al., 2015). An enactment model should practically stand in for the real system in a physical environment through the manifestation of its expected characteristics.

1.2 How to Specify It?

It is important to see the relationship between modeling formalisms and analysis methodologies. The latter are concerned with the overall process that makes use of models, while the former are tools used among others within this process to describe models. Therefore, the use of a unique formalism or multiple formalisms (which refers to multiformalism modeling) is guided by the objectives of the underlying analysis. We focus on unified modeling approaches in that context. We classify the unified modeling approaches in the literature into two categories based on whether they fall within one of the hyperplanes of Fig. 1 or are integrating two or three hyperplanes. They are:

- Unified analysis-specific modeling approaches, which aim at facilitating analysis in one specific hyperplane (i.e., simulation, formal methods, or enactment)

- Unified multianalysis modeling approaches, which combine methods from different hyperplanes of Fig. 1 (e.g., simulation combined with FM, or simulation combined with enactment, or enactment combined with FM, or all three methods combined).

In the remaining part of this chapter, we review contributions to unified modeling approaches, with this distinction in mind. Section 2 presents the methodology-specific unified modeling approaches, while Section 3 presents the multianalysis unified modeling approaches. Section 4 concludes the chapter.

2 UNIFIED ANALYSIS-SPECIFIC MODELING APPROACHES

Unifying approaches under this category reside in isolation in one of the three orthogonal hyperplanes of Fig. 1. The fundamental objective, which is shared by most approaches in this category, is usually to alleviate the complexity and rigor of direct system specification with the underlying formalism through high-level modeling interfaces. It is particularly meant to address the lack of requisite logic and mathematical skills to deal with most formalisms. Moreover, these approaches are generally motivated, inter alia, by the possibility of making the underlying formalisms and their capabilities accessible to wider communities including nonexpert users, ease of communication among stakeholders, and (semi) automated synthesis of executable analysis codes from the high-level models.

Usually, the development of unified modeling formalisms in this category follows one of the following prevalent styles: (1) interfacing, (2) federation, and (3) subsumption. Even if frontiers between these styles are thin, each has salient lineaments that help to identify whether a given approach falls under it or not.

2.1 Formalism Interfacing

A practical way to approach the unification of heterogeneous specifications is to interface them; that is, to use a third party as a glue to combine them. In his seminal work identified as multimodeling (Fishwick, 1995), Paul Fishwick proposes to interface different models by means of the object-oriented paradigm, where each model is encapsulated in an object, and intermodel communication is ensured by method calls.

Multimodeling brings an interesting light to the question of multiformalism, by proposing a classification of abstractions and indicating, for each class, a family of adequate formalisms:

- Conceptual models (CMs) describe, at the highest level of abstraction, the general knowledge about the system under study. Their purpose is to highlight system entities (or classes) as well as the relationships between them. They therefore constitute a knowledge base for subsequent abstractions. UML or natural languages are examples of adequate formalism for this abstraction level.
- Declarative models provide systems description in terms of sequences of events that move the system from one state to another. This level of abstraction is adequate to represent the dynamics of objects identified in the CM. Sequential automata (for passive objects) or Petri nets (for active objects) are adequate formalisms for this abstraction level.

- Functional models provide system descriptions in terms of sequences of functions (the outputs of the ones feeding the inputs of the others), arranged such that the final treatment performed by the system is found downstream of this sequence. Such a level of abstraction specifies the flow of processing within the system. Queueing networks or functional block diagrams are examples of adequate formalism for this abstraction level.
- Constraint (or conservative) models provide system descriptions in terms of laws and constraints. Differential equations or difference equations are examples of adequate formalism for this abstraction level.
- Spatial models focus on describing the decision rules for systems operation. The main idea, at this level of abstraction, is to represent global decisions as the fruit of the interactions of multiple local decisions. Cellular automata or multiagent systems are examples of adequate formalism for this abstraction level.

Fishwick suggests first building a CM of the system, and then breaking this model into declarative/functional/constrained/spatial hierarchies. The operational semantics of the resulting multimodel is obtained by the object-oriented implementation and integration of time management mechanisms (i.e., sequential/parallel simulation algorithms), or the use of simulation tools/environments such as SIMPACK (Fishwick, 1995). One can notice that even if Fishwick has developed his approach in the context of simulation, the methodology still stands from both the enactment and the formal analysis perspectives. A notable example in formal analysis is µSZ (Bussow et al., 1997).

2.2 Formalism Federation

Formalism federation consists of adding the existing formalisms and consecrating this sum as a new formalism. This comes from the desire to unify different (and possibly overlapping) views of the same system (i.e., different sets of interrelated concepts) that are traditionally expressed in different formalisms. A classic categorization of views is the separation between static, dynamic, and functional views. Another classification, more system-theoretically oriented is the distinction between structure and behavior. Prominent federated formalisms are UML and SysML.

UML, for Unified Modeling Language (Rumbaugh et al., 2004), is a standard object-oriented notation for software systems design, with the underlying unifying approach is to integrate static, dynamic, and functional views of a system, independent of any specific software development process (e.g., XP, RUP) or technology (e.g., Java, .NET). It also involves a profiling mechanism, which allows building generic extensions for customizing UML for particular domains and platforms. Class diagrams, component diagrams, and package diagrams mainly provide the static view in UML. Activity diagrams, sequence diagrams, state machine diagrams, and collaboration diagrams mainly provide the dynamic view. Use case diagrams mainly provide the functional view. It is known that these diagrams do not have formal semantics and no precise operational semantics is defined to make them executable. For example, activity diagrams semantics are described by informal texts. Each existing tool for modeling activity diagrams implements its own operational semantics. This implementation freedom is intentional since the goal of UML is to provide a unified framework for various application domains (like embedded systems, real time systems, or computer networks) with different needs. This freedom makes the validation problem of UML models dependent form semantic choices of available tools. Some tools with limited capabilities exist for validating OCL and UML specifications (Martin et al., 2007). The adequate use of OCL constraints helps a lot to avoid ambiguities, but even then tool support is lacking. A more precise subset of UML, called Foundational UML (fUML) (Object Management Group, 2013), is defined for specification of executable UML models. The fUML subset defines a semantics called base semantics. However, a small subset may make the formal verification of models feasible or decidable, but at the same time, it restricts the expressiveness of the language. In the opposite, a large subset is very expressive in modeling; however, it is difficult or impossible to do formal verification. The way of combining UML diagrams, OCL, and action semantics to deliver complete implementation code (mostly in Java, C#, and C++) for a system is not part of the UML standard leading to individual solutions that fail to consistently keep the link between the diagrams and the generated code.

SysML, for Systems Modeling Language (Object Management Group, 2010), is a standard graphical modeling language that customizes UML (by the profile mechanism) for the specification, analysis, design, verification and validation of systems including hardware, software, and processes. SysML is more expressive and flexible than UML. SysML reuses a subset of UML (activity diagrams, use-case diagrams, sequence diagrams, state diagrams, and package diagrams), and adapt class diagrams to be block definition diagrams. SysML also adds new diagrams (requirement diagrams, parametric diagrams). A broad range of UML tools supports SysML, and its diagrams can be exchanged between tools by using the standard XMI format. The advantages of SysML over UML for systems engineering is that SysML requirement diagrams can be used to efficiently capture functional, performance, and interface requirements, whereas UML is subject to the limitations of use-case diagrams to define high-level functional requirements. Also, SysML parametric diagrams can be used to define performance and quantitative constraints precisely, while UML provides no straightforward mechanism to capture this sort of essential performance and quantitative information.

2.3 Formalism Subsumption

Formalism subsumption is the act of defining a model of computation (rather than a language) that subsumes the semantics of different formalisms. In the context of simulation, an important step was taken in this approach with the advent of the DEVS (for Discrete Event Systems Specification) paradigm, which laid the foundation of a theory of M&S (Zeigler, 1976). Rooted in systems theory, DEVS proposes a unifying framework for all discrete-event systems specification formalisms (Vangheluwe, 2000). Various DEVS extensions have been proposed to deal with other kinds of systems, including the DEV&DESS hybrid systems modeling formalism (Zeigler et al., 2000), the DSDEVS variable structure systems modeling formalism (Barros, 1997), and a host of others.

A prominent DEVS-based methodology is DUNIP (Mittal, 2007; Mittal and Martín, 2013). DUNIP, for DEVS Unified Process, aims at exploring the integration of various concepts that had been developed through decades of research in DEVS-based simulation methodology, and applying it to the design and analysis of Systems of Systems in a full systems engineering life cycle. The overall objective of the framework is to harness the benefit of automated transformations in MDE to bind different phases of a rigorous MBSE process backed by the DEVS theory for a transparent simulation of Systems of Systems in a net-centric M&S setup. The fundamental visions of and MDSE processes in DUNIP are comprehensively captured in the DEVSML 2.0 stack (Mittal and Martín, 2013), which is a multilayered architecture with infrastructures at the different layers that work together to realize the M&S-based verification of discrete-event systems on a stand-alone or net-centric simulation platform. The top layer of DEVSML 2.0 stack contains the DEVS Modeling Language version 2.0 (DEVSML 2.0), a textual DSL for expressing systems' structure and behavior based on DEVS theory. A model written in DEVSML 2.0 is persisted in XML and considered as a Platform-Independent Model (PIM) to make it compatible with DEVS-based simulators implemented in multiple platforms. The PIM in the DEVSML 2.0 layer is transmitted through some DEVS-compliant middleware and APIs (in the lower layers) to a net-centric infrastructure that generates and deploy platform-specific simulation codes on a distributed multiplatform federation of simulation engines based on Java, C++, etc. at the bottom of the stack. DEVSML 2.0 also serves as the interface to integrate DUNIP with Domain-Specific Languages (DSL) such as Business Process Modeling Notations (White, 2004), UML, and SysML, so that the domain experts can create system models in the problem domains and transform them (semi) automatically to DEVSML 2.0-compatible format. In essence, DUNIP fosters the federation of diverse DEVS-based simulation engines to provide a transparent simulation support for DSLs via a net-centric virtual machine. Hence, it shields the modeler from the rigor of direct system specification with raw DEVS constructs through high-level concrete syntax for DEVSML 2.0 and the possibility of transforming domain-specific models into the XML format of DEVSML 2.0 for onward transformation to DEVS-based simulation codes.

3 UNIFIED MULTIANALYSIS MODELING APPROACHES

One significant benefit of the approaches presented in the previous section is the separation of analysis concerns between hyperplanes of Fig. 1. This means that the modeler in each hyperplane creates a specification of the system under study from the point of view of the hyperplane while abstracting away from the peculiarities of the methodologies in the other two hyperplanes, thereby leading to a considerably simplified and focused model in each hyperplane. However, since a comprehensive study of a system often requires combinations of multiple analysis methodologies, one would likely need to create separate models of the same system in each of the three orthogonal hyperplanes. The implication of this is that disparate models of the same system would be needed. Dealing with multiple disconnected views of the system in the different hyperplanes can be susceptible to miscommunication among domain experts (Bajaj et al., 2012). Moreover, the creation and, more importantly, the repeated update of the different models during the iterations of analysis processes can be hard since any change in the system variable may require that all models in the different hyperplanes be manually updated.

The required synergy of different analysis approaches can be achieved through a disciplined combination of them, so that they provide complimentary, rather than competitive, answers to evolving design questions. Recent publications such as Tolk and Hughes (2014) and Bocciarelli and D'Ambrogio (2014) suggest that there are growing interests, both in academia and industry, in the systematic combinations of disparate MBSE approaches to maximize the synergy between the different disciplines. The practical adoption of this collaborative approach to computational analysis of systems is, however, being constrained directly or indirectly by the same forces inhibiting the wide adoption of individual analysis methodologies especially by nonexpert users. A number of reasons have been identified in the literature for this shortfall. Chief among them are the following:

- Computational analysis methodologies often rely on some mathematics- and/or logic-based formalisms for the specification and manipulation of systems. This is necessary to ensure formal reasoning with models with precise semantics and devoid of ambiguities and inconsistencies. Domain users, however, seldom have the requisite skills to deal with such formalisms; they are considered as low-level expressions that do not match with the high-level artifacts which domain users are often accustomed to. Therefore, high-level modeling interfaces are required on top of families of formalisms, to make them accessible to nonexpert users. Surveys of some of such interfaces for discrete event simulation (Franceschini et al., 2014) and FM (Kefalas et al., 2003) highlight their features. It is important to note, however, that the tools vary in their capabilities to express different aspects of complex systems. Hence, accessibility to nonexpert users is still open to further research.
- Another source of concern which directly constrains the study of a system using multiple analysis methodologies is that there are usually little chances of portability of models between methodologies. This usually requires manual, or at best semiautomated, creation and updating of separate models, in different formalisms, of yet the same system for different kinds of analysis. This task can be time consuming and error prone.

These problems have been continuously acknowledged, and addressed through unifying modeling approaches, including the following two prevalent styles: (1) formalism transformation and (2) formalism weaving.

3.1 Formalism Transformation

Arguably, because of the limited chances of sharing system models among disparate analysis methodologies, different techniques rarely coexist in the same environment. Formalism transformation addresses the pairwise integration of system models and MBSE processes in the different hyperplanes of Fig. 1. This is usually done by identifying correspondences between elements of chosen formalisms in the different hyperplanes, in order to define the mapping rules between them, so that a model in one hyperplane may be used to drive the (semi) automated synthesis of models in other hyperplanes. Examples of such combined use of simulation and formal analysis include Kuhn et al. (2003), Traoré (2006), Trojet and Berradia (2015), and Yacoub et al. (2014). Similarly, several proposals have been made to bridge the gap between enactment and FM (Lano et al., 2004; Lilius and Paltor, 1999; Shah et al., 2009; Laleau et al., 2010), while some efforts to achieve pairwise integration of simulation and enactment are reported in Kapos et al. (2014), Risco-Martín et al. (2009), Schamai et al. (2009), and Shaikh and Vangheluwe (2011).

Noteworthy is the model-driven development for modeling and simulation (MDD4MS) approach proposed by Çetinkaya in her doctoral thesis (Çetinkaya, 2013). She argued that though the importance of simulation CM to the accurate development of M&S models was widely acknowledged in the literature, the systematic transformation of CMs, through intermediate models, to executable simulation models had not been sufficiently studied. She described the CM and two other models required in an M&S process as follows: (1) a CM, which is a nonexecutable higher-level abstraction of the system under study, that represents the structure of the system and what will be modeled in the future executable simulation model; (2) a platform-independent simulation model (PISM), which is a mathematical description of the processes and activities in the CM so that mathematical or computational analyses can be manually conducted; (3) a platform-specific simulation model (PSSM), which is derived from the PISM and the details of a specific execution platform toward the synthesis of an executable model that can allow the simulation to be carried out on a machine. Çetinkaya argued that nonconsumption of the CM in any development iteration involving other models creates a semantics gap between CM and PISM that may lead to lack of model continuity in all stages of model development. To address this problem, she proposed the MDD4MS framework to manage the simulation process that encompasses the three stages of model development. Fundamentally, MDD4MS mirrors the layered architecture of the OMG's model-driven architecture (MDA) framework (Mellor, 2004), with CM, PISM, and PSSM in the place of MDA's CIM, PIM, and PSM respectively. This generic framework was concretized with the CM, PISM, and PSSM created based on BPMN, DEVS, and the Java-based DEVS Distributed Simulation Object Library—DEVSDSOL (Seck and Verbraeck, 2009), respectively. With transformation rules written in ATLAS Transformation Language, ATL (Jouault et al., 2006), substantial parts of the PISM—which is manually refined—can be obtained from the CM through a partial model transformation process. Though not explicitly stated, the transformation is "partial" apparently because the mapping of the BPMN (source formalism) to DEVS (target formalism) is not surjective, that is, there is no guarantee that every element of the target formalism has the corresponding element(s) from which it can be derived in source formalism. The refined PISM is used in another partial model transformation process to generate a PSSM based on DEVSDSOL. Though not explicitly stated or claimed by Çetinkaya, the BPMN-based CM can arguably be independently refined to an enactment model. If it is considered in this sense, then we can reasonably say that an enactment model is being transformed into a DEVS-based model for simulation.

3.2 Formalism Weaving

Formalism weaving is the creation of a pivotal formalism by weaving the metamodels of existing formalisms to create a domain-specific language (DSL). The DSL's abstract syntax is specified to capture the concepts described in the underlying formalisms, while the concrete syntax is developed with high-level cognitive notations to shield the DSL's users from the complexity of the underlying formalisms. Intuitively, the operational/execution/logical frameworks of the underlying formalisms provide semantics domains for the DSL. As regards unification, formalism weaving differs from formalism transformation for the following three main reasons:

- While formalism transformation bridges two methodologies by establishing the rules of mapping one existing formalism into another existing one, formalism weaving creates a third formalism as the language of unification.
- To achieve the complete unification with formalisms weaving, there is still need to make use of formalisms transformation to define the semantics of the newly created formalism.
- The essence of formalism transformation is to bridge two languages, therefore realizing pairwise integration of their corresponding supported methodologies. Under usual circumstances, transformation rules are defined such that one model expressed in the source formalism is given as the input and a model specified in the target formalism is obtained as the output. Contrariwise, formalism weaving has the potential to integrate more than two formalisms, therefore opening the way to integration of methodologies, each residing in one of the three hyperplanes of Fig. 1. The high-level language for system specification (HiLLS) has been introduced to achieve this goal.

HiLLS is a system modeling language for constructing multianalysis system models, which helps domain experts express knowledge from various analysis perspectives in one single model (Aliyu et al., 2016). It can be formalized as a structure $\langle A, C, M^{AC}, D, \{S_i\}_{i \in D}, \{M_i^{AS}\}_{i \in D}\rangle$, in which A is the abstract syntax, C is the concrete syntax, and M^{AC} is the function mapping C to A. A special feature of HiLLS is that instead of having one single semantic domain, a family D of semantic domains is defined, each of which being represented by its abstract syntax S_i ($i \in D$). A set $\{M_i^{AS}\}_{i \in D}$ of corresponding semantic mappings is then established, each of which maps A to one S_i. By defining D as {DEVS, Object-Z, UML}, HiLLS enables its multiple analysis capabilities including simulation, formal analysis, and system enactment. Indeed, while the mapping to DEVS provides the operational semantics for

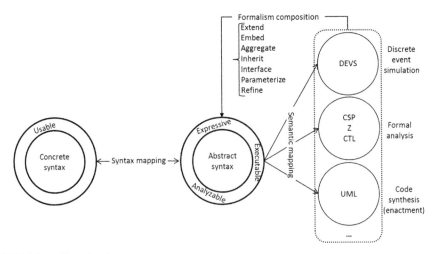

FIG. 2   HiLLS formalism structure.

simulation, the mapping to Object-Z (Smith, 2012) makes HiLLS models amenable to formal analysis, and the mapping to UML allows defining the operational semantics for enactment. These mapping operations can be seen as filtering activities, since A is built by weaving concepts from $\{S_i\}_{i \in D}$, using metamodel integration techniques proposed by Emerson and Sztipanovits (2006), as shown by Fig. 2. At the other hand, weaving operations imply semantic alignment be done between concepts coming from different sources. Therefore, the semantic unity of HiLLS-specified models is ensured by construction. Moreover, a concrete syntax is defined for graphically representing HiLLS models, which makes it easy for domain experts to learn, share, and discuss modeled systems. HiLLS visual representations are inspired by UML class diagram, system control-oriented transition diagrams, and Z schemas. Furthermore, the domain set D can be extended if need be to integrate formalism support for other analysis methodologies, provided the new concepts are aligned with the ones already existing in A, and that additional semantic mapping functions are defined to filter HiLLS toward target formalisms.

4 CONCLUSION

This chapter has explored unifying modeling approaches in the context of model engineering. The model-based design and development of a complex system may require an iterative process of modeling, performance evaluation, logical analysis for requirement verifications, and prototype implementation for runtime testing. Such iterations of analysis processes are often necessary for early revelations of subtle knowledge about the systems, which are, in most cases, beyond intuition. An undesired behavior discovered in the analysis of a system can be a signal of a fundamental flaw in the system's design; such discovery must be made at an early stage of development to forestall costly errors in the final system. Questions about different aspects of a system are usually best answered using some specific analysis methodologies; for instance, system's performance and behavior in some specified EFs can be efficiently studied using appropriate simulation methodologies. Similarly, verification of properties such as liveness, safeness, and fairness are better studied with appropriate formal methods while enactment methodologies may be used to verify assumptions about some time-based and human-in-the-loop activities and behaviors. Therefore, an exhaustive study of a complex (or even seemingly simple) system often requires the use of different analysis methodologies to produce complementary answers to likely questions.

There is no gainsaying that a combination of multiple analysis methodologies offers more powerful capabilities and rigor to test system designs than can be accomplished with any of the methodologies applied alone. While this exercise will provide (near) complete knowledge of complex systems and helps analysts to make reliable assumptions and forecasts about their properties, its practical adoption is not commensurate with the theoretical advancements, and evolving formalisms and algorithms, resulting from decades of research by practitioners of different methodologies. This shortfall has been linked to the prerequisite mathematical skills for dealing with most formalisms, which is compounded by little portability of models between tools of different methodologies that makes it mostly necessary to bear the hard task of creating and managing several models of the same system in different formalisms. Another contributing factor is that most of the existing computational analysis environments are dedicated to specific analysis methodologies (i.e., simulation, or formal analysis, or enactment) and are usually difficult to extend to accommodate other approaches.

Thus, one must learn all the formalisms underlining the various methods to create models and go round to update all of them whenever certain system variables change. The aim of unifying the modeling approaches is to alleviate the burdens on users of multiple formalisms that support these analysis methodologies. This chapter has reviewed major unification strategies along the three dimensions of simulation, formal methods, and enactment.

References

Aliyu, H.O., Maïga, O., Traoré, M.K., 2015. In: A framework for discrete event systems enactment. Proceedings of 29th European Simulation and Modeling Conference. EUROSIS-ETI, ISBN: 978-9077381-908, pp. 149–156.

Aliyu, H.O., Maïga, O., Traoré, M.K., 2016. The high level language for systems specification: a model-driven approach to systems engineering. Int. J. Model. Simul. Sci. Comput. 7(1):1641003.

Bajaj, M., Scott, A., Deming, D., Wickstrom, G., Spain, M.D., Zwemer, D., Peak, R., 2012. Maestro—a model-based systems engineering environment for complex electronic systems. INCOSE Int. Symp. 22 (1), 1999–2015.

Barros, F.J., 1997. Modelling formalisms for dynamic structure systems. ACM Trans. Model. Comput. Simul. 7 (4), 501–515.

Bocciarelli, P., D'Ambrogio, A., 2014. Model-driven method to enable simulation-based analysis of complex systems. In: Gianni, D., D'Ambrogio, A., Tolk, A. (Eds.), Modeling and Simulation-Based Systems Engineering Handbook. CRC Press, Boca Raton, FL, pp. 119–148.

Bruno, G., Agarwal, R., 1995. Validating software requirements using operational models. In: Objective Software Quality. Springer, Berlin, Heidelberg, pp. 78–93.

Bussow, R., Geisler, R., Grieskamp, W., Klar, M., 1997. The μSZ Notation. Version 1. Technical Report, TU Berlin, Germany.

Carson II, J.S., 2004. Introduction to modeling and simulation. Proceedings of the 36th Conference on Winter Simulation. IEEE Computer Society, pp. 9–16.

Çetinkaya, D.K., 2013. Model Driven Development of Simulation Models: Defining and Transforming Conceptual Models into Simulation Models by Using Metamodels and Model Transformations (Doctoral Dissertation). Delft University of Technology, Delft.

Clarke, E.M., Wing, J.M., 1996. Formal methods: state of the art and future directions. ACM Comput. Surv. 28 (4), 626–643.

Dowson, M., Fernström, C., 1994. Towards requirements for enactment mechanisms. In: Software Process Technology. Springer, Berlin, Heidelberg, pp. 90–106.

Emerson, M., Sztipanovits, J., 2006. In: Techniques for metamodel composition.OOPSLA 6th Workshop on Domain Specific Modeling, pp. 123–139.

Estefan, J.A., 2007. Survey of model-based systems engineering (MBSE) methodologies. INCOSE MBSE Focus Group. 25(8).

Finkelstein, A., Kramer, J., Nuseibeh, B., Finkelstein, L., Goedicke, M., 1992. Viewpoints: a framework for integrating multiple perspectives in system development. Int. J. Softw. Eng. Knowl. Eng. 2 (1), 31–57.

Fishwick, P., 1995. Simulation Model Design and Execution. Building Digital Worlds. Prentice Hall, Upper Saddle River, NJ.

Franceschini, R., Bisgambiglia, P.A., Touraille, L., Bisgambiglia, P., Hill, D., 2014. A Survey of Modelling and Simulation Software Frameworks using Discrete Event System Specification. OASIcs-Open Access Series in Informatics, vol. 43. Schloss Dagstuhl-Leibniz-Zentrum fuer Informatik.

Hall, A., 2007. Realizing the benefits of formal methods. J. Univers. Comput. Sci. 13 (5), 669–678.

Holmlid, S., Evenson, S., 2007. Prototyping and enacting services: lessons learned from human-centered methods. Proceedings of the 10th Quality in Services Conferencevol. 10.

Holmquist, L.E., 2005. Prototyping: generating ideas or cargo cult design? Interactions 12 (2), 48–54.

Hong, K.J., Kim, T.G., 2006. DEVSpecL: DEVS specification language for modeling, simulation and analysis of discrete event systems. Inf. Softw. Technol. 48 (4), 221–234.

Jeston, J., Nelis, J., 2014. Business Process Management. Routledge, London.

Jouault, F., Allilaire, F., Bézivin, J., Kurtev, I., Valduriez, P., 2006. ATL: a QVT-like transformation language. Companion to the 21st ACM SIGPLAN Symposium on Object-Oriented Programming Systems, Languages, and Applications, pp. 719–720.

Kapos, G.-D., Dalakas, V., Nikolaidou, M., Anagnostopoulos, D., 2014. An integrated framework for automated simulation of SysML models using DEVS. Simulation 90 (6), 717–744.

Kefalas, P., Eleftherakis, G., Sotiriadou, A., 2003. Developing tools for formal methods.9th Panhellenic Conference in Informaticspp. 625–639.

Kouvas, G., Grefen, P., Juan, A., 2010. Business process enactment. In: Dynamic Business Process Formation for Instant Virtual Enterprises. Springer, London, pp. 113–132.

Kuhn, D.R., Craigen, D., Saaltink, M., 2003. In: Practical application of formal methods in modeling and simulation. Summer Computer Simulation Conference. Society for Computer Simulation International, pp. 726–731.

Kurpjuweit, S., Winter, R., 2007. In: Viewpoint-based meta model engineering.EMISA 2007pp. 143–159.

Laleau, R., Semmak, F., Matoussi, A., Petit, D., Hammad, A., Tatibouet, B., 2010. A first attempt to combine SysML requirements diagrams and B. Innov. Syst. Softw. Eng. 6 (1–2), 47–54.

Lamsweerde, A.V., 2000. In: Formal specification: a roadmap.ACM Conference on the Future of Software Engineering, pp. 147–159.

Lano, K., Clark, D., Androutsopoulos, K., 2004. UML to B: formal verification of object-oriented models. In: Integrated Formal Methods. Springer, Berlin, Heidelberg, pp. 187–206.

Lilius, J., Paltor, I.P., 1999. vUML: a tool for verifying UML models.14th IEEE International Conference on Automated Software Engineering, pp. 255–258.

Maria, A., 1997. In: Introduction to modeling and simulation. Proceedings of the 29th Conference on Winter Simulation. IEEE Computer Society, pp. 7–13.

Martin, G., Büttner, F., Richters, M., 2007. USE: a UML-based specification environment for validating UML and OCL. Sci. Comput. Program. 69 (1), 27–34.

Mellor, S.J., 2004. MDA Distilled: Principles of Model-Driven Architecture. Addison-Wesley Professional, Reading, MA.

Mittal, S., 2007. DEVS Unified Process for Integrated Development and Testing of Service Oriented Architectures. (Doctoral dissertation). University of Arizona.

Mittal, S., Martín, J.L.R., 2013. Model-driven systems engineering for net-centric system of systems with DEVS unified process. In: Proceedings of the 2013 Winter Simulation Conference. IEEE Press, pp. 1140–1151.

Object Management Group, 2010. System Modeling Language TM (SysML) Version 1.2. June. Available from: http://www.omg.org/spec/SysML/1.2/. [(Accessed 4 October 2017)].

Object Management Group, 2013. A Foundational Subset for Executable UML. Available from: http://www.omg.org/spec/FUML/1.1. [(Accessed 4 October 2017)].

Ottensooser, A., Fekete, A., 2007. An enactment-engine based on use-cases. In: Business Process Management. Springer, Berlin, Heidelberg, pp. 230–245.

Risco-Martín, J.L., Jesús, M., Mittal, S., Zeigler, B.P., 2009. eUDEVS: executable UML with DEVS theory of modeling and simulation. Simulation 85 (11 – 12), 750–777.

Rumbaugh, J., Jacobson, I., Booch, G., 2004. Unified Modeling Language Reference Manual. The Pearson Higher Education, London.

Schamai, W., Fritzson, P., Paredis, C., Pop, A., 2009. In: Towards unified system modeling and simulation with modelicaML: modeling of executable behavior using graphical notations. 7th International Modelica Conference. Linköping University Electronic Press, pp. 612–621.

Seck, M., Verbraeck, A., 2009. In: DEVS in DSOL: adding DEVS operational semantics to a generic event-scheduling simulation environment. Summer Computer Simulation Conference. Society for Modeling & Simulation International, pp. 261–266.

Shah, S.M., Anastasakis, K., Bordbar, B., 2009. From UML to alloy and back again. In: Models in Software Engineering. Springer, Berlin, Heidelberg, pp. 158–171.

Shaikh, R., Vangheluwe, H., 2011. In: Transforming UML2.0 class diagrams and statecharts to atomic DEVS. Symposium on Theory of Modeling & Simulation: DEVS Integrative M&S Symposium. Society for Computer Simulation International, pp. 205–212.

Smith, G., 2012. The Object-Z Specification Language. vol. 1 Springer Science & Business Media, New York.

Tolk, A., Hughes, T.K., 2014. Systems engineering, architecture and simulation. In: Gianni, D., D'Ambrogio, A., Tolk, A. (Eds.), Modeling and Simulation-Based Systems Engineering Handbook. CRC Press, Boca Raton, FL, pp. 11–41.

Traoré, M.K., 2006. Making DEVS models amenable to formal analysis. Symposium on Theory of Modeling & Simulation: DEVS Integrative M&S Symposium, April 2–6. Society for Computer Simulation International, Huntsville, AL.

Traoré, M.K., Muzy, A., 2006. Capturing the dual relationship between simulation models and their context. Simul. Model. Pract. Theory 14 (2), 126–142.

Trojet, W., Berradia, T., 2015. System reliability using simulation models and formal methods. Int. J. Comput. Appl. 132 (17), 1–8.

Van Der Aalst, W.M., Ter Hofstede, A.H., Weske, M., 2003. Business Process Management: A Survey. Business Process Management. Springer, Berlin, Heidelberg, pp. 1–12.

Vangheluwe, H.L., 2000. In: DEVS as a common denominator for multi-formalism hybrid systems modelling. IEEE International Symposium on Computer-Aided Control System Design. IEEE, pp. 129–134.

White, S.A., 2004. Introduction to BPMN. vol. 2 IBM Cooperation, Armonk.

Wing, J.M., 1990. A specifier's introduction to formal methods. Computer 23 (9), 8–22.

Yacoub, A., Hamri, M., Frydman, C., 2014. In: A method for improving the verification and validation of systems by the combined use of simulation and formal methods.IEEE/ACM 18th International Symposium on Distributed Simulation and Real Time Applications, pp. 155–162.

Zeigler, B.P., 1976. Theory of Modeling and Simulation. John Wiley & Sons, Inc., New York, NY.

Zeigler, B.P., Nutaro, J.J., 2016. Towards a framework for more robust validation and verification of simulation models for systems of systems. J. Def. Model. Simul. 13(1). https://dx.doi.org/10.1177/1548512914568657.

Zeigler, B.P., Praehofer, H., Kim, T.G., 2000. Theory of Modeling and Simulation: Integrating Discrete Event and Continuous Complex Dynamic Systems. Academic Press, San Diego, CA.

Further Reading

Malik, P., Utting, M., 2005. In: CZT: a framework for Z tools. International Conference of B and Z Users. Springer, Berlin, Heidelberg, pp. 65–84.

Miller, T., Freitas, L., Malik, P., Utting, M., 2005. In: CZT support for Z extensions. International Conference on Integrated Formal Methods. Springer, Berlin, Heidelberg, pp. 227–245.

Object Management Group, 2011. Business Process Model and Notation (BPMN) Version 2.0. Document number: formal/2011-01-03.

CHAPTER

4

Model Composition and Reuse

Mikel D. Petty*, Eric W. Weisel†

*University of Alabama in Huntsville, Huntsville, AL, United States †Old Dominion University, Norfolk, VA, United States

1 INTRODUCTION

Model *composition* is combining or connecting separate models into an integrated composite model. The composite model is typically of a system, phenomenon, process, or scenario that is itself composed of the systems, phenomena, processes, or scenarios modeled by the component models. *Composability* is the capability to select and compose models in various combinations into simulation systems to satisfy specific user requirements. Its defining characteristic is that different simulation systems can be composed from different sets of models, each suited to some distinct purpose, and the different possible model compositions will be usefully valid. For nearly two decades composability has been an important research objective, especially in the defense-related modeling and simulation community, but general composability remains a challenging goal. Composability research has included both theoretical investigation of the limits and potential of model composition and practical implementation of frameworks and standards intended to provide useful composability capabilities.

In modeling and simulation, reusing an existing model for a new application has the potential to reduce the time, effort, and expense of development and testing. Moreover, model reuse may add credibility to the new application if the model was validated for its previous use or is familiar to the stakeholder community because of long-standing use. Because the models to be reused are usually implemented as software components, general software reuse techniques are often applicable, albeit with special considerations arising from the fact that the software components are implementations of models.

Model composition and reuse are not the same idea, but they are closely related. Models that are not composable can be reused, but composability is an important enabler for reuse, and reuse of a software implementation of a model may require that it be composable with other models. This chapter will define and explain model composition and reuse, survey selected research (both theoretical and applied) relevant to them, describe practical implementations of frameworks and standards that have been developed to support them, and identify several open research questions relevant to them.

2 MODEL COMPOSITION

This section begins by providing definitions of key terminology relating to model composition, beginning with *composability* itself. Model composition is then explained from a theoretical perspective, including a review of prior work in composability theory. Finally, practical aspects of model composition are discussed including discussion of the different levels at which models may be composed and different types of standards and frameworks that support model composition.

2.1 Definitions and Concepts[1]

Composability has been a goal of developers of simulation systems for at least two decades, especially in the US defense-related modeling and simulation community. For example, composability was mentioned multiple times in the requirements document for the US Army's one semiautomated forces (OneSAF) simulation system (US Army, 1998). Earlier published calls for composability research include Harkrider and Lunceford (1999) and Castro et al. (2002); more recent calls include Tolk et al. (2015) and Fujimoto (2016).

As composability became an increasingly important issue in simulation system development, different definitions of the term appeared in the simulation research literature. They were generally similar in concept but often differed in emphasis or level. The fact that composability could mean different things in different contexts had been noted earlier (Page and Opper, 1998). The following definitions of composability from the literature illustrate the variations on the theme:

> The ability to rapidly configure, initialize, and test an exercise by logically assembling a simulation from a pool of reusable components. *JSIMS (1997)*

> The ability to create, configure, initialize, test, and validate an exercise by logically assembling a unique simulation execution from a pool of reusable system components in order to meet a specific set of objectives. *Harkrider and Lunceford (1999)*

> The ability to build new things from existing pieces. *Pratt et al. (1999)*

> The ability to compose models/modules across a variety of application domains, levels of resolution and time scales. *Kasputis and Ng (2000)*

The following standard definition of composability was proposed: *Composability* is the capability to select and assemble simulation components in various combinations into valid

---

[1] Portions of this section were adapted and updated from Petty and Weisel (2003a).

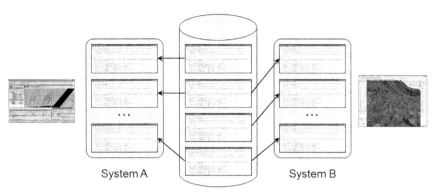

FIG. 1   Composability concept.

simulation systems to satisfy the specific user requirements (Petty and Weisel, 2003a).[2] The defining characteristic of composability is that different simulation systems can be composed in a variety of ways, each suited to a specific purpose, and the different compositions will be valid.[3] Composability is more than just the ability to put simulations together from components; it is the ability to combine and recombine, to configure and reconfigure, sets of components selected from those available into different simulation systems to meet different needs.

When assembling simulation components into a simulation system, the components to be composed may be drawn from a library or repository of components, as suggested in Fig. 1.[4] That library might include multiple network interfaces, different user interfaces, a range of classes of implemented entity models, a variety of implemented physical models at different levels of fidelity, and so on. Different sets of components from the repository may be composed into different simulation systems. The components may be reused in multiple simulation systems.

Discussions on composability often differ on the question as to what is being composed and what is formed by the composition. A number of different answers can be found in the literature; they will be referred to as *levels* of composability. Nine levels of composability are documented here.[5] These levels have been drawn from various sources, some of which

---

[2] At least one source adds "meaningfully" at the end of this definition of composability (Davis and Anderson, 2003).

[3] In modeling and simulation terminology, a *valid* model recreates the modeled characteristics of the real-world system it models with sufficient fidelity or accuracy to be useful (Balci, 1998; Petty, 2010).

[4] The screen images in the figure are two semiautomated forces systems (Petty, 1995), OneSAF (left) and VR-Forces (right). The images are intended only to suggest two similar but different simulation models resulting from different compositions of components. No assertion is intended that the actual OneSAF or VR-Forces software have any software components in common.

[5] In this list the levels are named after the unit of composition, that is, the components being composed. Another method of naming the levels is after the result of composition, that is, what is produced from the components. Some sources in the literature have adopted the former convention (Page and Opper, 1998; JSIMS, 1997), others the latter (Post, 2002).

explicitly or implicitly include several of the levels defined here in composability (e.g., Biddle and Perry, 2000). Composability levels from different sources that were essentially the same have been combined. Those listed here have different meanings and implications, but there may be some overlap in component and scale between them.

- *Application*. Applications such as simulations, real C4I systems, networks, communications equipment, and auxiliary software components are composed into simulation events, exercises, or experiments (Post, 2002). For this to be a level of composability, rather than simply software integration, the composition must be done in a way that allows combining and recombining the applications into different systems and events. (The distinction between composition and integration will be discussed in more detail later.) This level of composability has also been called "event-level" (Post, 2002).
- *Federate*. Federates are composed into persistent federations.[6] A federation is persistent if it is reused for a number of different purposes (such as events, exercises, or experiments), though possibly with some changes to the set of federates that have been composed. The composition may be supported by an interoperability protocol, such as high-level architecture (HLA), aggregate level simulation protocol (ALSP), or distributed interactive simulation (DIS).[7] Examples of this level of composability include the Joint Training Confederation and the Combat Trauma Patient Simulation (Petty and Windyga, 1999). This level of composability has also been called "federation-level" (Post, 2002).
- *Package*. Preassembled packages comprising sets of models that form a consistent subset of the battlespace are composed into simulations (Page and Opper, 1998; JSIMS, 1997).
- *Parameter*. Parameters are used to configure preexisting simulations (Page and Opper, 1998; JSIMS, 1997). Some sources also include in this level of composability, which may be called the "simulation" level, the idea that a limited pool of packages may be composed into simulations. Here that is included in package level.
- *Module*. Software modules[8] are composed into software executables. The executables may be federates in a federation or stand-alone simulation systems. The OneSAF family of software products has this level of composability (US Army, 1998; Courtemanche and Burch, 2000; Courtemanche and Wittman, 2002).
- *Model*. Separate models of smaller-scale processes or objects[9] are composed into composite models of larger-scale processes or objects. For example, models of platform/entity

---

[6]The terms "federate" and "federation" have specific meanings in the context of the High-Level Architecture (HLA) distributed simulation interoperability protocol. Here they are being used with more generic meanings analogous to their HLA meanings to denote simulation applications linked together, but not necessarily with the HLA distributed simulation interoperability protocol.

[7]DIS is Distributed Interactive Simulation, ALSP is Aggregrate Level Simulation Protocol, and HLA is High Level Architecture.

[8]The term "software component" is also used with this meaning, and if used here, would make this the "component" level of composability. However, in this report the term "component" is used in a general sense as the unit of composition at any level.

[9]Here "object" means simulated real-world object, not software object. The former is not always implemented as the latter and assuming so is an oversimplification. Even if a real-world object class, such as an aircraft type, is implemented as a software object class, it is not correct to assume that the subcomponents of that real-world object class, such as sensors and weapons, are implemented as subclasses of the software object class, that is, a mixing of is-a and part-of relationships.

subsystems, such as sensors and weapons, may be composed into composite models of platforms/entities, such as aircraft (Post, 2002). Models of physical processes, such as wind and rain, may be composed into composite models of larger-scale physical phenomena, such as weather. The various weather models, in turn, may be composed into a climate model. The composite models may be implemented as modules or federates. This level of composability was a design goal of both ModSAF (Ceranowicz, 1994) and OneSAF (Henderson and Rodriquez, 2002). This level of composability has also been called "object-level" (Post, 2002), "component" (JSIMS, 1997), and "reconfigurable models" (Diaz-Calderon et al., 2000).

- *Data*. Sets of data are composed into databases. The data sets may be initially separate because they describe different entities, are from different sources, or represent different aspects of some common phenomena. For example, different data sets were composed to represent electronic warfare entities and actions in DIS (Wood and Petty, 1995).
- *Entity*. Platforms/entities are composed into groupings such as military units, force structures, and scenario orders of battle (Post, 2002). This level of composition may be hierarchical, with several layers of groupings having composed into higher-level groupings. This level of composition is typically done with data, rather than with software, as in ModSAF and WARSIM. This level of composition has also been called "federate-level" (Post, 2002).
- *Behavior*. Low-level atomic behaviors are composed into high-level composite behaviors, which are to be executed by autonomous simulation entities in a computer generated forces system or constructive simulation. The behaviors may be expressed in a variety of forms. Examples include hierarchically organized finite state machines as used in ModSAF and its variants (Calder et al., 1993) and process flow diagrams (Peters et al., 2002).

Composability can be understood from both *syntactic* and *semantic* perspectives (Pratt et al., 1999; Ceranowicz, 2002).[10] Syntactic composability, that is, the actual implementation of composed models, requires that the components that implement the models be constructed so that their implementation details, such as parameter passing mechanisms, external data accesses, and timing assumptions are compatible for all of the different configurations that might be composed. The question in syntactic composability is whether the components can be connected and effectively exchange data. In contrast, semantic composability is the question as to whether the models that make up the composed simulation system can be meaningfully composed, that is, if their combined computation is semantically valid. Syntactic composability is necessary but not sufficient to produce semantic composability.

3 THEORETICAL LIMITS AND POTENTIAL FOR MODEL COMPOSITION[11]

From the beginning of the modeling and simulation community's interest in composability, the need for a theoretical understanding of the issues involved was recognized

---

[10]Some early sources use the terms *engineering composability* and *modeling composability* as equivalent to *syntactic composability* and *semantic composability*, respectively.

[11]Portions of this section were adapted and updated from Petty and Weisel (2003b), Petty et al. (2003b,, 2005). A similar review of composability theory results appeared in Balci et al. (2017).

(Davis et al., 2000; Kasputis and Ng, 2000). The latter source is clear: "We are discovering that unless models are designed to work together, they don't (at least not easily and cost effectively). Without a robust, theoretically grounded framework for design, we are consigned to repeat this problem for the foreseeable future" (Kasputis and Ng, 2000).

Beginning in the late 1990s, several research efforts developed important elements of a theory of composability. That work, which was based on mathematical logic and computability theory, was intended to establish a firm theoretical basis for understanding composability in general, and in particular, the formal characteristics of a composition of models. The significant outcomes of that work include the following:

- Standard definitions of composability and related terms (Petty and Weisel, 2003a), which have subsequently been widely used (e.g., Davis and Anderson, 2003; Mielke and Phillips, 2003; Davis and Anderson, 2004; Porcarelli et al., 2005; Waziruddin et al., 2003; Szabo and Teo, 2012; Benali and Bellamine-Ben Saoud, 2011; Fujimoto, 2016; Peng et al., 2017).
- Formal (i.e., theoretical) definitions of *model*, *simulation*, and *validity* that are consistent with their common informal meanings but are precise enough to support formal reasoning (Petty and Weisel, 2003b; Petty et al., 2003b).
- Determination of the computational complexity of algorithmic processes for selecting models to be composed from a repository (Page and Opper, 1999; Petty et al., 2003a).
- Determination that a simple form of composition is theoretically sufficient to assemble any composite model (Petty, 2004).
- Formal resolution of the central question as to whether models that are separately valid can be assumed to remain valid when composed (Weisel et al., 2003; Weisel, 2004).
- Surveys and assessments of practical software engineering approaches to achieving composability (Weisel et al., 2004; Petty et al., 2014, 2016).

Concepts closely related to composability include *interoperability* and *integratability*; compared to composability, they focus less on modeling semantics and more on technical connectivity at the software and hardware levels, respectively (Page et al., 2004). All of these concepts were integrated in a single overarching conceptual framework, known as the *Levels of Conceptual Interoperability Model* (Tolk and Muguira, 2003; Turnitsa, 2005), which defines levels or degrees to which models can exchange and consistently interpret data.

Verifying that a composition of components satisfies its requirements specification was the subject of Mahmood (2013). A comprehensive model verification framework was proposed and three theoretical verification approaches (algebraic analysis using Petri nets, state-space analysis using colored Petri nets, and model checking using communicating sequential processes) were investigated within that framework. Stemming from the same work, the related concept of *pragmatic composability* takes the context within which a composite model executes into consideration. The computational costs of validating a composition of models using multiple validation approaches were compared in Szabo and Teo (2012).

A process calculus for reasoning about software components (not necessarily models) and patterns of composition was introduced in Achermann and Nierstrasz (2005). Other existing simulation-oriented theories and methodologies that address some of the same ideas as composability theory include discrete event system specification (DEVS) (Zeigler et al., 2000), semantic descriptors (Kasputis et al., 2004), base object models (BOM) (SISO, 2006), denotational semantics (Mosses, 1990), model-based systems engineering (MBSE)

(Wymore, 1993), ontological descriptions of the simulation domain (Collins and Clark, 2004; Collins, 2004), and interoperability analysis (Harmon, 1996). The common goal of the comparisons was to improve composability theory with the insights of these theories. The two most closely related to composability theory were DEVS and MBSE; comparisons can be found in Weisel et al. (2005) for DEVS and Mielke et al. (2005) for MBSE.

3.1 Theoretical Limits of Model Composition

Of course, all simulations that are executed on computers are subject to the theoretical limits of computation, as embodied in computational complexity theory (Garey and Johnson, 1979) and computability theory (Davis et al., 1994). These limits are encountered infrequently in the day-to-day practice of modeling and simulation.[12] However, composability seems to push the boundaries of computation's theoretical limits; several operations associated with or implied by the definition of composability are constrained by the theory of computation.

In practice, model developers often assume, whether explicitly or implicitly, that model validity is preserved under composition, that is, if two models have been determined separately to be valid, then those models (and equivalently, software components implementing those models) may be composed and the resulting composite model will necessarily or automatically be valid. Therefore, a key question for composability theory is whether validity actually is preserved by composition. In other words, if two models are separately valid, can it be assumed that their composition is necessarily valid?

To address this specific question a series of formal theorems were developed and proved in the early 2000s. They considered several increasingly inclusive classes of models, starting from a very simple class and ending with a "computable" class that includes all models that can be executed on a digital computer. They also considered several increasingly sophisticated validity metrics, including a "trajectory metric" that formalizes a simulation practitioner's intuitive notion of error accumulating over the course of a simulation execution. Ultimately it was shown that in general, with the exception of some unrealistically simple combinations of model classes and validity metrics, the composition of two separately valid models cannot be assumed to be valid (Weisel et al., 2003; Weisel, 2004). Of course, this result does not imply that separately valid models are never valid when composed; rather it means that they cannot be assumed to be valid.

Experienced simulation practitioners are generally aware of this at a practical level. In practical terms, two separately valid models may nevertheless produce results that are invalid when they are composed. A composite model must be validated after it is composed even if its components are understood to be valid. For example, see Bunus and Fritzson (2004) for a focused discussion of structural and numerical inconsistencies that may arise when composing components that implement physics-based models based on differential equations.

Note that because this result is based on computability theory, it is general to all computer-based simulations and applies regardless of whether the specific mechanism or framework is

---

[12] The exception to this assertion is the inherently discrete nature of computation and the consequent finite numerical precision of digital computers, which leads inescapably to discretization errors when numerically integrating the continuous differential equations of physics-based models (Colley, 2010). This issue is considered regularly for some simulation practitioners.

used to compose the models. No software framework or theory of simulation that operates on digital computers can guarantee the validity of a composition of models.

From a computability theory viewpoint, any algorithm that executes on a digital computer, including any model to be composed and the component that implements it, is a computable function (Davis et al., 1994; Petty, 2004). Therefore model composition may be understood as a special case of function composition, where the functions are restricted to be computable. Multiple seemingly different mathematical forms of function composition have been defined. It might be assumed that these different forms of function composition would have some effect on what can and cannot be achievable by model composition. It turns out that they do not, at least at a theoretical level. Using induction on the number of models to be composed, it was proven that simple composition, that is, composition of the form of $f(g(x))$, or more generally, $f_1(f_2(...f_n(x)...)$, is sufficient to assemble any composite model (Petty, 2004). Other nonsimple forms of composition add no additional compositional or computational power. Consequently, theoretical analysis of model composition may rely on simple composition without loss of generality.

Given a library or repository of composable simulation components, assembling a simulation system requires selecting a subset of those components that will collectively meet the user's objectives (Clark et al., 2004). The process of selecting a set of components to be composed from a repository of available components that will satisfy a given set of simulation system requirements is known as *component selection*. This deceptively simple problem is a general software engineering issue, applicable to any repository of components, regardless of whether or not those components are implementations of models (Kaur et al., 2014). In the software engineering context, component selection is sometimes also known as *specification matching* (Beizer, 1995).

There are two computational problems subsumed in component selection. The first problem is determining which requirements each component satisfies. This may be done either prior to component selection for a simulation ("What are the requirements this component satisfies?") or as needed when a set of requirements for a simulation system are provided to the repository ("Does this component satisfy any of the system's requirements?"). Given an answer to the first problem, the second problem is then to select a set of components that collectively meet all of the system requirements. Both these problems are well known in the software engineering context; they are summarized, respectively, as "How do we describe software components in unambiguous, classifiable terms?" and "[H]ow do you find the [components] that you need?" in Pressman and Maxim (2015).

As can be easily seen the first of the two component selection problems is problematic. Suppose a (perfectly reasonable) requirement for a component is that its execution terminates, rather than entering a nonterminating loop, for all inputs. Determining if a computation will terminate is known as the "halting problem" (Turing, 1936), and has been proven uncomputable in general. Even for system requirements that are in principle algorithmically decidable, those determination may require superpolynomial computation time and thus be infeasible in practice (Page and Opper, 1999). The consequence of these results is that determination of the requirements satisfied by a component have to be done by nonalgorithmic methods, such as heuristic assessment or manual (i.e., human) labeling; the requirements so identified could be recorded in metadata associated with the component (Fox et al., 2004).

The second of the two component selection problems is computationally difficult even if the first has been solved, that is, even if the requirements satisfied by each of the components in the repository have somehow been determined. In initial work on this problem, four variants of the second component selection problem based on two forms of objectives decidability (bounded and unbounded) and two forms of composition (emergent and nonemergent) were defined, and the bounded nonemergent variant of the component selection problem was proven to be NP complete (Page and Opper, 1999). In subsequent work, an additional form of composition (antiemergent) was defined, leading to two additional variants of the problem, and then a general form of the second component selection problem that subsumes all six variants was defined. That general form separates the problem of determining which objectives a component or composition satisfies from the problem of selecting a set of components to meet the objectives by invoking an oracle for the former problem. The general form was proven to be NP complete (intractable in general) even if the requirements satisfied by a component or composition are known (Petty et al., 2003a; Petty, 2006).[13]

Component selection is not the only aspect of composability subject to theoretical limitations; three others are mentioned here. First, the testing of components to be composed or reused is sometimes done with automated software design checkers. Determining if the design of a software component, expressed as a set of logical constraints, can reach an unacceptable state is NP complete (Jackson, 2006). Second, a library or repository of composable components, as with any software component library, will require configuration management. The operation of finding a sequence of configuration management operations to transform a set of components from an initial state to an end state that meets a given set of requirements is NP complete (Sun, 2006; Sun and Couch, 2007). Finally, a common objective of composability research and engineering is to develop a form of specification for a component sufficiently powerful to allow algorithmic processing of the component based on its specification, including selection for composition, validity determination within a composition, and testing (Morse et al., 2004). Such specifications area sometimes known as metadata or metamodels. Some component specification formalisms have been developed and proposed (Achermann and Nierstrasz, 2005). However, given a component specification expressed in a formalism powerful enough to specify a model, determining if an execution of the component will halt is uncomputable (Overstreet, 1982; Overstreet and Nance, 1985).

Of course, simulation practitioners faced with a theoretical limit on a computational task cannot simply abandon the enterprise of modeling and simulation. Instead, to accomplish their objectives they must seek a heuristic solution, that is, a method that works well in most situations, even if it is not theoretically complete or perfect. Several heuristic solutions to the theoretical problems of model composition have been developed. An independent study of heuristics to select a set of components to compose models have shown that such selection is possible in situations where simplifying assumptions can be made regarding the property of emergence, which is the question as to whether a composition of components satisfies some objective that none of the components satisfies individually (Fox et al., 2004). The selection heuristic studied used a greedy approach, a standard algorithmic method (Cormen et al., 2001).

---

[13] In computational complexity theory, an "oracle" is a notional process that can perform any arbitrary computation in a single operation. Oracles are used to separate different parts of computation in order to study the parts' individual complexity.

3.2 Frameworks and Standards for Practical Model Composition[14]

Although a theoretical understanding of model composition and composability is important, much composability research and development has been aimed at developing methods to achieve composability in practical implementations of simulation systems. The overall problem of developing components and a framework for those components so that they can be assembled and reassembled effectively and efficiently, which will be referred to as *composability engineering*, requires that the components be constructed so that their implementation details are compatible for the different compositions of components that might be composed.

In early composability engineering, some module level composability was achieved using dynamically loadable modules (Franceschini et al., 1999) and special purpose architectures that supported forms of composability were designed (Biddle and Perry, 2000). A study of the degree to which object frameworks could achieve a composable semiautomated forces architecture concluded that composability based on object frameworks is "an implementation issue" (Courtemanche and Burch, 2000). Multiple efforts worked toward allowing the composition of autonomous behaviors into more complex composite behaviors (von der Lippe et al., 2000; Peters et al., 2002). These efforts have achieved varying degrees of success with respect to theoretical composability. However, in practical situations the degree to which the theory is satisfied is not as important as usefulness of the composability capabilities present in the system to the developers and users of the system.

A range of technologies, tools, protocols, standards, control mechanisms, interfaces, and processes have been developed with the intention of enabling rapid, efficient, and flexible assembly of simulation systems from components in a practical setting. Broadly speaking these technologies may be classified into five approaches: common library, product line, interoperability protocol, object model, and formal. Table 1 summarizes these approaches; each is then described in turn.

The *common library* approach depends on an organizing framework for a library of reusable software modules or components. The library may include components with varying levels of composability, that is, some of the modules may be composable with all or most of the components in the library, whereas other components may work with only a small subset of the other components. Ordinarily, none of the components in a common library is a stand-alone model or simulation system that can be executed individually, although there can be exceptions. The components in the library are made composable, or reusable, through compliance with a shared common interface. The modules interoperate with the other modules in the library, or a subset of them, through the common interface. In addition to the common interface, the organizing framework may include tools, services, and standards. The components may share a common set of assumptions about the simuland[15] and how it should be modeled. The components may be developed collectively as part of the common library from the outset, or they may be developed independently and integrated later. A notable example of a

---

[14]Portions of this section were adapted and updated from Petty et al. (2014).

[15]The simuland is the system, phenomenon, entity, or process to be simulated, that is, it is the subject of a model.

TABLE 1   Approaches to Composability Engineering

Approach	Level(s) of Composability	Example(s)
Common library	Package Module Model	JMASS (Handley et al., 2000; Meyer, 2001)
Product line	Module Model Entity	OneSAF (Wittman and Harrison, 2001; Courtemanche and Wittman, 2002)
Interoperability protocol	Federate	CTPS (Petty and Windyga, 1999) CATT (Marshall, 1999) CCTT (Marshall, 1999) MATREX (RDECOM, 2008) Joint Training Confederation (Fischer, 1996) JSIMS (Carlisle et al., 2003)
Object model	Model	Base object models (SISO, 2003, 2006)
Formal	Model	MBSE (Wymore, 1993) DEVS (Zeigler et al., 2000) ForSyDe (Attarzadeh-Niaki and Sander, 2016)

Adapted from Weisel, E.W., Petty, M.D., Mielke, R.R., 2004. A survey of engineering approaches to composability. In: Proceedings of the Spring 2004 Simulation Interoperability Workshop, Arlington VA, April 18–23, pp. 722–731.

simulation system developed using the common library approach was the Joint Modeling and Simulation System (JMASS) (Handley et al., 2000; Meyer, 2001).

The *product line* approach is based on a self-contained software development system that utilizes a library of simulation components and an automated (or semiautomated) process for combining those components into specific simulation "products." The intent is to produce multiple related simulation products with as much sharing of components as possible; this is accomplished through careful preplanning of the overall product line. The simulation development system also allows modification and reuse of the components in new products. It ensures that the appropriate data transfer protocol is used for intercomponent and interproduct communication. The components may be written by different teams and still work together in various combinations as long as they comply with the product line specification, which may be quite detailed. This approach often utilizes behavior, entity, model, or module levels of composability. Composition of behavior, entity, model, or module level components results in composites that are composite behaviors, military units, composite models, or executables.

Several simulation systems use the product line approach. Of special interest is the OneSAF system. OneSAF is a constructive entity-level combat model and simulation system; it is widely used by the US Army (Wittman and Harrison, 2001; Courtemanche and Wittman, 2002; Parsons et al., 2005; Petty, 2009). The OneSAF product line architecture framework is used to define software components, including their services and interface, and the products they compose. OneSAF products are made up of one or more components. These products include a military scenario planner, a model composer, a simulation generator, an exercise configuration and setup manager, a simulation core, a simulation controller, an interface to real-world command and control systems, an after action review tool, and a software maintenance environment. OneSAF is well documented; useful sources include US Army (1998), Courtemanche and Burch (2000), Franceschini et al. (2003), Henderson and Rodriquez (2002), Henderson (2003), Grainger and Henderson (2003), Parsons et al. (2005), Reece et al. (2005), and Petty (2009), among many others.

The *interoperability protocol* approach is based on the run-time exchange of simulation data or services among independently executing simulation applications (Tolk, 2012). This approach is typically based on using a standard distributed simulation interoperability protocol such as DIS (Hofer and Loper, 1995; Loper, 1995), the HLA (Dahmann et al., 1998; Petty and Gustavson, 2012), or the Test and Training Enabling Architecture (TENA) (Noseworthy, 2008). Independent simulation models and support utilities, executing on multiple computational platforms connected by a network, are composed into distributed simulation systems. During the execution of a simulation, the applications exchange data about the state of the portion of the overall simulation each is responsible. In this approach, the components that are composed can execute independently, but they interoperate by sending and receiving data via the protocol.

A large number of simulation systems use the interoperability protocol approach. Examples include the Combat Trauma Patient Simulation (CTPS) (Pettitt et al., 1998, 2009; Petty and Windyga, 1999), the Combined Arms Tactical Trainer (CATT) (Marshall, 1999), the Close Combat Tactical Trainer (CCTT) (Marshall, 1999), the Modeling Architecture for Technology, Research, and Experimentation (MATREX) (RDECOM, 2008) the Joint Training Confederation (Fischer, 1996; Tufarolo and Page, 1996), and the Joint Simulation System (JSIMS) (Carlisle et al., 2003). These are just examples; there are many others.

The *object model* approach to composability engineering depends on a standard for model or component specifications, not implementations as with the interoperability protocol approach. Compliance with the object model standard is intended to facilitate interoperability among compliant components, and thus by extension reusability in a multiple systems. The components are not individually stand-alone simulation systems; they are instead intended to be composed with other compliant components. The object model standard enables interaction of the models with supporting tools and services as well, primarily through distributed simulation interoperability protocols. This approach utilizes the model level of composability.

An implementation of this approach is the BOM standard (SISO, 2003; Petty and Gustavson, 2012). According to that standard, a BOM is meant to be a "reusable package of information representing an independent pattern of simulation interplay" that will improve "interoperability, reuse, and composability, by providing 'patterns' and 'components' of simulation interplay that can be used as building blocks in the assembly of simulations and enterprises of simulations" (SISO, 2003). The BOM specification provides a simulation standard that allows combat model developers and simulation engineers to create modular conceptual models and composable object models, which can be used as the basis for a simulation or simulation environment (SISO, 2006). The BOM concept is based on the assumption that components of models, simulations, and federations can be either decomposed or newly developed, and then reused as building blocks in the development of a new simulation or a federation.

Two types of BOMs have been defined (SISO, 2003). Interface BOMs have messages and triggers related to one or more class of objects and provide a reusable pattern of interplay. An encapsulated BOM includes additional information, such as behaviors to be modeled. A simulation system is constructed by the composition of individual BOMs. Composite BOMs can be converted to a High-Level Architecture Federation Object Model to support interoperability via the HLA protocol (Möller et al., 2007).

In the *formal* approach, models and compositions of models are specified using a specialized formal mathematical notation. The formal approach to composability is motivated by a desire to unambiguously specify the structure and behavior of a composite model. This approach often utilizes the model level of composability. The DEVS (Zeigler et al., 2000) is an example of a simulation formalism which supports composability through the use of "coupled" (i.e., composite) models using "ports" (i.e., interfaces). The MBSE (Wymore, 1993) is another formal notation, syntactically quite different from but semantically very similar to DEVS, that has analogous properties and limits (Mielke et al., 2005). The Formal System Design (ForSyDe) methodology is a more recent formal modeling methodology offering composability and model composition capabilities; it was designed specifically for embedded and cyber-physical systems (Attarzadeh-Niaki and Sander, 2016).

3.3 Model Reuse[16]

Composability and reusability are distinct concepts and properties (Balci et al., 2011; Mahmood, 2013), but they are closely related. Models that are not composable can be reused, but composability is an important enabler for reuse (Igarza and Sautereau, 2001). The effective reuse of previously developed simulation software components is both one of the motivating goals and defining characteristics of composability.[17] In this section, various aspects of reuse, related to both software in general and simulation software in particular are surveyed.

In both software engineering in general and modeling and simulation in particular, reusing a module or component has the potential to save time, effort, and expense for development or testing. Moreover, in modeling and simulation reuse may add credibility to the new application if the component underwent verification and validation for its previous uses. Unfortunately, the reuse of software components, data sets, and other assets in modeling and simulation development is neither as frequent nor as effective as it could be, and as a consequence, the potential benefits of reuse are not being fully realized (Petty et al., 2010).

3.4 Definitions and Concepts

The concept of a *component* is fundamental in the context of both general software reuse and simulation frameworks. A component can be variously defined as (1) a reusable software module that implements and encapsulates a set of related functionality and potentially

---

[16]Portions of this section were adapted and updated from Petty et al. (2010, 2014).

[17]Arguably, composability is a special case of software reuse, in that it is possible to reuse software in ways that do not fit the definition of composability, but all composability involves reuse by definition. Furthermore, any type of software may be reused, but composability is only concerned with simulation software.

communicates with other components via an interface; (2) a unit of executable or source code that is available for reuse (Mili et al., 2002); and (3) an interchangeable unit of software that conforms to a specification and has a known set of inputs and expected output behavior but with implementation details that may be hidden or unknown (Morse et al., 2004).

In the context of modeling and simulation software, software components may implement simulation models or simulation support utilities. The former, which when a distinction is necessary will be referred to as *model components*, implement a model and thus are capable of simulating all or part of some real-world system of interest, such as physics-based model of aircraft flight dynamics (e.g., Cimini et al., 1992). The latter, which will be referred to as *support components*, provide nonmodeling support functionality specific to a simulation implementation, such as a future event list in a discrete event simulation (e.g., Banks et al., 2010). Nevertheless, both types of components are software components, and as such have all the properties and attributes of software components in general.

In software engineering, *reuse* is "using a previously developed asset again, either for the purpose for which it was originally developed or for a new purpose or in a new context" (Petty et al., 2010). In this context, an *asset* may be software components, software design diagrams, data sets, software documentation, or other artifacts of software development process. For brevity and simplicity, hereinafter assets will be assumed to be primarily software components, unless otherwise stated. *Reusability* is "the degree to which an artifact, method, or strategy is capable of being used again or repeatedly" (Balci et al., 2011). *Repeated use* is using a previously developed asset for substantially the same purpose or in the same context as previous uses; for example, running another training exercise using the same federation as the last training exercise. Repeated use is considered to be a special case of reuse; it may not require the use of specialized reuse mechanisms.

Supplemental information associated with a component that helps software developers to select and use the component is known as *metadata*. Metadata, in general, is "structured, encoded data that describe characteristics of information-bearing entities to aid in the identification, discovery, assessment, and management of the described entities" (ALCTS, 2000). In modeling and simulation, metadata associated with a software component that implements a model can describe the model's functionality, intended uses, underlying assumptions, and modeling uncertainties. The goal of such metadata is to facilitate appropriate reuse and reduce inappropriate reuse of the component (Morse et al., 2004; Taylor et al., 2015).

3.5 Software Reuse in General

As already suggested, simulation software reuse is, in large part, a special case of general software reuse. Reuse is a major issue in general software engineering and is documented in an extensive literature that includes both emerging results for researchers and practical advice for developers. At considerable peril of oversimplification, the software reuse literature can be partitioned into three broad topical areas, each with its own large literature, of which only representative examples can be mentioned:

- *Methods for implementing reusable software components*. These methods include code-level software engineering practices such as structured programming (Jensen, 1979), object-oriented development (Meyer, 1988) [both of which enables reuse by producing reusable classes and depends on reuse in the form of inheritance (Cox, 1986)], and software testing (Deutsch, 1979; Beizer, 1983). It has been argued that simulation composability and software engineering component-based software development are fundamentally the same (Bartholet et al., 2004).

- *Technical capabilities enabling software reuse.* These capabilities include software architectures (Jacobson et al., 1997), component repositories (dos Santos et al., 2009), class libraries (Mili et al., 2002), and discovery mechanisms (Hummel and Atkinson, 2004).
- *Management practices enabling and exploiting software reuse.* These practices include software development processes (Royce, 1998; Mili et al., 2002) and business case incentives (Jacobson et al., 1997).

3.6 Special Considerations for Model Reuse

Reuse has, of course, been a major goal within simulation software development as well. The work on general software reuse applies to simulation software development essentially because simulation software development is, for the most part, a special case of general software development. In modeling and simulation the components that are being composed, and the way in which these components are composed, may vary (Petty and Weisel, 2003a). The same is true at a more general level in software reuse. Modeling and simulation assets of different types, such as models or data, may be reused. Even within a single type of asset, the scope or size of the reused asset is quite variable. Consider reusable source code assets; such assets may be classes, modules, libraries, federates, and more, and any of these can be usefully reused in a suitable context.

We define a *unit* of reuse as the size and nature of the reusable asset. Table 2 provides definitions and examples of units of reuse among reusable assets. In the table, units of reuse are organized into two types and listed. For each unit, the form in which that unit is expressed is stated, examples are identified, and explanatory comments are provided.

Reusable modeling and simulation assets may be reused in different ways. This may depend on the unit of reuse, that is, a source code module will be reused in a way different than a terrain file. However, in some cases even a single unit of reuse may be reused in different ways in different circumstances. We define a *mode* of reuse as way, or method, a reusable asset is reused. Table 3 provides definitions and examples of modes of reuse among reusable assets. In the table, modes of reuse are related to when in the system development process that mode is used. Table 4 cross references the modes of reuse with the units of reuse to which the modes apply.

There are modeling and simulation-specific aspects of simulation software development that relate to reuse. Four examples are given, as follows:

- *Interoperability protocols.* The development of distributed simulation interoperability protocols, such as DIS, HLA, and TENA, have all been motivated in part by the express expectation that they would increase reuse (Hofer and Loper, 1995; Dahmann et al., 1998; Noseworthy, 2008).
- *Modeling and simulation standards.* In addition to interoperability protocols, standards of various types have been defined with the intent of increasing reuse of the standardized assets (Henninger et al., 2009); examples include natural environment

TABLE 2   Units of Reuse

Type	Unit	Expressed as	Example(s)	Comment(s)
Model	Component	Source code Object code	Network interface Event queue class Java library classes	Reusable software package that encapsulates a set of related functionality and communicates with other components via an interface. Encapsulated unit with a known set of inputs and expected output behavior where the details may be hidden or unknown. An interchangeable element of a system that conforms to a specification. AKA package or module, but compare to *Module*. Within one but not necessarily all federates in a federation; compare to *Middleware*.
	Module	Source code	Search algorithm Coordinate conversion	Reusable "chunk" of code that does not satisfy definition of Component, i.e., not encapsulated, with no defined interface. May be reused via "copy and paste." Compare to *Component*.
	Middleware	Source code Object code	RTI TENA middleware MATREX	Within all federates in a federation; compare to *Component*.
	Stand-alone model	Source code Object code	Workforce model Most CFD models	Complete implemented model that will execute as-is, e.g., an Arena model of an assembly line. Analogous to a federate but not interoperable.
	Federate	Source code Object code	Gateway OneSAF JTLS	Complete federate, reusable without modification, though it may be modified.
	Federation	Source code Object code	EnviroFed	Existing Federation rerun for new exercise or experiment. Data (e.g., scenario) may change from earlier uses.
	Service	Source code Object code	Web validation service	Similar to a component, with encapsulated functionality and interface, but not available for integration; rather invoked with RPC, web, SOA, GIG, etc.
	Modeling method (or paradigm)	Text UML	Discrete event Monte Carlo Lanchester equations Finite state machines	Set or organizing principles and common algorithms and data structures for a class of models. Category of models with common basis. Concepts and structure reused, but model(s) reimplemented.
	Model specification	Text UML	Dead reckoning models Radar return equation	Precise specification for a model that, if implemented properly, will produce anticipatable results.

TABLE 2   Unitsof Reuse—cont'd

Type	Unit	Expressed as	Example(s)	Comment(s)
	Conceptual model	Text UML	Various	Single conceptual model may have multiple implementations. Conceptual models may be modified and/or composed.
Data	Terrain file	Custom binary XML	OTF JNTC CTDB Ft. Knox	The unit here is the specific terrain file, not the file format.
	Performance data file	Custom binary Text XML	Ph/Pk tables	The unit here is the specific performance data file, not the file format.
	Other data files	Custom binary Text XML	Various	Potentially reusable data files come in a variety of categories: Configuration file, Scenario file, Visual model file, Symbol/Icon file, etc.
	Data model	HLA OMT ER Diagrams UML	RPR FOM Various BOMs TENA data model	The unit here is the data model, i.e., the structure of the data, not the actual data values. Could be categorized as type **Model**, rather than type **Data**.

TABLE 3   Modes of Reuse

Mode	When	Description
Use method	Design	Implement model using concepts and conventions of modeling method.
Follow specification	Implementation	Implement model using details of model specification.
Integrate source	Compile	Integrate source code component/module/middleware unchanged into a body of source code.
Link object	Link	Link object code component/middleware into a body of object code.
Modify source	Implementation	Make modifications to a source code reuse unit, then reuse as appropriate for the unit.
Modify data	Execution	Make modification to a data reuse unit, then reuse as appropriate for the unit.
Use as-is	Execution	Reuse unit unchanged.
Invoke service	Execution	Invoke or call reuse unit offered as service via RPC, web, SOA, GIG, etc.
Use method	Design	Implement model using concepts and conventions of modeling method.

TABLE 4   Reuse Units and Modes

	Modes							
Units	Use Method	Follow Spec	Integrate Source	Link Object	Modify Source	Modify Data	Use As-Is	Invoke Service
Component			X	X	X		X	
Module			X		X		X	
Middleware			X	X	X		X	
Stand-alone model					X		X	
Federate					X		X	
Federation					X		X	
Service								X
Modeling method	X							
Model specification		X						
Conceptual model		X						
Terrain file						X	X	
Performance data file						X	X	
Other data files						X	X	
Data model		X					X	

data (Mamaghani, 1999), simulation data exchange models (SISO, 1999), asset discovery metadata (MSCO, 2009), and aerodynamic models (Hildreth and Jackson, 2009).

- *Systems engineering processes.* Systems engineering processes for development of simulation systems have been produced that explicitly encourage reuse at various steps within the process [e.g., the Federation Development and Execution Process (FEDEP) (IEEE, 2003) and the Distributed Simulation Engineering and Execution Process (DSEEP) (IEEE, 2010)].
- *Conceptual modeling.* A simulation conceptual model has been defined as "a repository of high-level conceptual constructs and knowledge specified in a variety of communicative forms … intended to assist in the design of any type of large-scale complete [modeling and simulation] application" (Balci et al., 2011) and as documentation of "those aspects of the simuland that are to be represented and those that are to be omitted" (Petty, 2010). Extensions to the Balci-Ormsby approach for designing large-scale simulations using conceptual models (Balci and Ormsby, 2007) specifically intended to enable reuse and composability have been developed (Balci et al., 2011). A number of benefits are expected to follow from a properly constructed conceptual model, several of which could enable reuse and composability (Seo et al., 2017).

Reuse has been used, supported, and advocated in a variety of ways in simulation software development; some selected interesting examples are mentioned. Reuse of modeling and simulation assets is seen as an essential technology (Bizub et al., 2009) in developing integrated Live-Virtual-Constructive simulation systems. Successful reuse and schedule risk mitigation were demonstrated using a configuration-controlled repository in a large simulation experiment (Kleinhample, 2009). Open architecture, object-oriented development, and product line techniques produced a library of reusable assets in the context of developing naval training applications (Belanger et al., 2009). Recent research into applications of conceptual models has considered reuse of conceptual models, which are claimed to be more easily reused than implemented models (Asaduzzaman, 2009). Finally, for composability and reuse to succeed in practical applications, business issues such as the protection of intellectual property and the economics of developing reusable software must be resolved (Joshi and Winters, 2009).

4 IMPLEMENTATION CONSIDERATIONS

This section discusses two important considerations regarding the implementation of reusable or composability components for modeling and simulation software. Such components must be stored and later found to be composed or reused, and before they are stored they must be implemented in a manner that supports composition and reuse.

4.1 Implementing Composable and Reusable Components

The goals of reusability and composability affect the implementation of models as components. The additional cost of producing reusable, compared to single-use, software has long been recognized (Brooks, 1975; Royce, 1998). Of course, subsequent development is made easier by the reuse of previously developed components. A contributing cause of the additional development effort for a simulation component that implements a model is the need to document the modeling assumptions and validity limits of the model. Good modeling practices, such as using parameters instead of constants, checking inputs for validity limits, and striving for clarity in model implementation, are beneficial to the development of reusable and composable components that embody those models. Not as obvious is that the reverse relationship is also present; the goals of reusability and composability support good modeling practices, for example, the requirement to document a model's assumptions and limits of validity would help the modeler to consider his/her models more carefully at the outset.

The scope, in terms of what the component models, of components that implement models of the real world can vary considerably, from a single model of a narrowly defined domain-specific object (e.g., a model of a radio transmitter) to a complex model of many objects and their interactions (e.g., a campaign-level combat model), but the former is more common and more amenable to description in component metadata. The proper scope or functionality of a component cannot be readily given in rules that are applicable in all situations, but the composability definition of "satisfies user requirements" provides a guideline; a component should have capabilities that are useful to potential users as a unit, neither too small nor too large. The criteria for making that determination will necessarily depend on the application.

There is often an assumption by developers and potential users that any component that has been placed in a repository is valid, at least within the bounds stated in its component specification. A component's validity constraints are the limits or bounds within which the component's model is deemed valid, and may be defined at a low level in terms of physical parameter values, time step sizes, and so on, or at a high level in terms of valid applications for the component. As already noted, the validity of individual components does not mean a composition of them can be assumed to be valid (Weisel et al., 2003), and thus compositions of components must be validated. Validating a composition could use traditional validation methods, such as comparing model output data to simuland observation data (Balci, 1998; Petty, 2010). In addition, composition validation could exploit a composition's component structure, such as automatically comparing the domains of validity for each component with the data they are receiving from other components in the composition.

Components can be integrated and used with other components only through well-defined interfaces associated with each component and documented in the component specification. A component interface should define "… a set of properties, methods, and events through which external entities can connect to, and communicate with, the component" (Krieger and Adler, 1998). A broadly interpreted interface is essentially synonymous with the component specification, or metadata. A narrower, and perhaps more useful, definition of component interface focuses on the application programming interface that is required for components within the framework. In some cases components would have customizable aspects that could be modified at runtime through the interface.

4.2 Repositories and Metadata

Implementation of a library of components, and using those components to develop simulation systems, will likely occur in a collaborative team context. Automated infrastructures to support collaboration composable development will be important. Reuse and composability depend on a component-oriented organizing framework for the dynamic registration, discovery, and composition of components that include models, algorithms, services, and systems, to simplify the development of new simulations from the currently existing components. Some useful characteristics of such a framework are as follows:

- Dynamic component registration and discovery, supported by a directory (or directories) of registered components and repositories.
- Semantic query, search, and reasoning capabilities for component selection, supported by component specifications (i.e., metadata).
- Distributed processing across multiple platforms, systems, services, and domains.
- Support for intelligent and polymorphic proxies for components.
- Automated composition processes to combine components.
- Virtual repositories that include version control.
- Ability to save compositions and composition templates.
- Compliance with relevant standards.
- Software authentication and information exchange services.

Components, by definition, conform to a specification. That specification is the basis for the component's metadata. Component metadata can be used to guide the processes of selecting a set of components to be reused for a specific purpose and potentially for determining if a selected set of components can be composed. Component metadata could potentially contain a wide range of information; categories of simulation component metadata are as follows:

- *Infrastructure*: name, version, dependencies.
- *Technology*: implementation language, operating system, compiler version.
- *Interfaces*: syntax, data definitions, standards.
- *Applications*: run modes, performance, intended uses.
- *Modeling*: assumptions, time model, range of validity.
- *Provenance*: developers, prior uses, validation history.

Component "provenance" or "pedigrees," histories of a component's development, validation, and previous use are useful to support future use of a component. Documentation of the systems in which a component was used and the other components it had been composed with can provide valuable insight into when and how that component should be used in new systems. The provenance should include a record of design decisions built into the component.

5 SUMMARY AND FUTURE RESEARCH

At a conceptual level, the content of this chapter can be summarized as shown in Fig. 2. In the figure, "Composability and interoperability" and "Reuse" are capabilities which are considered to be valuable, and "frameworks" and "standards" are technologies that enable those capabilities.[18] The arrows' labels summarize the means through which the technologies and goals support each other; their direction shows that the "from" technology or goal supports the "to" technology or goal.

Model reuse has multiple motivations, including the reduction of development costs by avoiding redundant development efforts and the increased confidence of reusing models that may have already been validated. These motivations have made model reuse a major research area. Nevertheless, although progress has been made, model reuse continues to be less frequent and less effective than it could be, and arguably, should be (Petty et al., 2010). Indeed, composability was recently described as "still our biggest simulation challenge" (Taylor et al., 2015).

Additional research, both theoretical and practical, is needed to increase and enhance model composition and reuse. Selected research topics, organized into three categories (composability theory, metadata and reuse, and reuse automation) are listed as follows (Balci et al., 2017).

---

[18] As defined earlier, "composability" and "interoperability" are not the same thing. However, they are both concepts that involve models and components working together, and so are combined for this figure.

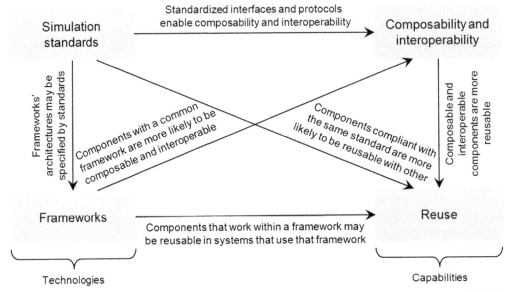

FIG. 2   Relationships among reuse-related topics. *Adapted from Petty, M.D., Kim, J., Barbosa, S.E., Pyun, J., 2014. Software frameworks for model composition. Model. Simul. Eng. 2014, 18 pages. doi:10.1155/2014/492737.*

5.1.1 Composability Theory

- Theoretical implications and consequences of composing models, understood as a special class of computable functions, in terms of the composition's combined computation.
- Formalisms and notations for expressing and abstracting model compositions.
- Validity effects of composing models at differing levels of abstraction (Fujimoto, 2016).
- Applications or insights from related theories, including model theory, category theory, and algorithmic information theory, for model composition.
- Models structures and modes of model composition that may preserve validity, or aspects of validity, under composition (Tolk et al., 2015).

5.1.2 Model Metadata

- Formalisms and formal languages for expressing model metadata, such as predicate calculus or ontology languages.
- Identification and enumeration of model characteristics, assumptions, and abstractions that should be represented in metadata.
- Standardized vocabularies or lexicons for model metadata.
- Algorithms to automatically generate metadata from a model's implementation as a component, or to verify metadata against a component.

5.1.3 Reuse Automation and Frameworks

- Algorithms or heuristics to automate or semi-automate the selection and composition of models (or their component implementations) (Fujimoto, 2016).

- Effective practical model reuse patterns, analogous to practical software design patterns (Gamma et al., 1995).
- Automated support for verification and validation of compositions, perhaps based on the composed components' metadata.
- Software frameworks designed to support and facilitate models (Petty et al., 2014).
- Model development and implementation standards specifically designed to enhance model reusability (Fujimoto, 2016).

Acknowledgments

Some portions of this chapter are based on research conducted in collaboration with the Johns Hopkins University Applied Physics Laboratory and supported by the US Department of Defense; this work is described in Petty et al. (2010). Other portions are based on research performed in collaboration with REALTIMEVISUAL Inc. and supported by the Republic of Korea's Agency for Defense Development; this work is described in Petty et al. (2014). The Simulation Interoperability Standards Organization gave permission for the adaptation of portions of the authors' prior Simulation Interoperability Workshop publications. Earlier versions of some portions of this chapter were developed from the same original research mentioned above and were documented in the final report of a workshop organized by the US National Science Foundation, later published in Balci et al. (2017). The US Army and VT MÄK gave permission to the use images on the left and right screen, respectively, in Fig. 1, which is gratefully acknowledged.

References

Achermann, F., Nierstrasz, O., 2005. A calculus for reasoning about software composition. Theor. Comput. Sci. 331 (2–3), 367–396.

Asaduzzaman, A., 2009. Conceptual modeling of multicore high performance computing systems.Proceedings of the 2009 Huntsville Simulation Conference, Huntsville, AL, October 28–29.

Association for Library Collections and Technical Services, Committee on Cataloging: Description and Access, Task Force on Metadata, 2000. Final Report, June 16. Online at: http://www.libraries.psu.edu/tas/jca/ccda/tf-meta6.html.

Attarzadeh-Niaki, S., Sander, I., 2016. An extensible modeling methodology for embedded and cyber-physical system design. Simul.: Trans. Soc. Model. Simul. Int. 92 (8), 771–794.

Balci, O., 1998. Verification, validation, and testing. In: Banks, J. (Ed.), Handbook of Simulation: Principles, Methodology, Advances, Applications, and Practice. John Wiley & Sons, New York, NY, pp. 335–393.

Balci, O., Ormsby, W.F., 2007. Conceptual modelling for designing large-scale simulations. J. Simul. 1 (3), 175–186.

Balci, O., Arthur, J.D., Ormsby, W.F., 2011. Achieving reusability and composability with a simulation conceptual model. J. Simul. 5 (3), 157–165. https://dx.doi.org/10.1057/jos.2011.7.

Balci, O., Ball, G.L., Morse, K.L., Page, E., Petty, M.D., Tolk, A., Veautour, S.N., 2017. Model reuse, composition, and adaptation. In: Fujimoto, R., Bock, C., Chen, W., Page, E., Panchal, J.H. (Eds.), Research Challenges in Modeling and Simulation for Engineering Complex Systems. Springer-Verlag, Cham, Switzerland, pp. 87–116.

Banks, J., Carson, J.S., Nelson, B.L., Nicol, D.M., 2010. Discrete-Event System Simulation, fifth ed. Prentice Hall, Upper Saddle River, NJ.

Bartholet, R.G., Brogan, D.C., Reynolds, P.F., Carnahan, J.C., 2004. In search of the philosopher's stone: simulation composability versus component-based software design.Proceedings of the Fall 2004 Simulation Interoperability Workshop, Orlando, FL, September 19–24.

Beizer, B., 1983. Software Testing Techniques. Van Nostrand Reinhold, New York, NY.

Beizer, B., 1995. Black-Box Testing. John Wiley & Sons, New York, NY.

Belanger, W., McInnis, S., Beatty, W., 2009. Component reuse across multiple modeling and simulation programs. Proceedings of the 2009 Huntsville Simulation Conference, Huntsville, AL, October 28–29.

Benali, H., Bellamine-Ben Saoud, N., 2011. Towards a component-based framework for interoperability and composability in modeling and simulation. Simul.: Trans. Soc. Model. Simul. Int. 87 (1–2), 133–148.

Biddle, M., Perry, C., 2000. An architecture for composable interoperability.Proceedings of the Fall 2000 Simulation Interoperability Workshop, Orlando, FL, September 17–22.

Bizub, W., Wallace, J., Ceranowicz, A., Powell, E., 2009. Next-generation live virtual constructive architecture framework.Proceedings of the 2009 Interservice/Industry Training, Simulation, and Education Conference, Orlando, FL, November 30–December 3 2009.

Brooks, F.P., 1975. The Mythical Man-Month: Essays on Software Engineering. Addison-Wesley, Reading, MA.

Bunus, P., Fritzson, P., 2004. Automated static analysis of equation-based components. Simul.: Trans. Soc. Model. Simul. Int. 80 (7–8), 321–345.

Calder, R.B., Smith, J.E., Courtemanche, A.J., Mar, J.M.F., Ceranowicz, A.Z., 1993. ModSAF behavior simulation and control.Procedings of the Third Conference on Computer Generated Forces and Behavioral Representation, Orlando, FL, March 17–19, pp. 347–356.

Carlisle, P., Babineau, W., Wuerfel, R., 2003. The Joint Simulation System (JSIMS) federation management toolbox. Proceedings of the Fall 2003 Simulation Interoperability Workshop, Orlando, FL, September 14–19.

Castro, P.E., Antonsson, E., Clements, D.T., Coolahan, J.E., Ho, Y., Horter, M.A., Khosla, P.K., Lee, J., Mitchiner, J.L., Petty, M.D., Starr, S., Wu, C.L., Zeigler, B.P., 2002. Modeling and Simulation in Manufacturing and Defense Systems Acquisition, Pathways to Success. National Research Council, Washington, DC.

Ceranowicz, A.Z., 1994. ModSAF capabilities.Proceedings of the Fourth Conference on Computer Generated Forces and Behavioral Representation, Orlando, FL, May 4–6, pp. 3–8.

Ceranowicz, A.Z., June 10, 2002. Composability Wrapup, Personal Communication.

Cimini, F.C., Campbell, C.E., Petty, M.D., 1992. A simple flight dynamics model for computer generated forces. Proceedings of the Southeastern Simulation Conference 1992, Pensacola, FL, October 22–23pp. 41–47.

Clark, J., Clarke, C., De Panfilis, S., Granatella, G., Predonzani, P., Sillitti, A., Succi, G., Vernazza, T., 2004. Selecting components in large COTS repositories. J. Syst. Softw. 73 (2), 323–331.

Colley, W.N., 2010. Modeling continuous systems. In: Sokolowski, J.A., Banks, C.M. (Eds.), Modeling and Simulation Fundamentals: Theoretical Underpinnings and Practical Domains. John Wiley & Sons, Hoboken, NJ, pp. 99–130.

Collins, J.B., 2004. Standardizing an ontology of physics for modeling and simulation.Proceedings of the Fall 2004 Simulation Interoperability Workshop, Orlando, FL, September 19–24.

Collins, J.B., Clark, D., 2004. Towards an ontology of physics.Proceedings of the 2004 European Simulation Interoperability Workshop, Edinburgh, Scotland, June 28–July 1.

Cormen, T.H., Leiserson, C.E., Rivest, R.L., Stein, C., 2001. Introduction to Algorithms, second ed. The MIT Press, Cambridge, MA.

Courtemanche, A.J., Burch, R.B., 2000. Using and developing object frameworks to achieve a composable CGF architecture.Proceedings of the Ninth Conference on Computer Generated Forces and Behavioral Representation, Orlando, FL, May 16–18, pp. 49–62.

Courtemanche, A.J., Wittman, R.L., 2002. OneSAF: a product line approach for a next-generation CGF.Proceedings of the Eleventh Conference on Computer Generated Forces and Behavioral Representation, Orlando, FL, May 7–9, pp. 349–361.

Cox, B.J., 1986. Object-Oriented Programming: An Evolutionary Approach. Addison-Wesley, Reading, MA.

Dahmann, J.S., Kuhl, F., Weatherly, R., 1998. Standards for simulation: as simple as possible but not simpler: the high level architecture for simulation. Simulation 71 (6), 378–387.

Davis, P.K., Anderson, R.H., 2003. Improving the Composability of Department of Defense Models and Simulations. RAND National Defense Research Institute, Santa Monica, CA.

Davis, P.K., Anderson, R.H., 2004. Improving the composability of DoD models and simulations. J. Def. Model. Simul. 1 (1), 5–17.

Davis, M.D., Sigal, R., Weyuker, E.J., 1994. Computability, complexity, and languages. In: Fundamentals of Theoretical Computer Science, second ed. Morgan Kaufmann, San Diego, CA.

Davis, P.C., Fishwick, P.A., Overstreet, C.M., Pegden, C.D., 2000. Model composability as a research investment: responses to the featured paper.Proceedings of the 2000 Winter Simulation Conference, Orlando, FL, December 10–13, pp. 1585–1591.

Department of Defense Modeling and Simulation Coordination Office, February 20, 2009. Modeling and Simulation (M&S) Community of Interest (COI) Discovery Metadata Specification (MSC-DMS), Version 1.2. On-line at: http://www.msco.mil/resource_discovery.html. [(Accessed 17 March 2009)].

Deutsch, M.S., 1979. Verification and validation. In: Jensen, R.W., Tonies, C.C. (Eds.), Software Engineering. Prentice-Hall, Englewood Cliffs, NJ, pp. 329–408.

Diaz-Calderon, A., Paredis, C.J.J., Khosla, P.K., 2000. Organization and selection of reconfigurable models. In: Proceedings of the 2000 Winter Simulation Conference, Orlando, FL, December 10–13, pp. 386–393.

dos Santos, R.P., Werner, C.M.L., da Silva, M.A., 2009. Incorporating information of value in a component repository to support a component marketplace infrastructure.Proceedings of the 10th IEEE International Conference on Information Reuse & Integration, Las Vegas, NV, August 10–12, pp. 266–271.

Fischer, M.C., 1996, Joint training confederation.Proceedings of the First International Simulation Technology and Training Conference, Melbourne, Australia, March 25–26.

Fox, M.R., Brogan, D.C., Reynolds, P.F., 2004. Approximating Component Selection.Proceedings of the 2004 Winter Simulation Conference, Washington, DC, December 5–8pp. 429–434.

Franceschini, D., Zimmerman, J., McCulley, G., 1999. CGF system composability through dynamically loadable modules.Proceedings of the Eighth Conference on Computer Generated Forces and Behavioral Representation, Orlando, FL, May 11–13pp. 341–347.

Franceschini, D.J., Hawkes, K.R., Graffuis, S., 2003. System composition in OneSAF.Proceedings of the Spring 2003 Simulation Interoperability Workshop, Kissimmee, FL, March 30–April 4.

Fujimoto, R.M., 2016. Research challenges in parallel and distributed simulation. ACM Trans. Model. Comput. Simul. 26 (4) (Article 22).

Gamma, E., Helm, R., Johnson, R., Vlissides, J., 1995. Design Patterns: Elements of Reusable Object-Oriented Software. Addison-Wesley, Upper Saddle River, NJ.

Garey, M.R., Johnson, D.S., 1979. Computers and Intractability, A Guide to the Theory of NP-Completeness. W.H. Freeman and Company, New York, NY.

Grainger, B., Henderson, C., 2003. Battlespace composition in the OneSAF objective system.Proceedings of the Spring 2003 Simulation Interoperability Workshop, Kissimmee, FL, March 30–April 4.

Handley, V.K., Shea, P.M., Morano, M., 2000. An introduction to the Joint Modeling and Simulation System (JMASS). Proceedings of the Fall 2000 Simulation Interoperability Workshop, Orlando, FL, September 17–22.

Harkrider, S.M., Lunceford, W.H., 1999. Modeling and simulation composability.Proceedings of the 1999 Interservice/Industry Training, Simulation and Education Conference, Orlando, FL, November 29 1999–December 2, pp. 876–881.

Harmon, S.Y., 1996. Interoperability between distributed simulations I: interacting models of physical processes. Proceedings of the 15th Workshop on the Interoperability of Distributed Interactive Simulation, Orlando, FL, September 16–20.

Henderson, C., 2003. Model execution in the OneSAF objective system.Proceedings of the Spring 2003 Simulation Interoperability Workshop, Kissimmee, FL, March 30–April 4.

Henderson, C., Rodriquez, A., 2002. Modeling in OneSAF.Proceedings of the Eleventh Conference on Computer Generated Forces and Behavioral Representation, Orlando, FL, May 7–9pp. 337–347.

Henninger, A.E., Morse, K.L., Loper, M.L., Gibson, R.D., 2009. Developing a process for M&S standards management within DoD.Proceedings of the 2009 Interservice/Industry Training, Simulation, and Education Conference, Orlando, FL, November 30–December 3.

Hildreth, B., Jackson, E.B., 2009. Benefits to the simulation training community of a new ANSI standard for the exchange of aero simulation models.Proceedings of the 2009 Interservice/Industry Training, Simulation, and Education Conference, Orlando, FL, November 30–December 3.

Hofer, R.C., Loper, M.L., 1995. DIS Today. Proc. IEEE 83 (8), 1124–1137.

Hummel, O., Atkinson, C., 2004. Extreme harvesting: test driven discovery and reuse of software components.- Proceedings of the 2004 IEEE International Conference on Information Reuse and Integration, Las Vegas, NV, November 8–10, pp. 66–72.

Igarza, J., Sautereau, C., 2001. Distribution, use and reuse: questioning the cost effectiveness of re-using simulations with and without HLA.Proceedings of the Fall 2001 Simulation Interoperability Workshop, Orlando, FL, September 9–14.

Institute of Electrical and Electronics Engineers, 2003. IEEE Std 1516.3-2003—IEEE Recommended Practice for High Level Architecture (HLA) Federation Development and Execution Process (FEDEP), New York, NY.

Institute of Electrical and Electronics Engineers, 2010. IEEE Std 1730-2010 (Revision of IEEE Std 1516.3-2003)—IEEE Recommended Practice for Distributed Simulation Engineering and Execution Process (DSEEP), New York, NY.

Jackson, D., 2006. Dependable Software by Design. Scientific American, New York, NY, pp. 68–75.

Jacobson, I., Griss, M., Jonsson, P., 1997. Software Reuse: Architecture, Process, and Organization for Success. ACM Press, New York, NY.

Jensen, R.W., 1979. Structured programming. In: Jensen, R.W., Tonies, C.C. (Eds.), Software Engineering. Prentice-Hall, Englewood Cliffs, NJ, pp. 221–328.

Joshi, B., Winters, J.D., 2009. The legal fork in the OTD roadmap–what lies ahead? Proceedings of the 2009 Interservice/Industry Training, Simulation, and Education Conference, Orlando, FL, November 30–December 3.

JSIMS Composability Task Force, September 30, 1997. JSIMS Composability Task Force Final Report.

Kasputis, S., Ng, H.C., 2000. Composable simulations.Proceedings of the 2000 Winter Simulation Conference, Orlando, FL, December 10–13, pp. 1577–1584.

Kasputis, S., Osvalt, I., McKay, R., Barber, S., 2004. Semantic descriptors of models and simulations.Proceedings of the Spring 2004 Simulation Interoperability Workshop, Arlington, VA, April 18–23.

Kaur, P., Singh, J., Singh, H., 2014. Component selection repository with risk identification.Proceedings of the 2014 International Conference on Computing for Sustainable Global Development (INDIACom), New Delhi, India, March 5–7, pp. 524–531.

Kleinhample, R.C., 2009. Reuse—don't throw out the baby with the bathwater.Proceedings of the 2009 Interservice/Industry Training, Simulation, and Education Conference, Orlando, FL, November 30–December 3.

Krieger, D., Adler, R., 1998. The emergence of distributed component platforms. IEEE Comput, 43–53.

Loper, M.L., 1995. Introduction to distributed interactive simulation. In: Clarke, T.L. (Ed.), Distributed Interactive Simulation Systems for Simulation and Training in the Aerospace Environment. SPIE Optical Engineering Press, Bellingham, WA, pp. 3–15.

Mahmood, I., 2013. A Verification Framework for Component Based Modeling and Simulation: "Putting the Pieces Together" (Ph.D. Thesis). KTH School of Information and Communication Technology.

Mamaghani, F., 1999. SEDRIS as a standard for interchange virtual world data sets.Proceedings of IEEE Virtual Reality 1999, Houston, TX, March 13–17, p. 74.

Marshall, H.A., 1999. SAF in CATT training systems, update 1999.Proceedings of the Eighth Conference on Computer Generated Forces and Behavioral Representation, Orlando, FL, May 11–13, pp. 277–283.

Meyer, B., 1988. Object-Oriented Software Development. Prentice-Hall, New York, NY.

Meyer, R.J., 2001. Joint Modeling and Simulation System (JMASS): what it does, and what it doesn't! Proceedings of the Spring 2001 Simulation Interoperability Workshop, Orlando, FL, March 25–30.

Mielke, R.R., Phillips, M.A., 2003. Development and application of an academic battle lab.Proceedings of the 2003 Interservice/Industry Training, Simulation and Education Conference, Orlando, FL, December 1–4, pp. 1371–1380.

Mielke, R.R., Petty, M.D., Weisel, E.W., 2005. A comparison of model-based systems engineering and composability theory.Proceedings of the Huntsville Simulation Conference 2005, Huntsville, AL, October 25–27, pp. 300–308.

Mili, H., Mili, A., Yacoub, S., Addy, E., 2002. Reuse Based Software Engineering: Techniques, Organization, and Controls. John Wiley & Sons, New York, NY.

Möller, B., Gustavson, P., Lutz, B., Löfstrand, B., 2007. Making your BOMs and FOM modules play together.Proceeding of the 2007 Fall Simulation Interoperability Workshop, Orlando, FL, September 16–21.

Morse, K.L., Petty, M.D., Reynolds, P.F., Waite, W.F., Zimmerman, P.M., 2004. Findings and recommendations from the 2003 composable mission space environments workshop.Proceedings of the Spring 2004 Simulation Interoperability Workshop, Arlington, VA, April 18–23, pp. 313–323.

Mosses, P.D., 1990. Denotational semantics. In: van Leeuwen, J. (Ed.), Handbook of Theoretical Computer Science, Volume B—Formal Models and Semantics. Elsevier, Amsterdam, The Netherlands, pp. 573–631.

Noseworthy, J.R., 2008. The Test and Training Enabling Architecture (TENA)—supporting the decentralized development of distributed applications and LVC simulations.Proceedings of the 12th IEEE/ACM International Symposium on Distributed Simulation and Real-Time Applications, Vancouver, Canada, October 27–29, pp. 259–268.

Overstreet, C.M., 1982. Model Specification and Analysis for Discrete Event Simulation (Ph.D. Dissertation). Virginia Polytechnic Institute and State University.

Overstreet, C.M., Nance, R.E., 1985. A specification language to assist in analysis of discrete event simulation models. Commun. ACM 28 (2), 190–201.

Page, E.H., Opper, J.M., 1998. Theory and practice in user-composable simulation systems.Presentation for DARPA Advanced Simulation Technology Thrust, October 30.

Page, E.H., Opper, J.M., 1999. Observations on the complexity of composable simulation.Proceedings of the 1999 Winter Simulation Conference, Phoenix, AZ, December 5–8, pp. 553–560.

Page, E.H., Briggs, R., Tufarolo, J.A., 2004. Toward a family of maturity models for the simulation interconnection problem.Proceedings of the Spring 2004 Simulation Interoperability Workshop, Arlington, VA, April 18–23, pp. 1059–1069.

Parsons, D., Surdu, J., Jordan, B., 2005. OneSAF: a next generation simulation modeling the contemporary operating environment.Proceedings of the 2005 European Simulation Interoperability Workshop, Toulouse, France, June 27–29.

Peng, D., Warnke, T., Haack, F., Uhrmacher, A.M., 2017. Reusing simulation experiment specifications in developing models by successive composition—a case study of the Wnt/β-catenin signaling pathway. Simul.: Trans. Soc. Model. Simul. Int. 93 (8), 659–677.

Peters, S.D., LaVine, N.D., Napravnik, L., Lyons, D.M., 2002. Composable behaviors in an entity based simulation. Proceedings of the Spring 2002 Simulation Interoperability Workshop, Orlando, FL, March 10–15.

Pettitt, M.B., Goldiez, B.F., Petty, M.D., Rajput, S., Tu, H., 1998. The combat trauma patient simulator.Proceedings of the 1998 Spring Simulation Interoperability Workshop, Orlando, FL, March 9–13, pp. 936–946.

Pettitt, M.B., Mayo, M., Norfleet, J., 2009. Medical simulation training simulations. In: Cohn, J., Nicholson, D., Schmorrow, D. (Eds.), The PSI Handbook of Virtual Environment Training and Education: Developments for the Military and Beyond, Volume 3: Integrated Systems, Training Evaluations, and Future Directions. Praeger Security International, Westport, CT, pp. 99–106.

Petty, M.D., 1995. Computer generated forces in distributed interactive simulation. In: Clarke, T.L. (Ed.), Distributed Interactive Simulation Systems for Simulation and Training in the Aerospace Environment, SPIE Critical Reviews of Optical Science and Technology. In: vol. CR58. SPIE Press, Bellingham, WA, pp. 251–280.

Petty, M.D., 2004. Simple composition suffices to assemble any composite model.Proceedings of the Spring 2004 Simulation Interoperability Workshop, Arlington, VA, April 18–23, pp. 299–307.

Petty, M.D., 2006. Corrigendum to 'computational complexity of selecting components for composition'.Proceedings of the Fall 2006 Simulation Interoperability Workshop, Orlando, FL, September 10–15, pp. 489–490.

Petty, M.D., 2009. Behavior generation in semi-automated forces. In: Nicholson, D., Schmorrow, D., Cohn, J. (Eds.), The PSI Handbook of Virtual Environment Training and Education: Developments for the Military and Beyond, Volume 2: VE Components and Training Technologies. Praeger Security International, Westport, CT, pp. 189–204.

Petty, M.D., 2010. Verification, validation, and accreditation. In: Sokolowski, J.A., Banks, C.M. (Eds.), Modeling and Simulation Fundamental: Theoretical Underpinnings and Practical Domains. John Wiley & Sons, Hoboken, NJ, pp. 325–372.

Petty, M.D., Gustavson, P., 2012. Combat modeling with the high level architecture and base object models. In: Tolk, A. (Ed.), Engineering Principles of Combat Modeling and Distributed Simulation. John Wiley & Sons, Hoboken, NJ, pp. 413–448.

Petty, M.D., Weisel, E.W., 2003a. A composability lexicon.Proceedings of the Spring 2003 Simulation Interoperability Workshop, Orlando, FL, March 30–April 4, 2003, pp. 181–187.

Petty, M.D., Weisel, E.W., 2003b. A formal basis for a theory of semantic composability.Proceedings of the Spring 2003 Simulation Interoperability Workshop, Orlando, FL, March 30–April 4 2003, pp. 416–423.

Petty, M.D., Windyga, P.S., 1999. A high level architecture-based medical simulation. Simulation 73 (5), 279–285.

Petty, M.D., Weisel, E.W., Mielke, R.R., 2003a. Computational complexity of selecting components for composition. Proceedings of the Fall 2003 Simulation Interoperability Workshop, Orlando, FL, September 14–19, pp. 517–525.

Petty, M.D., Weisel, E.W., Mielke, R.R., 2003b. A formal approach to composability.Proceedings of the 2003 Interservice/Industry Training, Simulation and Education Conference, Orlando, FL, December 1–4, pp. 1763–1772.

Petty, M.D., Weisel, E.W., Mielke, R.R., 2005. Composability theory overview and update.Proceedings of the Spring 2005 Simulation Interoperability Workshop, San Diego, CA, April 3–8, pp. 431–437.

Petty, M.D., Morse, K.L., Riggs, W.C., Gustavson, P., Rutherford, H., 2010. A reuse lexicon: terms, units, and modes in M&S asset reuse.Proceedings of the Fall 2010 Simulation Interoperability Workshop, Orlando, FL, September 20–24.

Petty, M.D., Kim, J., Barbosa, S.E., Pyun, J., 2014. Software frameworks for model composition. Model. Simul. Eng. 2014, 18 pages, https://doi.org/10.1155/2014/492737.

Petty, M.D., Kim, J., Park, S., Lee, S., 2016. A methodology for quantitative assessment of the features and capabilities of software frameworks for model composition. Int. J. Model. Simul. Sci. Comput. 7 (1) 21 pages, https://doi.org/10.1142/S1793962315410020.

Porcarelli, S., Castaldi, M., Di Giandomenico, F., Bondavalli, A., Inverardi, P., 2005. A framework for reconfiguration-based fault-tolerance in distributed systems. In: de Lemos, R., Gacek, C., Romanovsky, A. (Eds.), Architecting Dependable Systems, Volume 2, Lecture Notes in Computer Science. Springer-Verlag, New York, NY.

Post, G.M., June 12, 2002. J9 Composability Summary Comments, Personal Communication.

Pratt, D.R., Ragusa, L.C., von der Lippe, S., 1999. Composability as an architecture driver.Proceedings of the 1999 Interservice/Industry Training, Simulation and Education Conference, Orlando, FL, November 29 1999–December 2, pp. 882–891.

Pressman, R.S., Maxim, B.R., 2015. Software Engineering: A Practitioner's Approach, eighth ed. McGraw Hill Education, New York, NY.

Reece, D.A., McCormack, J., Zhang, J., 2005. A case-based behavior design aid for OneSAF.Proceedings of the Fourteenth Conference on Behavior Representation in Modeling and Simulation, Universal City, CA, May 16–19, pp. 191–199.

Royce, W., 1998. Software Project Management: A Unified Framework. Addison-Wesley, Reading, MA.

Seo, K., Hong, W., Kim, T.G., 2017. Enhancing model composability and reusability for entity-level combat simulation: a conceptual modeling approach. Simul.: Trans. Soc. Model. Simul. Int. 93 (10), 825–840.

Simulation Interoperability Standards Organization, 1999. SISO-STD-001.1-1999: Real-time Platform Reference Federation Object Model (RPR FOM 1.0).

Simulation Interoperability Standards Organization, October 10, 2003. Base Object Model (BOM) Template Specification Volume I—Interface BOM, SISO-STD-003.1-2003-DRAFT-V0.7.

Simulation Interoperability Standards Organization, 2006. Base Object Model (BOM) Template Specification, SISO-STD-003-2006. Online at:www.sisostds.org. [(Accessed 7 May 2011)].

Sun, Y., August 2006. Complexity of System Configuration Management. (Ph.D. Dissertation). Tufts University.

Sun, Y., Couch, A., 2007. Complexity of system configuration management. In: Bergstra, J., Burgess, M. (Eds.), Handbook of Network and System Administration. Elsevier, Amsterdam, The Netherlands, pp. 623–652.

Szabo, C., Teo, Y.M., 2012. An analysis of the cost of validating semantic composability. J. Simul. 6 (3), 1–12.

Taylor, S.J.E., Khan, A., Morse, K.L., Tolk, A., Yilmaz, L., Zander, J., Mosterman, P.J., 2015. Grand challenges for modeling and simulation: simulation everywhere—from cyberinfrastructure to clouds to citizens. Simul.: Trans. Soc. Model. Simul. Int. 91 (7), 648–665.

Tolk, A., 2012. Standards for distributed simulation. In: Tolk, A. (Ed.), Engineering Principles of Combat Modeling and Distributed Simulation. John Wiley & Sons, Hoboken, NJ, pp. 209–241.

Tolk, A., Muguira, J.A., 2003. The Levels of Conceptual Interoperability Model (LCIM).Proceedings of the Fall 2003 Simulation Interoperability Workshop, Orlando, FL, September 14–19, pp. 53–62.

Tolk, A., Balci, O., Combs, C.D., Fujimoto, R., Macal, C.M., Nelson, B.L., Zimmerman, P., 2015. Do we need a national research agenda for modeling and simulation?.Proceedings of the 2015 Winter Simulation Conference, Huntington Beach, CA, December 6–9pp. 2571–2585.

Tufarolo, J.A., Page, E.H., 1996. Evolving the VV&A process for the ALSP joint training confederation.Proceedings of the 1996 Winter Simulation Conference, Coronado, CA, December 8–11, pp. 952–958.

Turing, A.M., 1936. On computable numbers, with an application to the entscheidungs-problem. Proc. Lond. Math. Soc. Ser. 2 42, 230–265 (Correction, ibid. 43, 544–546, 1937).

Turnitsa, C.D., 2005. Extending the Levels of Conceptual Interoperability Model.Proceedings 2005 Summer Simulation Multiconference, Cherry Hill, NJ, July 24–28, pp. 479–487.

U. S. Army Research, Development, and Engineering Command, 2008. MATREX simulation architecture.Proceedings of the 2008 Department of Defense Modeling and Simulation Conference, Orlando, FL, March 10–14 Online at:www.matrex.rdecom.army.mil. [(Accessed 7 May 2011)].

United States Army, August 21, 1998. One Semi-Automated Forces Operational Requirements Document, Version 1.1. Online Document at URL:http://www-leav.army.mil/nsc/stow/saf/onesaf/onesaf.htm/.

von der Lippe, S., McCormack, J.S., Kalphat, M., 2000. Embracing temporal relations and command and control in composable behavior technologies.Proceedings of the Ninth Conference on Computer Generated Forces and Behavioral Representation, Orlando, FL, May 16–18, pp. 183–192.

Waziruddin, S., Brogan, D.C., Reynolds, P.F., 2003. The process for coercing simulations.Proceedings of the Fall 2003 Simulation Interoperability Workshop, Orlando, FL, September 14–19.

Weisel, E.W., May 2004. Models, Composability, and Validity (Ph.D. Dissertation)Old Dominion University.

Weisel, E.W., Mielke, R.R., Petty, M.D., 2003. Validity of models and classes of models in semantic composability. Proceedings of the Fall 2003 Simulation Interoperability Workshop, Orlando, FL, September 14–19, pp. 526–536.

Weisel, E.W., Petty, M.D., Mielke, R.R., 2004. A survey of engineering approaches to composability.Proceedings of the Spring 2004 Simulation Interoperability Workshop, Arlington, VA, April 18–23, pp. 722–731.

Weisel, E.W., Petty, M.D., Mielke, R.R., 2005. A comparison of DEVS and semantic composability theory.Proceedings of the Spring 2005 Simulation Interoperability Workshop, San Diego, CA, April 3–8, pp. 956–964.

Wittman, R.L., Harrison, C.T., 2001. OneSAF: a product line approach to simulation development.Proceedings of the European 2001 Simulation Interoperability Workshop, London, UK, June 25–27.

Wood, D.D., Petty, M.D., 1995. Electronic warfare and distributed interactive simulation. In: Clarke, T.L. (Ed.), Distributed Interactive Simulation Systems for Simulation and Training in the Aerospace Environment, SPIE Critical Reviews of Optical Science and Technology. In: vol. CR58. SPIE Press, Bellingham, WA, pp. 179–194.

Wymore, A.W., 1993. Model-Based Systems Engineering. CRC Press, Boca Raton, FL.

Zeigler, B.P., Praehofer, H., Kim, T.G., 2000. Theory of Modeling and Simulation: Integrating Discrete Event and Continuous Complex Dynamic Systems, second ed. Academic Press, San Diego, CA.

Service-Agent-Based Model Composition

Lin Zhang, Chun Zhao, Feng Li

School of Automation Science and Electrical Engineering, Beihang University, Beijing, China

1 INTRODUCTION

Model composition is one of the key technologies of model engineering and an important way of constructing large-scale complex systems (e.g., systems of systems). Especially for complex simulation systems, different component models have to be selected with respect to different simulation scenarios to rapidly build a new simulation system. This places great demands on the degree of automation and intelligence of model composition. In recent years, it is becoming a trend to encapsulate models into services and realize model compositions with the help of service composition technologies.

The concept of cloud simulation was proposed by Li et al. (2009), in which modeling and simulation (M&S) resources were encapsulated into services to be shared by M&S users with the support of the cloud simulation platform. This idea was also named as modeling & simulation as a service (MSaaS), which is a concept that combines service-based approaches and cloud computing (Cayirci, 2013; Siegfried et al., 2014). NATO Modeling and Simulation Group MSG-131 defined M&Saas as: "MSaaS is a means of delivering value to customers to enable or support modeling and simulation user applications and capabilities as well as to provide associated data on demand without the ownership of specific costs and risks." As such, MSaaS is an architectural and organizational approach that promotes abstraction, loose coupling, reusability, composability, and discovery of M&S services. The objective of MSaaS is to effectively and efficiently support operational requirements and to improve development, operation, and maintenance of M&S applications. Cayirci (2013) pointed out that MSaaS is a model for provisioning M&S services on demand from a cloud service provider (CSP), which keeps the underlying infrastructure, platform, and software requirements/details hidden from the users. Siegfried et al. (2014) indicated that technical development in the area of cloud computing technology and service-oriented architectures (SOAs) offers opportunities to better utilize M&S capabilities.

Tolk et al. (2006) and Tolk (2013) presented the idea of composable M&S services. The levels of conceptual interoperability model (LCIM) was developed to manage the basic challenges of service interoperation. The model theory was used to investigate definitions of interoperability and composability and to provide the implications for verification and validation procedures, model-based approaches, and simulation interoperability standards. Zeigler et al. (2000) and Zeigler and Sarjoughian (2012) developed a set of generic SOA-based models based on the discrete event system specification (DEVS) framework so that "virtual build and test" of systems of systems comply with SOA standards. Wang and Wainer (2016) proposed a cloud-based simulation (CBS) architecture to deploy resources as services, named Cloud Architecture for Modeling and Simulation as a Service. Guo et al. (2011) presented an architecture and provided a specification for simulation software as a service and service-oriented experiment, which can support automatic deployment of simulation services for carrying out experiments. Tsai et al. (2011) introduced the concept of simulation software-as-a-service (SimSaaS) with a multitenancy architecture configuration model and a cloud-based runtime to support rapid simulation development in an elastic cloud environment. Walker et al. (2016) showed that how Galaxy tools can be used to realize the idea of models and simulations as a service (MaSS), in which models and simulations were delivered as easy to use, on-demand web services (WSs) accessed through the user's browser. Galaxy allows users to seamlessly customize and run simulations on cloud computing resources. Wang and Wainer (2014) proposed a simulation as a service methodology with application for crowd modeling, simulation, and visualization and presented a method based on a distributed architecture with simulation in the cloud, and composition using workflows.

In the process of model composition, there are two fundamental phases, namely model matching and model coupling. Many researchers are involved in the research of model matching which ranges from manual combination to fully automatic algorithms. Clarke and Walker (2002) proposed a standard design language for aspect-oriented software development which can provided a means to assess these languages and their incompatibilities. Fleurey et al. (2010) proposed a generic framework for composition. This framework is independent of a modeling language. A method that shows how the generic composition is specialized for class diagrams has been illustrated. Mandelin (2006) represented models and diagrams as graphs whose nodes have attributes such as name, type, connections, and containment relations. Based on probabilistic models, high-quality correspondences can be found using search algorithms. Nejati et al. (2007) described two operators for manipulating hierarchical state charts, which involve match and merge. Match operator is heuristic, which can improve the accuracy of matching based on both static and behavioral properties of the models. Meanwhile, merge operator preserves the hierarchical structure of the input models. In this way, merges that preserves the semantics of models were constructed automatically. Semantic composition methods have gained more attention in the past decade. The implementation of semantic composition was based on rule-based languages and framework. Aiming at the problem of lacking declarative semantic definition, Rubin et al. (2008) proposed a new declarative procedure for model matching and merging. Based on this procedure, property-driven framework for model composition was presented. Reddy et al. (2005) presented a more advanced composition procedure based on signatures rather than names and the conflicts that occurred during composition. Clarke and Walker (2001) presented a means to solve design problem by separating the design of cross-cutting requirements into composition patterns. Rubin et al. (2008) proposed a declarative approach for model composition, which could augment and strengthen existing structural and heuristic approaches. Li et al. (2018) proposed a service network-based method for service composition and scheduling in the cloud simulation environment, which can reflect the characteristic of uncertainty of composition paths in the cloud environment.

In addition, some researchers paid their attentions on model composition tools. In order to merge two or more models and implement composition in the context of the systems biology markup language, Randhawa et al. (2007) developed a Windows-based modeling tool. To make modelers build model easily, Randhawa et al. (Randhawa et al., 2009) presented a model aggregation process. In this process, models were defined in terms of components that were designed for the purpose of being combined. An online model composition tool for system biology models was proposed by Coskun et al. (2012, 2013), which was an all-in-one web-based solutions that supported advanced SBML functionalities. As a modeling and simulation environment, MS4 Modeling Environment (MS4 Me™) was developed (Zeigler and Sarjoughian, 2012), which supported model coupling based on DEVS concepts and theory.

Although SOA has been widely used in variety of areas, most service composition methods still lack intelligence and are difficult to realize automatic composition of services, especially for services in a dynamic and uncertain environment. To deal with service composition problems like the deficiency of intelligence and adaptability to complex and large-scale environments, the concept of SA was proposed by Si et al. (2009). The combination of service and agent will take advantages of both technologies so that a service has the ability to perceive and adapt to environments (Liu et al., 2014; Zhao et al., 2017). When a model is encapsulated into an SA, model composing will be more effective and adaptive. This chapter will introduce a model composition approach based on the SA.

2 CONCEPT OF SA

An agent is an entity that can make a response with sensors and adapt to environments. An agent is assumed to be rational and can achieve its goals through learning or knowledge.

Many researchers think that agent technology started a new paradigm for software development after the object-oriented technology (Chen and Cheng, 2010; Siegfried et al., 2014). The concept of agent has been used in many areas, such as manufacturing, real-time control systems, electrical commerce, network management, transportation systems, information management, scientific computing, health-care, entertainment, etc. Agent technology can significantly improve the analysis capability of systems with the following features (Mccoy and Natis, 2003):

(1) the system is generally composed of distributed component systems;
(2) the component systems exist in a dynamic environment;
(3) the component systems need to interact with each other flexibly.

As we know, SOA provides an architecture for new generation information systems, which is flexible, easy to integrate, and implement, but the service itself in SOA does not have intelligence or self-adaption capacity. By using the agent, a service in SOA can be changed from passive invocation to active searching, pushing, and adapting. In this sense, the agent can extend the service function, and make service intelligent so as to spontaneously respond to and adopt to the surrounding environments and understand the customers' needs better, and therefore a service becomes an SA (Zeigler and Sarjoughian, 2012).

Based on the basic agent model, this section proposes an SA modeling method. This method expands semantics of the BDI model (belief, desire, intention), and focuses on the description of agent's perception, scheme, behavior, coordination, cooperation, and finally forms a semantic BDI agent. Then this agent is used to model the service and form an SA.

Paper by Norvig and Russell (1995) put forward the simple agent model, which describes the interaction between an agent and an environment as well as the internal handling mechanism. In this model, a simple agent is only based on current perceptual behavior without considering other historical data. The function of the agent is based on the operation rule of agent function, that is, if→condition→then→action. This function can be realized only if the environment of the agent is visible. Some agents can also contain current status information.

An SA is an intelligent encapsulation of a service, or can be regarded as an agent that drives a service. An SA can be both a service provider and a service demander. In a simulation environment, an SA has two roles that can ensure that every SA has an opportunity to gain profit. Fig. 1 depicts a conceptual model of an SA.

In Fig. 1, we can see that an SA is composed of two parts: core services and intelligence module. Services contain basic service description, service interface, service status, data, etc., whereas the intelligence module includes a knowledge library, an environment sensor, a clock controller, and a processor. The knowledge library is used to store the status, basic data, rules, and function data of an SA. The environment sensor can be managed by a message queue. The clock controller generates motivation to trigger the function of the SA. The processor includes the functions of service providers and service demanders.

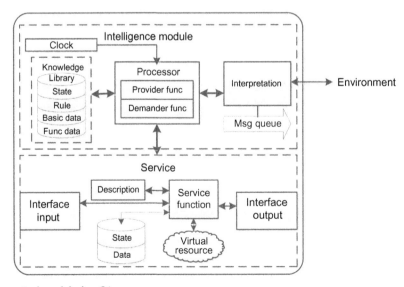

FIG. 1   Conceptual model of an SA.

Core services represent a basic business, a function, or a component. In order to make service matching easier, a detailed description of a service is needed. The tag is an ideal way of description. It can not only describe functions and attributes from different perspectives but also build a fast and flexible classification mechanism. The service interface is used to establish the inputs and outputs of a service. Each interface has more than one ports. A service composition can be established if interfaces of services match. Service status changes as time passes. Service status is like a small data library and it saves static and dynamic data until service is canceled. The service information belongs to basic information which can be described and recorded by Web Service Description Language (WSDL).

The clock controller is the excitation trigger of an SA. As the clock controller vibrates, the function of the SA is triggered to complete each response. The vibration frequency of the clock controller is synchronized with the service center, which can enable the service center to collect SA information. The service center is the management center of SAs, which provides the bulletin board, environment information, supply and demand information of SAs, and other related information.

The knowledge library is used to store data that is generated during the initialization and running processes of an SA which include status data, basic data, rule data, and function data. Status data is real-time information of the SA including busy, idle, etc. Basic data is used to record static data and other basic properties that describe the SA. An address book of the SA's friends (other SAs that have relationships with the SA) is also included in the basic date. Rule data is the formal description of behavior rules in the process of SA negotiation. Function data is the real-time data and history data in the process of SA execution.

The environment sensor is used for interaction between an SA and its environment, among which the environment includes service center and other SAs. Communications between them are based on messages. The messages are encapsulated into message packets and stored in message queues to be invoked by the SA. An SA realizes publishing and searching behavior through messages and the service center.

Processor is the core module of an SA. When an SA is in different roles, there are different functions to support the roles. The clock controller is used to trigger the execution of every function. Real-time data and history data are stored in data memory. For example, service providers possess the functions of service publishing and service transactions. Service users have the characteristics of demand publishing and demand bidding. Both service monitoring and service execution will operate core services.

3 SA CONSTRUCTION OF A MODEL

3.1 Service Description of a Model

The objective of service description is to enable models: (1) to be published and invoked; (2) to be found and matched with requirements; and (3) to be coupled through interface matching. As a result, service description of a model can be established as follows:

$$SE = <SID, RSID, Info_{State}, Interface, Info_{Basic}, Func_{Templ}, Func_{Data}, Func_{Call}, Func_{Order} >$$

where SID is the only identity of a service used to specify a service in the environment.

RSID is the only identity of a model. The mappings between services and models are many-to-many, namely, one model can be encapsulated into different services, while different models can also form one service by cooperation.

Info_state is current state information. There are five kinds of state information representing five status of a service:

$$Info_{state} = \{Check, Publish, Idle, Busy, Unavailable\}$$

Check represents a checking process for a service. In this process, the service cannot be used.

Publish is a status that the manager confirms services can be invoked. At this status, services are in an open situation but cannot be used.

Idle is the status that the service is published and ready to be invoked.

Busy is a status that a service is in use. It means that this service is busy now and it cannot accept other tasks.

Unavailable means a service is not available now. It occurs because of service maintenance or malfunctions.

Interface refers to service interface that is one important part of a service. Interfaces can be used to make a decision that if services can be invoked or composed. It can also be used to validate if service inputs satisfy requirements and if the outputs can be used. The structure of *Interface* is shown in Eq. (1).

$$Interface = < Input, Output > \tag{1}$$

$$Input = << Interface1, Vaule1 >, < Interface2, Vaule2 > \cdots < Interfacen, Vaulen >, Rule > \tag{2}$$

$$Output = << Interface1, Vaule1 >, < Interface2, Vaule2 > \cdots < Interfacen, Vaulen >, Rule > \tag{3}$$

Interfaces are divided into inputs and outputs. Their basic functions are displayed in Eqs. (2), (3).

The internal functions of a model consist of $Func_{Templ}$, $Func_{Data}$, $Func_{Call}$, $Func_{Order}$.

$Func_{Templ}$ plays the template function that can provide description templates for models.

$Func_{Data}$ is the data function, in which static data is a basic description of a model. It represents the data that cannot be accumulated, while, dynamic data is accumulated data over time.

$Func_{Call}$ is the invoking function, which can be described by WSDL and invoked by WEB.

$Func_{Order}$ is the command function, which is designed for models that cannot be physically connected to a network or a cloud. In this case, tasks can be completed by sending an order to a person or a team who can operate the model.

3.2 SA Description of a Service

An SA conducts service update and service encapsulation with the intelligent features of an agent. In a simulation platform, an SA can simulate users' behaviors and conduct cooperation and interaction among different services.

An SA serves not only as a service provider but also as a service demander. Hence, it has the characteristics of both providers and demanders at the same time.

An SA can be represented as follows:

$$SA = < SAID, SID, Info_{state}, Info_{basic}, Msg_{sa}, Clk_{SA}, Func >$$

Among which:

$SAID$ is the unique identification of an SA.

SID is the unique identification of a service, which is used to determine a specific service encapsulated by an SA.

$Info_{state}$ is the status information used to describe dynamic information of an SA. There are four information states: *working, prepared, searching*, and *waiting*.

$Info_{basic}$ is the basic information of an SA. Unlike $Info_{state}$, $Info_{basic}$ is more like a data library that can store basic descriptions, rules, and data.

Msg_{sa} is the message controller among different SAs. Messages are represented by message queues.

Clk_{SA} is the clock of an SA. Different SA clocks have different clock response frequencies. *Func* will be stimulated when clock is triggered.

Func represents the basic behaviors of an SA. The behaviors of a *Func* are the process of negotiation among SAs.

4 A MODEL COMPOSITION METHOD BASED ON SA

After a model has been encapsulated as an SA based on Section 3, the composition process of SAs will be given in this section. When SAs are used to support a composition process, their characteristics of self-organization can be used to negotiate with each other to establish a chain of SAs, and therefore obtain a composition of models accordingly.

The model composition process is shown in Fig. 2. The service demander submits an application for services; the client side receives the task and search for the appropriate SA to issue task demands to it. Upon receipt of the port notification, the SA will establish organizations as the organizer to complete service composition. The organizer will preserve a set of abstract service composition templates of cooperators required for the completion of this kind of tasks in its own knowledge library. Each of the SAs owns a set of address lists, of which one of the organizer includes individual information of the SAs that corresponds with the template. The organizer makes the decision on the characters on which to cooperate with through searching templates; and by searching the address list, it selects the appropriate SAs to form the candidate set, to which it will send requests for cooperation and then form the cooperative organization (composition) according to their responses. After all the roles of the template receive responses, the organization will be formed and the notification of start will be sent. If the current record in the organizer's address list falls short of cooperation for the completion, it could be expanded according to the expansion strategy. Since the organizer itself is a service provider as well, it has to fulfill its own portion of tasks and notify each other the beginning of work after completing its tasks according to the matching relation of the forward and the backward level. The last SA to complete the task shall send the result back to the Client. During the whole procedure, the organizer is able to save the information of cooperation for later updating after the end of cooperation.

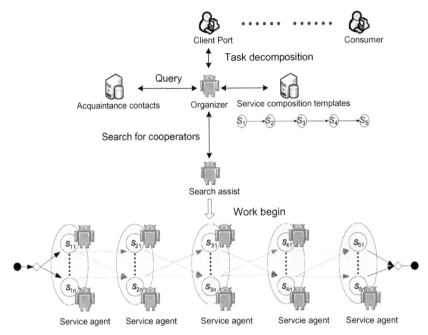

FIG. 2   SA-based model composition process.

In fact, in order to form the service composition, the organizator's work of searching for cooperator itself is one of the tasks of the SA. From this point of view, aside from being in the prepared state or the working state, the SA could be in the state of "organizer's searching for cooperator." However, the actual states of the roles as organizers and cooperators are different. The organizer needs to preserve a set of service composition template and its corresponding address list of acquaintances, with which it will check out acquaintances' states and complete organization work. Service composition template records the ways for matching between the forward and backward levels of different types of services; of course, it is limited to the types of services, the real SA is embodied in the address list of acquaintances. The address list and the service composition template constitute the framework of the cognitive world of SA together. These can only be achieved by the building ontology. As for the cooperator, it is required to respond to the requests of the organizer. Therefore, every SA bears two types of states—the searching state as organizer and the waiting state as cooperator. The sequence of the states is: the prepared state, the searching state, the waiting state, and the working state (Fig. 3).

Behaviors in each state are described as follows:

(1) In the working state, the SA provides services and receives outputs from its previous stage. It delivers the result to the next SA once its own work is finished. Information of other organization members is therefore necessary. Basic information contains names and addresses. The organizer releases such information with an "organization formed" notification.

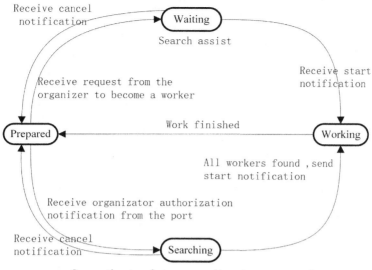

FIG. 3   Individual state transition model.

Both organizers and cooperators eventually evolve to working state and return to prepared state irrespective of whether their work is accomplishable or not.

(2) In the prepared state, all behaviors are blocked for a new message to come. The messages are classified into three categories. Firstly, if the message is from the client port, the SA either becomes an organizer to enter searching state or updates its library according to the message content. Secondly, if the message is a composition request from the organizer, it replies and waits for further acknowledgment. The organizer accepts its reply if the organization member role is still vacant. Agreement is reached as the SA will become a cooperator in waiting state and wait for start notification. At last, if other messages arrive, it will reenter the prepared state with nothing done. Prepared state is both the SA's initial state and the eventually returned state through a service composition experience.

(3) Waiting state is for the cooperators. In this state, the cooperator receives other organization members' ID information and in/out match information from organizer. When a work start notification is received, the cooperator exits this state and enters working state.

The cooperator is designed to help its organizer find other cooperators to continue the work of forming the organization. This is a complement of "organizers seek for cooperators" behavior. The SAs in prepared state are also permitted to assist search. This tiny behavior is of much importance for its wide range of involved SAs. The resulting effect would be discussed later. In waiting state, the cooperator may also receive cancellation and return to the prepared state.

(4) Searching state is for the organizers. There are different search modes for an organizer. Mode selection is due to the client port's requirements such as highest quality or minimal

resource assumption. Search mode determines how to choose appropriate cooperators. Templates provide information about the roles are needed. Contacts provide their links, including ID and address. Cooperators' ID information includes name, service type, service quality, and resource consumption, as well as agent credit level and agent benefits. By default, the organizer adopts "first come first serve" strategy to choose cooperators. For example, if the client port requires minimal resource consumption, the organizer will sort and search its acquaintances referring to resource information in its contacts. Acquaintances of historically less resource consumption will be added into the result set. If anyone in the set responds, it would be accepted to be a cooperator of the organization.

Secondly, once a model is published on a server with a WSDL file, it becomes a WS. Consumers or agents could use an application programming interface (API) to call it. An SA is wrapped and published within a service container after the states have been successfully designed. As a result, agents running in web servers become WSs.

5 A TOOL PROTOTYPE FOR MODEL COMPOSITION AND A CASE STUDY

A software prototype has been developed by using JADE4.1 (Java Agent Development Framework) platform to perform model composition with SA-based method. The JADE is a multiagent system (MAS) software development platform in JAVA language. This platform enables multiagent negotiation; WSs are published in a Tomcat container; WSs package by WSIG (JADE Web Service Integration Gateway) plug-in. The purpose of WSIG is to achieve the integration of MAS and WS architecture.

There are two key points throughout our development experience. Firstly, agent behavior sequence and communication language need to be taken into account. Secondly, well-defined domain ontology is necessary to support information exchange between SAs and between SAs and users.

A hotel booking model is used to demonstrate the system. A test page with basic service information is depicted in Fig. 4. After inputting a simple object access protocol (SOAP) text, we click on "send" button and wait until system responds. The first element is a functional request for M Service; the second element is nonfunctional for high quality. Success information will return in the end.

In the SA test system, "WX" represents a hotel-booking model. "SS" represents the online flight reservation model. "FS" represents a scenic site model. "LJD" represents shopping mall model. "YD" represents the return ticket model.

As shown in Fig. 5, the port notified WX service 1 to be the organizer. It discovered from its own template that the sequence of service composition was WX-SS-FS-LJD-YD. And its address list of acquaintances included FS service 1, LJD service 1, YD service 1, and YD service 2.

After WX service 1 enters into the searching state, it sent the request generated by the organization to these friends (acquaintances). YD service 1 has been performing another task already, it would not respond to the request now as it was in working state. Also, FS service 1 has formed an organization with the organizer WX service 2 already. Though it would not respond, it could help search for other cooperators. YD service 2 and LJD service 1 were both

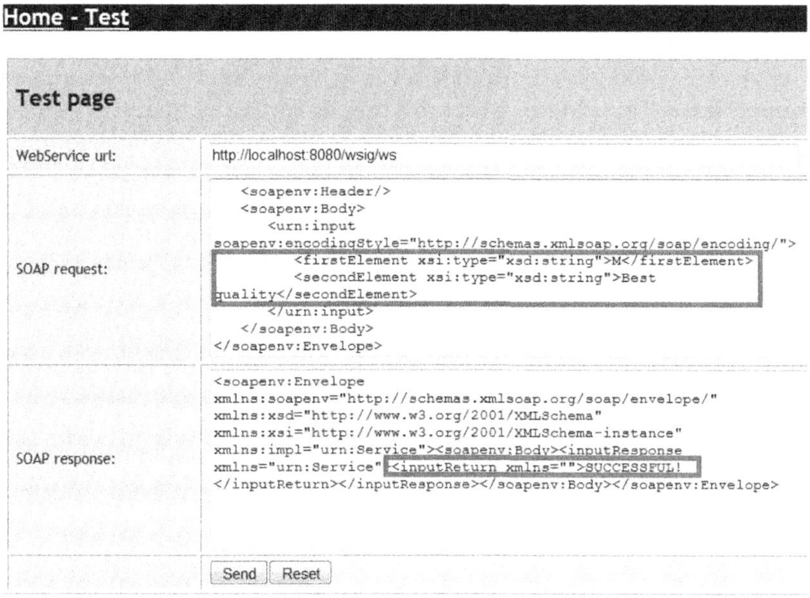

Home - Test
Test page
WebService url:
http://localhost:8080/wsig/ws
SOAP request:
<soapenv:Header/>
<soapenv:Body>
<urn:input
soapenv:encodingStyle="http://schemas.xmlsoap.org/soap/encoding/">
<firstElement xsi:type="xsd:string">M</firstElement>
<secondElement xsi:type="xsd:string">Best
quality</secondElement>
</urn:input>
</soapenv:Body>
</soapenv:Envelope>
SOAP response:
<soapenv:Envelope
xmlns:soapenv="http://schemas.xmlsoap.org/soap/envelope/"
xmlns:xsd="http://www.w3.org/2001/XMLSchema"
xmlns:xsi="http://www.w3.org/2001/XMLSchema-instance"
xmlns:impl="urn:Service"><soapenv:Body><inputResponse
xmlns="urn:Service"><inputReturn xmlns="">SUCCESSFUL!
</inputReturn></inputResponse></soapenv:Body></soapenv:Envelope>
Send
Reset

FIG. 4 System test page.

YD service 1
(working)
SS service 1
(prepared)
FS service 1
(waiting)
WX service 1
(searching)
WX service 2
(searching)
YD service 2
(prepared)
LJD service 1
(prepared)
FS service 2
(prepared)

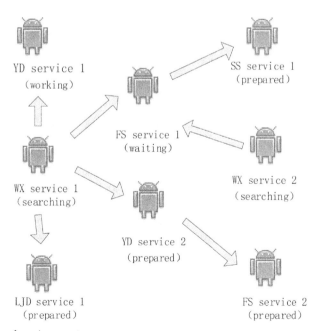

FIG. 5 Collaboration of service-agents.

in the prepared state; they agreed to form the organization, for which WX service 1 replied: wait for notification, at which they entered into the waiting state. For the reason that no appropriate character could be found, WX service 1 requested for a second-level search, which is required for the waiting or prepared friends (acquaintances) to search their own address lists. FS service 1 found its friend SS service 1; YD service 2 found its friend FS service 2. By then, all the types of roles in the template have found their cooperators—the organization was thus completed.

Fig. 6 depicts the view from Sniffer Agent. Sniffer Agent is a special agent to monitor other agents' communication in our system. Here the port YJ, the organizer WX2, and the last SA YD2 are listed to show service combination process.

The first purple arrow indicates user input is sent in message form to the port YJ. The second gray arrow means YJ informs WX2 to be an organizer. Four gray messages in lines 3, 4, 5, 7 indicate WS2 sends cooperation requests to four SAs with a high-quality limitation. Three blue messages in lines 6, 8, 9 are three responses. Response from YD2 is highlighted in red (line 11). Arrows in lines 10, 12, 13, 14 are "cooperation-accepted" notifications. Receivers enter waiting state. The following gray arrows in lines 15–18 are cancellation notifications to the

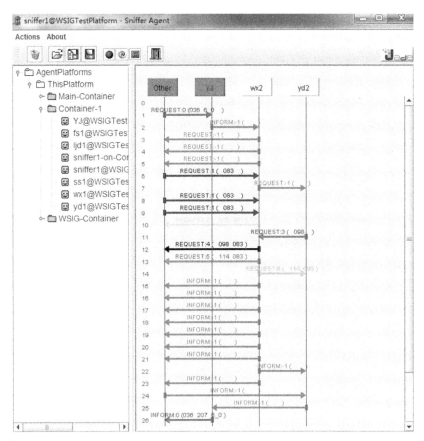

FIG. 6 Sniffer agent interface.

SS1, FS1, LJD1, and YD1. Arrows in lines 19–22 indicate that the organizer tells who will receive their work output. SAs work in sequential in this template. WX2 begins first and notifies the next SA in line 23 when it finishes. In the end, YD2 receives begin notification from its last cooperator (line 24) and sends the final result to YJ. YJ then displays a success notification in the UI web page.

6 CONCLUSIONS

The automatic and efficient model composition technology is the key for the construction of complicated system models. As of now, agent technologies and service-oriented approaches are playing increasingly important roles in terms of the construction and composition of complicated system models. This chapter probes into a model composition approach that is based on SA, involving the advantages of both agents and services. The concept of SA is introduced. The conceptual model of an SA is given and a method of encapsulating a model into an SA is developed. Based on the operating principles of an SA, this chapter introduces an approach of automatic model composition and develops a preliminary prototype system to realize the approach. However, the complexity of model composition means that the proposed approach of this chapter just represent a preliminary methodology framework, which is still far away from the implementation of the automatic model combination for a complex system. Many open problems still need to be carefully studied, including the QoS descriptions of models, the multiparameter model matching based on semantics, the clustering of models, the optimization of model composition pathway, the verification, validation and evaluation of composed models, etc.

References

Cayirci, E., 2013. In: Modeling and simulation as a cloud service: a survey.Winter Simulation Conference, Washington, DC, USA, December 8–11.

Chen, B., Cheng, H.H., 2010. A review of the applications of agent technology in traffic and transportation systems. IEEE Trans. Intell. Transp. Syst. 11 (2), 485–497.

Clarke, S., Walker, R.J., 2002. In: Towards a standard design language for AOSD.International Conference on Aspect-Oriented Software Development.

Clarke, S., Walker, R.J., 2001. Composition Patterns: An Approach to Designing Reusable Aspects. pp. 5–14.

Coskun, S.A., Qi, X., Cakmak, A., Cheng, E., Cicek, A.E., Yang, L., Jadeja, R., Dash, R.K., Lai, N., Ozsoyoglu, G., 2012. PathCase-SB: integrating data sources and providing tools for systems biology research. BMC Syst. Biol. 6 (1), 67.

Coskun, S.A., Cicek, A.E., Lai, N., Dash, R.K., Ozsoyoglu, Z.M., Ozsoyoglu, G., 2013. An online model composition tool for system biology models. BMC Syst. Biol. 7 (1), 88.

Fleurey, F., Baudry, B., France, R., Ghosh, S., 2010. In: A generic approach for automatic model composition.Proc. AOM at Models, p. 5002.

Guo, S., Bai, F., Hu, X., 2011. In: Simulation software as a service and service-oriented simulation experiment.IEEE International Conference on Information Reuse and Integration, Las Vegas, NV, USA, August 3–5.

Li, B.H., Chai, X.D., Hou, B.C., Tan, L.I., Zhang, Y.B., Hai-Yan, Y.U., Han, J., Yan-Qiang, D.I., Huang, J.J., Song, C.F., 2009. Networked modeling & simulation platform based on concept of cloud computing—cloud simulation platform. J. Syst. Simul. 21 (17), 5292–5299.

Li, F., LaiLi, Y., Zhang, L., Hu, X., Zeigler, B.P., 2018. In: Service composition and scheduling in cloud-based simulation environment.Proceedings of the 2018 Spinr Simulation Multi-Conference, Baltimore, MD, USA, April 15–18, 2018.

Liu, J., Zhang, L., Tao, F., 2014. In: Service-oriented model composition.Summer Simulation Multiconference.

Mandelin, D., 2006. In: A Bayesian approach to diagram matching with application to architectural models.International Conference on Software Engineering.

Mccoy, D.W., Natis, Y.V., 2003. Service-Oriented Architecture: Mainstream Straight Ahead. .

Nejati, S., Sabetzadeh, M., Chechik, M., Easterbrook, S., 2007. In: Matching and merging of Statecharts specifications. International Conference on Software Engineering.

Norvig, P., Russell, S.J., 1995. Artificial Intelligence: A Modern Approach. Prentice Hall, Englewood Cliffs, New Jersey.

Randhawa, R., Shaffer, C.A., Tyson, J.J., 2007. In: Fusing and composing macromolecular regulatory network models. Spring Simulation Multiconference.

Randhawa, R., Clifford, A.S., Tyson, J.J., 2009. Model aggregation: a building-block approach to creating large macromolecular regulatory networks. Bioinformatics 25 (24), 3289–3295.

Reddy, R., France, R., Ghosh, S., Fleurey, F., Baudry, B., 2005. Model composition—a signature-based approach. In: In: Aspect Oriented Modeling.

Rubin, J., Chechik, M., Easterbrook, S.M., 2008. In: Declarative approach for model composition.International Workshop on MODELS in Software Engineering.

Si, N., Zhang, L., Tao, F., Guo, H., 2009. In: Research on multi-agent system based service composition methodology in semantic SOA (in Chinese).Proceedings of the 5th Conference on Multi-Agent System and Control, Chongqing, China, September 19–20.

Siegfried, R., Tom, V.D.B., Cramp, A., Huiskamp, W., 2014. M&S as a Service: Expectations and Challenges. Siso.

Tolk, A., 2013. In: Interoperability, composability, and their implications for distributed simulation: towards mathematical foundations of simulation interoperability.IEEE/ACM International Symposium on Distributed Simulation and Real Time Applications.

Tolk, A., Turnitsa, C.D., Diallo, S.Y., Winters, L.S., 2006. Composable M&S web services for net-centric applications. J. Def. Model. Simul. 3 (1), 27–44.

Tsai, W.T., Li, W., Sarjoughian, H., Shao, Q., 2011. In: SimSaaS: simulation software-as-a-service.Spring Simulation Multi-Conference, Springsim '11. Volume 2: Proceedings of the Simulation Symposium, Boston, MA, USA, April 03–07, 2011.

Walker, M.A., Madduri, R., Rodriguez, A., Greenstein, J.L., Winslow, R.L., 2016. Models and simulations as a service: exploring the use of galaxy for delivering computational models. Biophys. J. 110 (5), 1038–1043.

Wang, S., Wainer, G., 2014. A simulation as a service methodology with application for crowd modeling, simulation and visualization. Simul. Trans. Soc. Model. Simul. Int. 91 (1), 71–95.

Wang, S., Wainer, G., 2016. Modeling and simulation as a service architecture for deploying resources in the cloud. Int. J. Model. Simul. Sci. Comput. 07, 1.

Zeigler, B.P., Sarjoughian, H.S., 2012. Guide to Modeling and Simulation of Systems of Systems. Springer, Germany.

Zeigler, B.P., Kim, T.G., Praehofer, H., 2000. Theory of Modeling and Simulation. Academic, USA.

Zhao, C., Zhang, L., Liu, Y.K., Li, B.H., 2017. Agent-based simulation platform for cloud manufacturing. Int. J. Model. Simul. Sci. Comput. 8, 3.

Further Reading

North Atlantic Treaty. Modelling and Simulation as a Service: New Concepts and Service-Oriented Architectures.

6

Verification, Validation, and Accreditation (VV&A)– Requirements, Standards, and Trends

Axel Lehmann*, Zhongshi Wang[†]

*Institut für Technische Informatik, Universität der Bundeswehr München, Neubiberg, Germany
[†]ITIS GmbH, Neubiberg, Germany

1 INTRODUCTION–THE NEED FOR QUALITY ASSURANCE OF COMPUTER-BASED SIMULATION

Generation of new knowledge as well as product innovations is mainly based on three pillars of enabling methods: application of fundamental axioms and natural laws, application of experiments in real scenarios, and application of computer-based modeling and simulation. The latter one, modeling and simulation (M&S) is of increasing importance as the most powerful and flexible approach that provides additional or even new opportunities for various kinds of application domains: M&S provides new learning and training capabilities, serves as tool for decision support as well as a tool for the analysis and evaluation of the existing or even planned systems and processes. Innovative M&S concepts—such as component-based modeling, parallel and distributed simulation, collaborative or agent-based simulation—as well as rapid advancements of computer and communication technologies are driving forces for increasing M&S performance and range of applications.

As a result, these advancements generally lead to an increasing complexity of application scenarios and corresponding M&S. A measure for M&S state complexity can be defined as the total numbers of input, output, and state variables spanning the whole state space of an M&S. This results in new challenges for model designs and simulation implementations as well as quite often to tremendous amounts of data. While in the past simulations were mostly used for systems analysis or training, meanwhile M&S technologies are also used for modeling and evaluation of socio-technical systems, including challenges of adequate representation of human behavior and uncertainties. Increasing systems and corresponding M&S complexities can also show emergent system behavior, for example, in agent-based simulation, which seem to be far from real behavior, even unrealistic, as it has never been thought about.

Besides benefits of M&S innovations and advances, increasing risks with respect to quality of M&S design, development, and operation have to be considered, especially regarding credibility and utility of input data used for simulation as well as output data and their interpretation. As demonstrated by real cases, the use of erroneous models, faulty simulations, invalid data, or interpretation errors (e.g., Challenger accident) can lead to severe situations, wrong decisions, erroneous and safety-critical training results, or to economic damages. Especially in the context of the increasing importance of multifaceted M&S applications efficient quality control, utility, and credibility assurance mechanisms have to be applied to avoid such safety-critical, expensive, and other unwanted side effects. This requires the application of standardized system engineering processes not only for M&S design, development, operation, and maintenance, but also for quality control. Quality control has to be performed at different levels—project level, product level, and at the technical level. Quality assurance measures should include measures to establish credibility by proving correctness and validity, as well as utility with respect to the aims and purposes of a specified M&S application. In summary, quality assurance measures include project-specific planning of an M&S verification, validation, and acceptance or accreditation (VV&A) process, tailoring of an adequate set of VV&A activities, and a corresponding selection of efficient verification and validation (V&V) techniques.

At first, this chapter presents a brief overview of terminology and basic concepts applied in the context of model engineering. Regarding M&S development and application as a specific systems engineering process, multiple M&S development phases have to be distinguished. According to classical engineering processes, each M&S phase delivers an intermediate product together with its documentation. Each of these phase products as well as the documentation can be subject of V&V. In this regard, integrated V&V strategies and activities, as well as several M&S quality- and efficiency-related concepts have been developed. These include, for example, international guidelines and standards for M&S documentation, for generalized V&V planning processes, for multistage tailoring, and for M&S use risk identification and analysis. Findings and lessons learned from the application of these guidelines and standards in several use cases are finally summarized.

2 BASIC TERMINOLOGY

Following J. Rothenberg´s definition, a model can be defined as a symbolic, abstract, cost-effective, and safe referent of something else for some specific cognitive purpose (Shannon, 1975). Most of the human´s everyday reasoning and decision-making processes are based on mental models. Mental models are being used for analyzing scenario-specific situations or for making decisions. The understanding and analysis of real-world problems require—at least in general—simplification and abstraction of reality as humans are only able to conclude interdependencies between about 5–7 parameters at the same time.

As mental models are built intuitively and based on experiences, they belong to the category of inductive models. Inductive modeling approaches (Bridewell et al., 2008) enable reasoning that concludes on similar situations or cases. In contrast, the majority of computer-based models and simulations used nowadays for analyses, for decision-making, or for education and training are designed and applied as deductive models expressing qualitative or quantitative dependencies between goal, input, and state parameters. Interdependencies between these parameters can be described by mathematical formulas, by numerical algorithms, or by logical algorithms. The following text focuses on the design and application principles of these kinds of deductive models.

Along these lines, and in accordance with Maisel and Gnugnoli (1972), simulation can be defined as a (numerical) technique for conducting experiments on a computer; this technique involves certain types of mathematical and logical models that describe real-world behavior or the behavior of a system over a period of time. This definition describes simulation explicitly as a technique for solving and using models by means of computers. This definition also indicates that (computer-based) simulation is a multistage process which differentiates between stages like conceptual model (CM) design, its formal description, technical solution, and experimental applications.

For collecting evidences of correctness, validity and utility of models, and simulations and data, some general principles and specific techniques for M&S V&V can be applied. An informal but clear distinction between these terms has been presented by Balci (1997): application of verification techniques provides evidences for M&S correctness ("Is the *model right*?") opposed to validation techniques which can provide evidences for validity and utility of an M&S regarding its specific application goals and constraints ("Is it *the right model*?"). Meanwhile, multiple and more precise definitions for V&V are provided by national and international standards, like in IEEE, NATO, SISO, or DoD standards.

In GM-VV-Volume 1 (SISO-Guide-GM-VV, 2012–2013), the following definitions are established:

- *Verification*: The process of providing evidence justifying the M&S system's correctness.
- *Validation*: The process of providing evidence justifying the M&S system's validity. Validity is the property of an M&S system's representation of the simuland to correspond sufficiently enough with the referent for the intended use. The simuland is the system or process that is simulated by a simulation while the referent is "… the codified body of knowledge about a thing being simulated" (IEEE STD 1516.4, 2007).
- *Acceptance*: The process that ascertains whether an M&S system is fit for its intended use.
- *Accreditation*: The official certification that a model, simulation, or federation of models and simulations are acceptable for use for a specific purpose.
- *Tailoring*: The modification of V&V processes, V&V organization, and V&V products to fit agreed risks, resources, and implementation constraints (SISO-Guide-GM-VV, 2012–2013). Project demands like use risk, time schedules, limited resources, or intellectual property rights can be constraints that have to be considered while planning M&S and VV&A activities. Therefore, project-specific tailoring of M&S activities including V&V planning is required in general.

These and some alternative but closely related definitions of international M&S-related terminology can be found in SISO-GUIDE-GM-VV-001.1-2012 (Volume 1, SISO-Guide-GM-VV, 2012–2013).

As mentioned above, more and more M&S applications incorporate significant uncertainties such as environmental or human behavior, especially in socio-technical models and simulations. In such cases, complete validation as defined above is often not feasible. Instead, to provide at least some confidence in M&S results, *Plausibility* checks should be performed as far as possible, for example, by face validation performed by domain experts.

3 M&S AS SYSTEMS ENGINEERING PROCESS

As already mentioned in the introductory subchapter, in the past M&S have been used primarily for the analysis and training purposes in natural sciences and engineering disciplines. With the increasing range of application domains, the complexity of systems, processes that are simulated as well as respective experiences have demonstrated that M&S developments and maintenance should be processed similar to other systems engineering processes (Kossiakoff et al., 2011): each M&S development and maintenance process should be phase-oriented structured in such a way that each process phase has a specific subgoal, delivers an (intermediate) phase or work product along with its documentation. The work product and its documentation can be seen as the specification for the processing of the next M&S phase, etc. (see Fig. 1). Similar to typical systems engineering processes, a generic M&S process can be instantiated and tailored according to the specific project goals, demands, and requirements.

This generic M&S development process is not limited to M&S in engineering disciplines but should be also considered for modeling of socio-technical systems and processes. In general, M&S developments are not following a straightforward development process like in waterfall models (Bell and Thayer, 1976). Instead, in practice most M&S developments are iterative processes caused by adaptations, refinements, or maintenance of intermediate work products. A revision of a work product in phase i may result in corresponding revision in phase $(i+1)$ as schematically indicated on the right-hand side of Fig. 1. As a result, each M&S phase of a distinct M&S project will offer a set of different intermediate work product

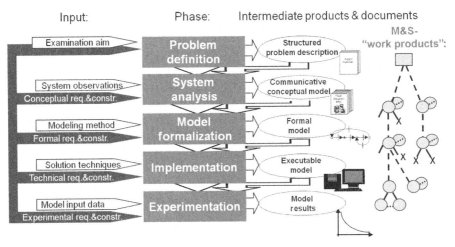

FIG. 1   M&S development as engineering process.

versions related to work product versions in foregoing and subsequent M&S phases. Major advantages of following such a generic multiphase development process including a graph-like version control structure of its work products and documentation are as follows:

- retrace of essential steps and decisions during M&S development and lifetime,
- backtracking of potential weaknesses or even errors regarding a specific M&S design, implementation, adaptations, or updated versions,
- providing source information for quality assurance, especially for phase-oriented V&V,
- enables opportunities for efficient reusability of intermediate work products,
- offering a basis for overall M&S process management.

4 A GENERIC GUIDELINE AND META-MODEL FOR M&S DOCUMENTATION AND QUALITY ASSURANCE

In accordance with a multiphase M&S process described in Fig. 1, an M&S documentation guideline has to define core requirements of structure and content of M&S development phases, work products, and corresponding documentation. These guidelines should provide templates for documenting an M&S project and its associated V&V efforts throughout the entire M&S development life cycle. As shown in Figs. 1 and 2, essential elements of documentation guidelines can be adapted from already existing international guidelines like the Generic Methodology for Verification and Validation (GM-VV) (SISO-Guide-GM-VV, 2012–2013).

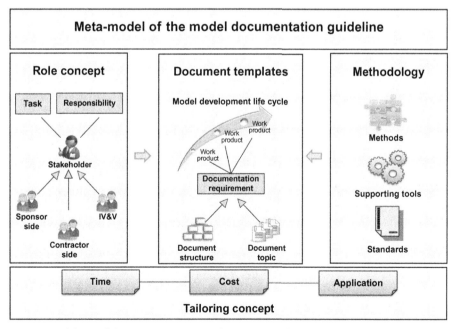

FIG. 2   Overview of the model documentation guideline (Wang and Lehmann, 2010).

As demonstrated in Fig. 2, a meta-model for M&S documentation includes major elements such as a role concept, documentation templates, a catalogue of basic modeling, implementation and V&V methods, tool support, as well as a tailoring concept (Wang et al., 2009).

The structure of a generic M&S development guideline can be based on a meta-model, which serves as basis to describe the core elements and their interrelations of M&S development. Such core elements are, for example, roles and responsibilities, work products and documentation templates, V&V activities, and a tailoring concept. The meta-model should specify semantics and relationships between these core elements. In addition, this meta-model for M&S development includes descriptions of different abstraction levels to meet the specific needs of each user group of this guideline.

4.1 Roles and Responsibilities

In order to define functional tasks and competencies for persons participating in M&S development, in its documentation and V&V activities, a well-structured role concept should be introduced. During the process of project planning, roles have to be assigned to individuals or organizational units including certain responsibilities, for example, contributory or decisive responsibilities. While several roles may support in the creation of a work product, exactly one responsible role can be assigned to each work product and its documentation. An M&S project will generally involve three parties, namely participants from the sponsor side, the developer side, and the independent V&V (IV&V) agent or institution (Arthur and Nance, 2000; Balci, 2010; Lewis, 1992). The role concept presented here has offered an organization-independent orientation for project management. Fig. 3 shows the basic relationships between roles, their responsibilities, and tasks as part of a meta-model for M&S development.

4.2 Documentation Templates

To facilitate the proposed model documentation activities and to avoid misinterpretations, concrete requirements regarding structure and content of each intermediate work product created during the M&S life cycle should be specified in the form of documentation templates (Fig. 4). A structured process for developing M&S applications and conducting their V&V should be applied as reference process, for example, as developed by Brade (2000) and Wang and Lehmann (2007a,b).

As shown in Fig. 4, a documentation guideline should also provide detailed documentation templates for intermediate work products in all M&S development phase, such as for the description of sponsor needs (SNs), for structured problem description (SPD), for specification of CM, formal model (FM), executable model (EM), and for simulation results (SRs). This documentation structure is in accordance with GM-VV and is basically compatible with other international M&S development concepts (such as Balci and Saadi, 2002; Banks et al., 2010; Sargent, 2015), M&S development guidelines and standards like REVVA (Brade and Jaquart, 2005), IEEE 1516.4 (2007), or DoD Standard Practice (DoD MIL-STD-3022, 2008). Basically, this template concept can be directly applied in an M&S project or project-specific adaptation as required.

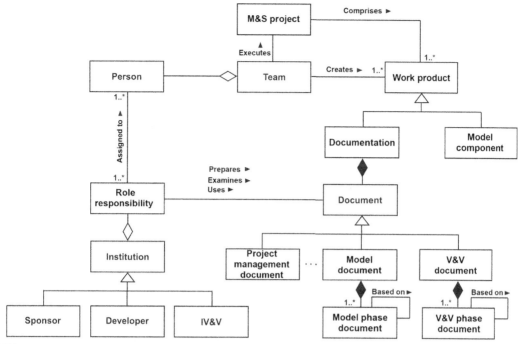

FIG. 3   Roles, responsibilities, and tasks (Wang and Lehmann, 2010).

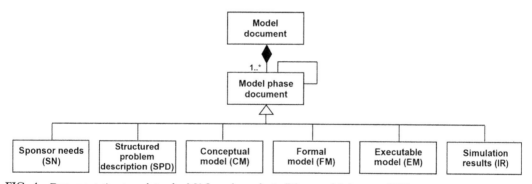

FIG. 4   Documentation templates for M&S work products (Wang and Lehmann, 2010).

4.3 Methodological Support

Generic M&S development and documentation guidelines as proposed in this subchapter should also include recommendations regarding the selection of design, implementation, and analysis methods, selection of supporting tools, as well hints to other standardized development processes, for example, the V-Modell XT, the official guideline for German government sponsored IT development processes like M&S (Kuhrmann et al., 2005; Wang and Lehmann, 2008). Like other M&S development and documentation guidelines and standards, the

V-Modell XT offers additional opportunities for developers and users to select efficient methods of preparing work products and documents, as well as for data and information exchange among all involved project participants. As proposed in the chapter "Generic Concept and Architecture for Efficient Model Management" of this book, an M&S management system can efficiently offer these supporting services.

Application of M&S Tailoring

Regarding different characteristics of organization structures, project requirements, resources and budgets, in general M&S development and documentation guidelines need some kind of project-specific adaptation, respectively, tailoring prior to its project-related application. For this purpose, consideration of a multistage tailoring concept is recommended, which enables the project-specific selection of essential intermediate work products, documents, development activities, and roles for M&S development and applications. In addition, tailoring decisions should be considered not only for V&V planning and for the execution of V&V activities with respect to primary M&S goals, but also in view of available resources and constraints, such as specified budget, project milestones, and deadlines, as well as specific application constraints. By applying these tailoring mechanisms described in detail in Section 6, all M&S phases, activities and roles, corresponding work products, and documentations relevant and significant for an intended M&S application purpose as well as expected range of M&S applications and M&S lifetime have to be taken into account. In most cases tailoring implies a reduction in project resources and potential overhead.

5 A GENERALIZED V&V CONCEPT—THE "V&V TRIANGLE"

As Fig. 5 illustrates, two closely related task groups of V&V activities (model V&V and data V&V) are specified for application as part of M&S development (Brade, 2000; Wang and Lehmann, 2007a). As an example in Fig. 5, this model development process defines five modeling phases (depicted by *black boxes*) and their related work products plus documentation (depicted by ellipses). These development phases symbolize sets of activities to transform one intermediate work product of phase (i) to its succeeding intermediate work product of phase ($i + 1$). For each transformation step, work product (i) has to be considered as specification for producing the succeeding work product of phase ($i + 1$). Tasks of V&V subphase activities are to check internal completeness and consistency of this work product. In addition, to check the transformation process from work product (i) to work product ($i + 1$), fulfillment and consistency with all previous work products have to be verified and validated. Each V&V activity as well as their results has to be documented in a V&V report (as depicted by the *gray boxes*). V&V report (ij) describes the results of the V&V activities performed by verifying and/or validating the transformation from work product of phase (i) to the work product of phase (j).

As described above, an extensive execution of a V&V process results in a triangle-like matrix of V&V reports. The columns of the matrix represent the V&V main phases, which are associated with the work products (also referred to as intermediate products); while

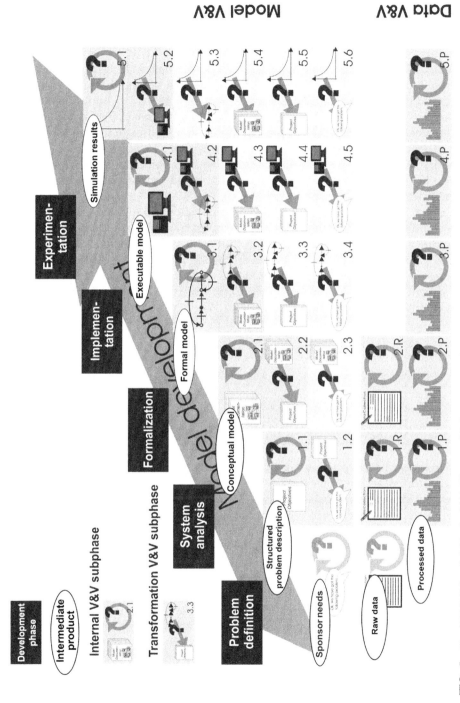

FIG. 5   The "V&V Triangle" (Brade, 2000; Wang and Lehmann, 2007a).

intersections between the columns and rows split the V&V main phases into V&V subphases. During V&V, each work product is examined for internal consistency and completeness with respect to the intended purpose of the model. Subsequently, the transformation consistency is checked by pairwise comparison of all work products. Regarding model V&V, one work product is input to a V&V phase, numbered 1–5. Each V&V phase is again split into subphases, each with a defined subaim to detect the internal defects or transformation defects. In each subphase numbered as $x.1$, the absence of internal defects in each particular work product should be demonstrated. For example, in subphase 1.1, it should be ensured that the problem description is free of misunderstandings and inconsistencies, and in subphase 3.1, a syntax check can be applied to the FM for comparison of the chosen formalism. In any other subphase, the pairwise comparison between the current work product and each previous work product can be performed to confirm the absence of transformation defects. For instance in subphases 3.2, 3.3, and 3.4, the FM could be compared with the CM, the SPD, and also with the SNs.

With respect to data V&V, two types of data should be distinguished: raw data and processed data. Raw data are obtained directly from different sources, which offer the required input data for an M&S application, but which are in general unstructured and unformatted data. Processed data are, however, created by analyzing, editing, transforming, and adapting raw data for or during the modeling process. Thus, data V&V involves credibility assessment of raw data and processed data, which are essential for creating a work product. It should be noted that raw data are usually only relevant for specifying work products in early M&S development phases, for example, for SPD and CM. But in general, raw data are not directly applicable for work products like FM and other succeeding work products. Therefore, the associated V&V of raw data are undefined for certain work products (Wang and Lehmann, 2007a).

6 TAILORING OF M&S, V&V ACTIVITIES

As obvious by the description of Fig. 5, an extensive implementation of a V&V process is very time consuming and resource intensive. In general, M&S tailoring can lead to reduction, extension, specialization, or to balancing of M&S activities, intermediate work products, and V&V activities as a consequence of project requirements and constraints. According to Fig. 6, tailoring actions are feasible at different M&S development levels: on process, product, subject, and/or role level (Wang et al., 2009). At the beginning of an M&S project, by *"static" tailoring*, a V&V plan should be discussed by all the three stakeholders of an M&S project (project sponsor, M&S developer, and independent V&V agent) and finally deployed: only M&S intermediate work products accessible and relevant for this project are determined, produced, documented, and are subject of V&V which implies a focus to a project's use risks, constraints, complexity, and required and available resources (Wang et al., 2009; SISO-Guide-GM-VV, 2012–2013). While during project execution certain project goals or requirements as well as resource constraints may change, the original V&V plan could be adapted (*"dynamic" tailoring*).

In the same way, but depending in addition on the availability and accessibility of M&S work products, V&V tailoring can be considered. For example, regarding availability of

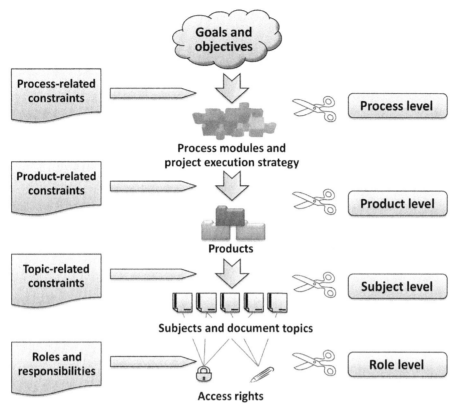

FIG. 6 Multistage tailoring process (Wang et al., 2009).

M&S resources and use risk constraints specified for a concrete project or M&S application—those tailoring actions can consider a reduction in the amount of V&V activities.

Fig. 7 demonstrates consequences of tailoring by reduction: if on the product level the work product "formal model" is not available or not accessible, this leads to a reduction in adequate V&V activities and missing V&V result reports 3.1, 3.2, and 3.3, as well as to missing V&V reports 4.2 and 5.3. If—in addition—some data are not available at the subject level, this can also lead to a reduction in V&V activities and missing V&V reports 2.3 and 4.5. A reason for tailoring an FM could be intellectual property rights held by a developer, his or her institution, so that this work product is not accessible for external review. Those tailoring actions can result not only in significant reduction in V&V efforts and costs but also in an increase in M&S use risks (Lehmann, 2014).

7 COST-BENEFIT-BASED SELECTION OF V&V TECHNIQUES

Quality and credibility assessment of M&S applications requires not only systematic planning and conducting VV&A activities as part of the V&V plan, but also selection and

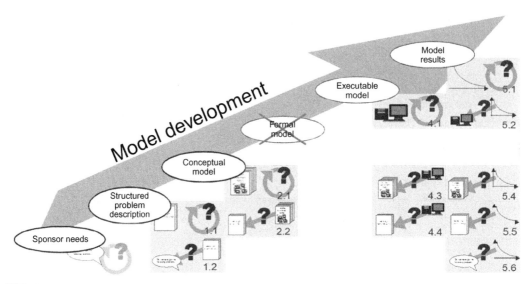

FIG. 7   Tailoring consequences of product and subject levels (Lehmann, 2014).

application of suitable V&V techniques for the purpose of gathering reliable evidences. Whether or not a V&V technique is appropriate to perform a particular V&V activity depends, on the one hand, on characteristics of this technique, and on the other hand, on work product and context where it will be applied. In practice, due to lack of information or knowledge about characteristic features of V&V techniques as well as their strength, weaknesses and application potentials, only a quite limited number of techniques are considered and repeatedly used in any V&V context and any M&S project constellation, although a variety of almost 100 V&V techniques are well known and documented in literature, for example, Balci's taxonomy (Balci, 1997, 1998). Therefore, mechanisms have to be developed supporting selection and application of V&V techniques depending on the project-specific requirements, each respective available work product, data, and related documentation in a systematic and efficient manner.

From the pool of V&V techniques documented in literature (e.g., Balci, 1998; Beizer, 1990; Sargent, 2011), we have defined a characterization scheme for the classification of the most common V&V techniques according to their scope of application, efficiency of use, and required skills and efforts for their application. Fig. 8 illustrates this characterization approach.

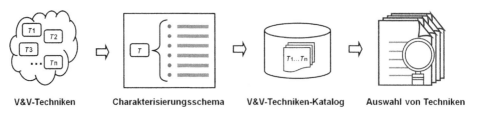

FIG. 8   Basic concept for selection of V&V techniques (Wang, 2013).

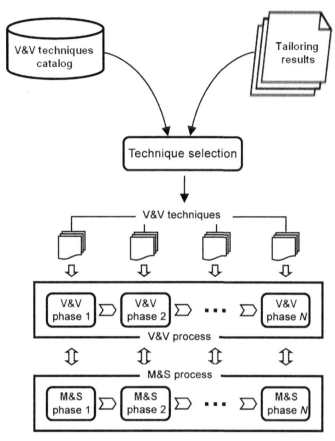

FIG. 9   V&V technique selection (Wang, 2013; Lehmann and Wang, 2017).

As shown in Fig. 9, the selection of suitable techniques for each phase in a V&V process is based on the analysis and evaluation of the:

- techniques collected and characterized in the V&V catalog and
- results of the above-mentioned multistage tailoring process which determines the relevant work products, activities, and documentation depending on specified project constraints, M&S goals, M&S requirements, and acceptance criteria.

Fig. 10A shows the proposed characterization scheme, which is used to build the V&V techniques catalog. The attributes are organized in two categories: Applicability and Cost of a V&V technique. The category Applicability includes attributes referring to:

- usability of a V&V technique to perform a V&V activity;
- operational conditions;
- relevance and dependency of V&V results.

On the other hand, the category Cost includes the information related to the effort and exposure time required for understanding and mastering a V&V technique, preparing test data, and executing test processes.

Category	Attribute	Description	Value example	Reviews	Inspections	Animation/visualization
Applicability	V and V activity	V and V stages associated to the M and S life cycle, in which the V and V technique can be applied, such as V and V phases defined in a V and V process	o V and V of M and S requirements o V and V of conceptual model o V and V of executable model o V and V of simulation results o and etc.	o Any	o Any	o V and V of executable model o V and V of simulation results
	V and V object	Artifacts that the V and V technique is able to examine, including work products, their documentation, and any other documents created in the course of a simulation project.	o M and S problem definitions o Requirements specification o Conceptual model o Formal model o Executable model o Data o Experiment design o Results presentation	o Project plans o Project reports o model-related work product and documentation	o model-related work product and documentation	Executable model: o Structure and internal behavior o Interaction of submodels o Runtime o Data Simulation results: o Design of experiments o Execution of experiments o Presentation of results
	M and S development paradigm	Development paradigm to which the technique is linked.	o Component-based o Object-oriented o Procedural paradigm o Others	o Any	o Any	o Any
	Modeling formalism	Modeling formalism to which the technique is linked.	o Discrete Event System specification (DEVS) o Petri nets o Process algebra o Others	o Any	o Any	o Any
	Simulation type	Simulation types to which a V and V technique can be used.	o Deterministic simulation o Stochastic simulation o Discrete event simulation o Continuous event simulation	o Any	o Any	o Any
	Implementation platform	Simulation language or tool to which the technique is linked, including general-purpose programming languages, general purpose simulation language and special purpose simulation packages.	o C, C++, C#, Java, etc. o General Purpose Simulation System (GPSS) o Arena, Simio, Simula, AnyLogic, etc.	o Any	o Any	o Any
	Availability of data of the real system	Whether or not observable, measurable data of the real system are necessarily required for applying the technique?	o Yes or o No	o Not necessary	o Not necessary	o Not necessary
	Static or dynamic	Whether static or dynamic behavior will be evaluated by using the technique?	o Static behavior o Dynamic behavior or o Both	o Both	o Both	o Both
	Quality of method	Is the technique subjective or objective?	o Subjecitve o Objective	o Subjective	o Subjective	o Subjective
	Dependency	Relationships of the technique with another.	o No o Yes, ...	o None	o None	o None
Cost	Data quality level	Requirements for identifying, preparing and applying test data.	o Low o Medium o High	o Low	o Low to medium	o Medium to high
	Formality level	Level of using formalized structure and process, formal logic, and mathematics in the technique.	o Low o Medium o High	o Very low to low	o Low to medium (5-phase process)	o Low
	Comprehensibility	Effort required for understanding the technique.	o Low o Medium o High	o High	o High	o High
	Human resource	Effort and time exposure required for applying the technique.	o Low o Medium o High	o Medium to high	o High	o Low to medium
	Technical resource	Special hardware or technical equipment required to be able to apply the technique.	o No o Yes, ...	o No	o No	o Display device
	Type of application	Manner how the technique is applied, team-based, self-organized, or both.	o Self-organized o Team-based o Both possible	o Teamwork	o Teamwork	o Teamwork or self-organized
	Participant	Role(s) involved in application of the technique.	o Sponsor o Model developer o Implementer o V and V agent o Others	o Sponsor o Review leader o Management Staff o Technical staff o Recorder	o Moderator o Reader o Model developer o Implementer o Inspector o Recorder	o Model developer o Sponsor (optional) o V and V agent (optional) o Subject matter expert (optional)
	Knowledge and experience	Knowledge, experience, skill, or qualification required to be able to apply the technique.	o No o Yes, ...	o No	o Training in inspection techniques	o No

(A) (B)

FIG. 10 (A) V&V technique characterization scheme (Wang, 2013; Lehmann and Wang, 2017) and (B) examples (Wang, 2013; Lehmann and Wang, 2017).

Applicability and Cost are similar to the two sides of a coin. The selection of a V&V technique with best effectiveness and lowest costs, as well, is hardly achievable in practice. For example, a certain objective V&V technique may appear more effective compared to a subjective one, but its application can be associated with high costs. Therefore, when selecting V&V techniques, both types of technique characteristics should be analyzed and evaluated taking into consideration the specified project goals, requirements, and its environment. Consequently, a reasonable balance between cost and benefit should be achieved.

Considering V&V technique characteristics (Balci, 1998; IEEE STD 1028, 1997; Gilb and Graham, 1993; Schulmeyer and Mackenzie, 2000), Fig. 10B shows the comparison of the three V&V techniques reviews, inspections, and animation/visualization.

8 INTERNATIONAL STANDARDS AND GUIDELINES

To show that these generic M&S development as well as VV&A process models described in the previous sections are in accordance with well-established international guidelines and standards, those guidelines and their characteristics most relevant in this context will be briefly summarized in this section. Nowadays relevant for quality assurance of M&S by VV&A are: SISO-Guideline-GM-VV (2012–2013), Federation Development and Execution Process (FEDEP) (IEEE STD 1516, 2010), VV&A overlay to FEDEP (IEEE STD 1516.4, 2007), and risk-based tailoring of VV&A (RTO-TR-MSG-054, 2012).

Generic Methodology for Verification and Validation (GM-VV)

The GM-VV includes a guideline for the development of a V&V plan, V&V activities, and corresponding selection of V&V techniques. This GM-VV guideline can be considered for every kind and application category of modeling and simulation. In GM-VV documentation, a generic approach of the most relevant decisions, processes, and activities is presented which should be considered as part of an M&S project.

In GM-VV-Volume 1.1, terminology and basic approaches for performing VV&A is comprehensively described. The proposed methodology recommends the definition of the project- and application-specific acceptance goal defined at beginning of the project by the project sponsor. In a next step, this acceptance goal has to be broken down into a set of acceptability criteria which—if fulfilled—will confirm that the acceptance goal of this M&S application can be achieved. In the following step, V&V planning assessment has to be performed which indicates how the acceptability criteria can be verified and validated by selection and execution of adequate V&V activities and techniques. Based on this a V&V plan can be established at begin of the project which might be revised during execution due to unexpected project-related constraints. The execution of this V&V plan will provide items of evidence which could be used to check if the fulfillment of acceptability criteria can be confirmed.

From an organizational and management point of view, V&V activities as part of a V&V plan should be performed on several levels:

- on the technical level concerning all engineering activities that have to be processed for receiving recommendation for acceptance;
- on project level concerning all management activities required to enable the technical efforts for performing the planned V&V activities;
- on the enterprise level, enabling all required support to ensure feasibility of V&V activities.

In GM-VV-Volume 1.2, descriptions of implementation options for this generic methodology, V&V processes, etc., are described in detail. Volume 1.3 is a reference guide linking this methodology to literature relevant in this context, to similar or connected M&S approaches, development guidelines, and standards.

In addition to this GM-VV guideline it is recommended to check the VV&A documentation guideline (DoD MIL-STD-3022, 2008) issued by the US Department of Defense which is also served as valuable input for the GM-VV.

9 FEDERATION DEVELOPMENT AND EXECUTION PROCESS (FEDEP) AND VV&A OVERLAY TO FEDEP

Regarding quality assurance measures for distributed simulation and their applications as federations of simulation federates, two other guidelines or standards (FEDEP and VV&A Overlay to FEDEP) should be considered. Both are related to HLA federations, their correctness, validity, and utility with respect to the project goals.

According to IEEE STD 1516 (2010), the FEDEP distinguishes seven work products as intermediate products for federation development and its documentation. Corresponding to these seven M&S development phases, in IEEE STD 1516.4 (2007) corresponding V&V checks and reports are recommended in the VV&A overlay (see Fig. 11).

FEDEP work products:	VV&A — overlay (activities):
• Definition of federation objectives	• Verification of federation objectives
• Conceptual analysis	• V&V of federation concept model
• Federation design	• Verification of federation design
• Federation implementation	• Verification of implementation products
• Federation integration and test	• Federation validation and acceptance
• Federation execution results	• V&V of federation output results
• Data analysis and evaluation	• Final consolidated VV&A

FIG. 11   FEDEP and VV&A overlay.

Risk-Based Tailoring of VV&A

Beside detailed descriptions regarding the activities and documentation of the VV&A overlay, another guideline describes precisely how identification and determination of M&S use risks can be performed, which is issued by the RTO technical report (RTO-TR-MSG-054, 2012).

This is a detailed guideline for risk-based tailoring of VV&A processes as an Overlay to simulation federations according HLA. Foundational work is documented in this guideline on applying use risk as a tailoring mechanism for the VV&A overlay. This report includes components of tailoring guidance. The resulting product is the V&V Composite Model (see Appendix 1 in this report) which describes the components of the V&V processes (i.e., phases, activities, and tasks) which can be selected in order to match the risk and resource constraints of the V&V efforts in the context of other relevant policies, standards, and guidelines. The V&V Composite Model is a superset of the possible activities and the context in which these activities can be tailored into working V&V processes. In Section 3 of this report the V&V Composite Model is described in detail.

10 PRACTICAL APPLICATIONS AND EXPERIENCES

As described in the introductory section of this chapter, there is an increasing awareness among M&S sponsors and developers, that quality assurance of M&S developments and applications becomes an essential part of M&s projects to avoid or at least minimize M&S use risks—risks that might heavily influence training results or rational decision-making. This section describes major experiences gained by several case studies in which the proposed quality and credibility assurance strategies according to the recommendations and guidelines presented in Sections 3–8 were performed and evaluated.

10.1 Case Studies

All the case studies were sponsored by either government agencies or industrial institutions. M&S development had been performed by professional simulation companies and V&V activities executed by independent institutions as V&V agent. Two case studies were concerned with developments of training simulators while the others concerned M&S projects with constructive simulation (e.g., agent-based simulation) used as decision supporting tools.

These studies were primarily focused on the analysis of feasibility and effectiveness of M&S development processes and documentation guidelines in accordance with international guidelines. In addition, three case studies were especially useful for evaluating the applicability and effectiveness of the "V&V Triangle" concept and the proposed multistage tailoring (see Wang and Lehmann, 2010; Wang et al., 2009) regarding M&S correctness, utility, and acceptability criteria.

Major purpose of the third case study—evaluation of decision options regarding scenario-based composition and operation of military convoys—was an overall evaluation of efficiency, effectiveness, and resource requirements of the proposed M&S-/V&V guidelines in the case of agent-based constructive simulation. V&V techniques applied were primarily desk checking, walk through, inspections, visualization, and face validation of subject matter experts (SMEs) to assess consistency and completeness of work products, as well as evidences for the fulfillment of acceptability criteria (Wang, 2011). Desk checking was selected to investigate the created model specifications. This technique is also known as self-inspection and can be performed by several V&V agents independently. The benefits of selecting this technique are that desk checking is easy and cost-effective. In addition, it is particularly useful in the initial phases of M&S development processes (Balci, 1998).

In many case studies, detailed information required for model design is missing, vague, and unclear. Quite often information required about the real system and its dynamic behavior is incomplete or even missing. Accordingly, tailoring decisions have to be determined concerning the selection and intensity of V&V activities to be applied. All project partners—sponsor, developer, and V&V agent—should be involved in this tailoring process. In this context it has been shown that the involvement of SMEs in M&S specification, development, and V&V planning and tailoring is beneficial. These studies indicate that V&V techniques like inspections, face validation, and visualization/animation are very useful. As described in the V&V techniques catalog, techniques like inspections, walkthrough are team-based and their execution requires active involvement of several different project roles, such as sponsor, model designer, model implementer, and V&V agent. Nevertheless it has to be taken into account that planning and organizing such V&V activities are both complex and time consuming.

10.2 Major Lessons Learned

This section presents major findings obtained by the application of the proposed M&S development, VV&A planning and tailoring concepts and guidelines in the case studies mentioned above. Major findings are as follows:

- The proposed M&S development and documentation guideline was perceived as beneficial by project sponsors as well as by M&S developers. Coaching of guidelines prior to their application turned out as important and effective effort to support novice M&S developers. A tool-supported M&S development and documentation guideline in the form of "living documents" should be available, allowing permanent updates and adjustments (Wang and Lehmann, 2008, 2010).
- Following these experiences, M&S developments, V&V, and documentation guidelines require additional resources (time, budget, experts). By repeated application, however, it has become a very effective way of improving M&S development, operation, maintenance, reuse, and refinement as demonstrated by two case studies performed by the same M&S developers.
- The proposed "V&V Triangle" concept can be efficiently used for performing management and technical activities. (According to the IEEE standard 1059, a V&V effort consists of two types of tasks: management tasks and technical tasks.) While management tasks refer to activities such as planning, organizing, and monitoring of V&V efforts, technical tasks refer to the quality assurance procedures, such as analyzing, evaluating, reviewing, and testing the M&S development processes and work products.

Results of the case studies indicate that the "V&V Triangle" is a comprehensible and efficient tool for planning V&V activities and for executing the examinations.

- Both tailoring approaches—static as well as dynamic tailoring—are required and beneficial. Static tailoring has to be arranged at the begin of an M&S project so that all efforts of M&S development, V&V, and documentation can be determined and planned at an early stage. However, experiences from two case study projects indicate that usually still a considerable amount of information, data, resources, etc., are unavailable at this stage, or certain requirements have changed over time. Therefore, dynamic tailoring performed during the M&S life cycle has to be considered besides static tailoring. Thus, a refined mechanism of dynamic adaptation in the course of a project was developed and integrated in an overall tailoring concept (Wang and Lehmann, 2008, 2010).

- Special attention should be given to intellectual property rights (IPR) protection. An approach to solving this problem was proposed as follows: the V&V agent specifies detailed V&V requirements. An inspector from the quality assurance department of the developer performs the specified V&V and documents V&V process and results of each test case. Final acceptability assessment of the V&V agent takes into account also these V&V protocols and results from the developer side.

- The expenditure of M&S and V&V documentation and V&V has to be considered and planned at the tendering stage of an M&S project. In order to prevent model documentation and V&V from becoming a burden or even useless, because of time and cost pressures during project execution, project sponsors should request calculation of planned documentation and V&V efforts in their project proposal and budget accepted by the project sponsor.

11 CONCLUSIONS

This chapter summarizes current status, demands, and perspectives for quality assurance of M&S and their data developments and applications over their life cycle. Main focus in this context is the importance of performing adequate VV&A activities accompanying the overall M&S life cycle. At first and for clarification, basic terminology of M&S and VV&A are briefly summarized followed by description of basic M&S development processes in accordance with systems engineering principles. The main focus of this chapter concerns the description of the proposed V&V planning process, resulting V&V activities, and the selection criteria for V&V techniques. It is demonstrated that the proposed generic M&S and V&V processes are in accordance with already existing IEEE and SISO standards like GM-VV or the VV&A overlay for the FEDEP. Finally, some general experiences and lessons learned by the application of the proposed M&S and V&V processes in case studies are shared.

Regarding permanently increasing complexity and uncertainties within the wide range of M&S applications, such guidelines are urgently required to assure the required quality, credibility, and utility of its results. M&S acceptability criteria and use risks have to be driving factors for the determination of the amount of M&S documentation, tests, and V&V activities to be performed. For selecting appropriate V&V techniques, the information about the characteristics of techniques and their potential applications is of crucial importance. The characterization approach in this chapter proposes an integrated process to identify, gather, and to use (reuse) all relevant information, so that uncertainties regarding the selection of V&V techniques can be mitigated and consequently more reliable V&V evidence can be achieved. The project-specific technique selection is also a conceptual extension of the existing multistage tailoring process.

Experience and feedback from all stakeholders—M&S sponsors, developers, and V&V agents—applying these concepts and its related guidelines in the case studies were very positive. A consequent next challenge requires research of effective methods for M&S use risk identification, evaluation, and management. Professional international organizations like IEEE or SISO are supporting current research and standardization efforts in regard to this challenge.

References

Arthur, J., Nance, R., 2000. In: Verification and validation without independence: a recipe for failure.Proceedings of the 2000 Winter Simulation Conference, Orlando, USA.

Balci, O., 1997. In: Verification, validation, and accreditation of simulation models.Proceedings of the 1997 Winter Simulation Conference, Atlanta, USA.

Balci, O., 1998. Verification, validation, and testing. In: Banks, J. (Ed.), Handbook of Simulation. John Wiley & Sons, New York, pp. 335–393.

Balci, O., 2010. Golden rules of verification, validation, testing, and certification of modeling and simulation applications. SCS M&S Mag., Issue 4. The Society for Modeling and Simulation International (SCS).

Balci, O., Saadi, S., 2002. In: Proposed standard processes for certification of modeling and simulation applications. Proceedings of the 2002 Winter Simulation Conference, San Diego, USA.

Banks, J., Carson II, J., Nelson, B., Nicol, D., 2010. Discrete-Event System Simulation, fifth ed. Prentice Hall, Upper Saddle River, NJ.

Beizer, B., 1990. Software Testing Techniques, second ed. Van Nostrand Reinhold, New York.

Bell, T., Thayer, T., 1976. In: Software requirements: are they really a problem? Proceedings of the 2nd International Conference on Software Engineering. IEEE Computer Society Press.

Brade, D., 2000. Enhancing modeling and simulation accreditation by structuring verification and validation results. Proceedings of the 2000 Winter Simulation Conference, Orlando, USA.

Brade, D., Jaquart, R., 2005. Final state of the REVVA methodology.Proceedings of the 2005 Spring Simulation Interoperability Workshop, San Diego, USA.

Bridewell, W., Langley, P., Todorovski, L., Džeroski, S., 2008. Inductive process modeling. Mach. Learn. 71 (1), 1–32.

DoD MIL-STD-3022, 2008. Documentation of Verification, Validation, and Accreditation (VV&A) for Models and Simulations. Department of Defense, USA.

Gilb, T., Graham, D., 1993. Software Inspection. Addison-Wesley, Boston, MA.

IEEE STD 1028-1997, 1997. IEEE Standard for Software Reviews. IEEE Computer Society.

IEEE STD 1516.4-2007, 2007. IEEE Recommended Practice for Verification, Validation, and Accreditation of a Federation—An Overlay to the High Level Architecture (HLA) Federation Development and Execution Process (FEDEP). IEEE Computer Society.

IEEE STD 1516-2010, 2010. IEEE Standard for Modeling and Simulation (M&S) High Level Architecture (HLA)— Framework and Rules. Revision of IEEE STD 1516-2000, IEEE Computer Society.

Kossiakoff, A., Sweet, W., Seymour, S., Biemer, S., 2011. Systems Engineering Principles and Practice. John Wiley & Sons, Hoboken, NJ.

Kuhrmann, M., Niebuhr, D., Rausch, A., 2005. Application of the V-Modell XT—report from a pilot project. In: Li, M., Boehm, B., Osterweil, L. (Eds.), Unifying the Software Process Spectrum, International Software Process Workshop. Springer, Heidelberg, Germany.

Lehmann, A., 2014. Verification and validation (V&V) of models and simulations (M&S)—past, present and future. In: NATO-Lecture Series on Application of the GM-VV- the Generic Methodology for Verification & Validation of Models, Simulations and Data. NATO STO-EN-MSG-123, The North Atlantic Treaty Organization (NATO).

Lehmann, A., Wang, Z., 2017. Efficient use of V&V techniques for quality and credibility assurance of complex modeling and simulation (M&S) applications. Int. J. Ind. Eng. Theory Appl. Pract. 24 (2), 220–228.

Lewis, R., 1992. Independent Verification and Validation: A Life Cycle Engineering Process for Quality Software. John Wiley & Sons Inc, New York.

Maisel, H., Gnugnoli, G., 1972. Simulation of Discrete Stochastic Systems. Science Research Associates, Inc., Chicago, USA.

RTO-TR-MSG-054, 2012. Risk-Based Tailoring of the Verification, Validation, and Accreditation/Acceptance Processes. (NMSG-054/TG-037 Final Report). Research and Technology Organization (RTO) of NATO.

Sargent, R., 2011. In: Verification and validation of simulation models.Proceedings of the 2011 Winter Simulation Conference, AZ, USA.

Sargent, R., 2015. In: An introductory tutorial on verification and validation of simulation models.Proceedings of the 2015 Winter Simulation Conference, Huntington Beach, CA.

Schulmeyer, G., Mackenzie, G., 2000. Verification and Validation of Modern Software-Intensive Systems. Prentice Hall, Upper Saddle River, NJ.

Shannon, R., 1975. Systems Simulation—The Art and Science. Prentice Hall Inc, Englewood Cliffs.

SISO-Guide-GM-VV, 2012–2013. Generic Methodology for Verification and Validation (GM-VV), Volume 1.1, Volume 1.2, Volume 1.3. Simulation Interoperability Standardization Organization (SISO).

Wang, Z., 2011. In: Towards a measurement tool for verification and validation of simulation models.Proceedings of 2011 Winter Simulation Conference, Phoenix, AZ.

Wang, Z., 2013. In: Selecting verification and validation techniques for simulation projects: a planning and tailoring strategy.Proceedings of the 2013 Winter Simulation Conference, Washington, DC.

Wang, Z., Lehmann, A., 2007a. Verification and validation of simulation models and applications: a methodological approach. In: Ince, A. et al., (Ed.), Recent Advances in Modeling and Simulation Tools for Communication Networks and Services. Springer, New York.

Wang, Z., Lehmann, A., 2007b. A framework for verification and validation of simulation models and applications. In: Park, J. et al., (Ed.), AsiaSim 2007—Proceedings of Asia Simulation Conference. Springer-Verlag, Berlin Heidelberg, p. 2007.

Wang, Z., Lehmann, A., 2008. In: Expanding the V-Modell® XT for verification and validation of modelling and simulation applications. Proceedings of the 7th International Conference on System Simulation and Scientific Computing (ICSC'2008). IEEE.

Wang, Z., Lehmann, A., 2010. Quality assurance of models and simulation applications. Int. J. Model. Simul. Sci. Comput. 1 (1), 27–45.

Wang, Z., Lehmann, A., Karagkasidis, A., 2009. A multistage approach for quality- and efficiency-related tailoring of modelling and simulation processes. Simul. News Europe 19 (2), 12–20.

Further Reading

IEEE STD 1059-1993, 1993. IEEE Guide for Software Verification and Validation Plans. IEEE Computer Society.

NY Times, 1995. Cruise ship carrying 1,500 runs aground. The New York Times (12 June).

7

A Practical Approach to Model Validation

Ke Fang, Ming Yang

Control & Simulation Center, School of Astronautics, Harbin Institute of Technology, Harbin, China

1 INTRODUCTION

1.1 Background

Nowadays, simulation users and modelers have the common sense that a simulation has to be proved to possess acceptable similarity to the real-world origin in the application domain before use. The concept of "credibility," which represents an acceptability measure, is well known in the simulation community. However, as we all know, credibility assessment is not an easy job.

Model is the core of a simulation system. Its credibility influences the whole system's credibility to a great extent. In the verification, validation, and accreditation (VV&A) framework (IEEE, 1997), the model credibility is mainly achieved by the work of "validation." Thus model validation is actually essential for determining the credibility of the simulation.

Simulation model is not a simple object. It usually contains multiple inputs, outputs, and a sophisticated processing mechanism (Zhang, 2011; Zhang et al., 2014). The credibility of a simulation model is influenced by many factors. Naturally it is well known that the model credibility can be measured as the similarity between model outputs and real-world outputs by the same input. However, this approach remains an unsolved problem. How to compare two series of data in the application domain? How to calculate the similarity between a single output and its real-world reference through a bunch of simulation datasets? How to make an aggregation of those individual output similarity to represent total credibility? How to take an insight into the model and find the cause of credibility deficiency? We have to answer these questions before getting a convincing result.

Although a lot of earlier study has been conducted to realize model validation, we still need a thorough solution to determine the model credibility in practical VV&A process. This chapter presents "what to do" and "how to do" with achieving the goal. Furthermore, the chapter provides a case study to explain the use of the methods introduced in the earlier sections.

1.2 Scope and Fundamentals

Model validation aims at achieving model credibility. Because the primary principle is comparing the simulation outputs and "real" world outputs, the model under validation must be runable by some means. The other prerequisite of model validation is that the outputs of the "real" world can be observable, or at least there is adequate expert knowledge on them.

When the outputs are valid, we need appropriate validation techniques to compare them, and need acceptability criteria to judge whether the model is acceptable. Although there is no regulation on qualitative or quantitative results, some simulation users and validation agencies prefer to get a quantitative credibility which is able to make a direct comparison between similar models.

Model validation is a comprehensive work. We can deduce how many specific jobs have to be involved in model validation. Fig. 1 shows the essentials of model validation, which are indicated by the subjects marked with a round-corner rectangle.

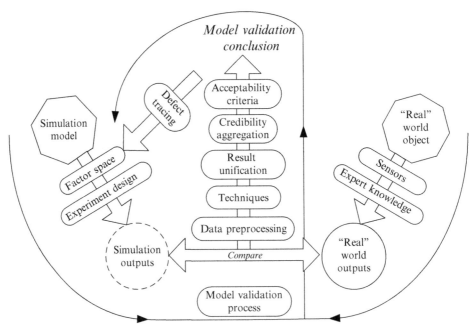

FIG. 1   Fundamentals of model validation.

(1) Factor space

The simulation model needs a factor space (Wang, 1992; Li et al., 2000) to describe and organize the credibility indicators. In early days the AHP (analytic hierarchical process) (Saaty, 1997) hierarchy is widely used. However, researchers and VV&A engineers prefer to use a network instead nowadays.

(2) Experiment design

Various initial states and uncertain factors make the model outputs different at each running. In order to gather the outputs as thoroughly as possible, and meanwhile reduce the simulation times, we must perform an experiment design, such as the orthogonal design (Su et al., 2016), Latin hypercube design (Shields and Zhang, 2016), etc.

(3) Sensors and expert knowledge

We need sensors deployed on the "real-" world object to get the observed outputs. However, many physical quantities are hard or even impossible to measure. In this case, we have to use expert knowledge instead.

(4) Data preprocessing

The data has to be preprocessed to meet the requirements of certain techniques. For example, the Theil inequality coefficient (TIC) (Kheir and Holmes, 1978) method requires that two data series are aligned with the horizontal axis attribute (usually time). In this case, we have to perform interpolation methods, etc.

(5) Techniques (similarity analysis methods)

Appropriate techniques are used to perform similarity analysis between simulation and observed data. There are three major categories of techniques: statistics analysis methods, time domain analysis methods, and frequency domain analysis.

(6) Result unification

There has to be a unified description of the similarity analysis results so that they can be compared and aggregated. The conversion must be based on the mechanism of each technique, and results in the range of credibility, monotonically increasing, and same difference amplitude.

(7) Credibility aggregation

We need a way to aggregate the partial credibility to the total credibility, if a quantitative result is required. The aggregation method, or the calculation function, is selective from taking the minimum, weighted average, neural network composition, etc.

(8) Acceptability criteria

Once the credibility is achieved, we need the acceptability criteria to determine whether the model is acceptable or not. Acceptability criteria must be measurable. The value of it must be carefully calculated according to the tolerance of simulation.

(9) Defect tracing

If the credibility result is negative, it is natural for the user to wonder which part of the model causes credibility deficiency. The factor space can be used to search and locate the "fault node" by assessing the impact of partial credibility to the total one.

(10) The model validation process

All the essential work above forms a model validation process, which usually needs the workflow (WfMC, 1995) technique to drive. Because the model validation process is synchronized with the model development process, the workflow must have the ability of coupling (WfMC, 1996).

1.3 Relation to VV&A

The concept of VV&A is well known by the IEEE 1278.4 standard, Recommended Practice Guide for Distributed Interactive Simulation − Verification, Validation, and Accreditation (IEEE, 1997). It integrates all jobs which are involved in simulation credibility assessment, and categorizes them into three kinds of work. VV&A covers a much wider range of work than model validation, but holds the same purpose of assuring that the simulation has an acceptable credibility before use. Fig. 2 shows the relation between model validation and VV&A.

1.4 Approach of Model Validation

A model validation approach must fulfil the fundamentals of credibility assessment, and has the ability to achieve a convincing credibility result. Fig. 3 shows an example of the approach of model validation.

The execution procedures of the approach are explained as below:

(1) Build a model validation process

Use workflow or OA (office automation) tools (Ke et al., 2005) to build a model validation process, which is able to be facilitated automatically, or plan a practical schedule of model validation calendar.

(2) Build a factor space

Collect all indicators that influenced model credibility, reveal their relationship, and use a modeling method (AHP, MADN (Fang et al., 2011; Fang et al., 2012), etc.) to build a factor space of model validation.

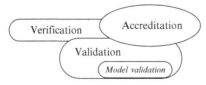

FIG. 2   Relation between model validation and VV&A.

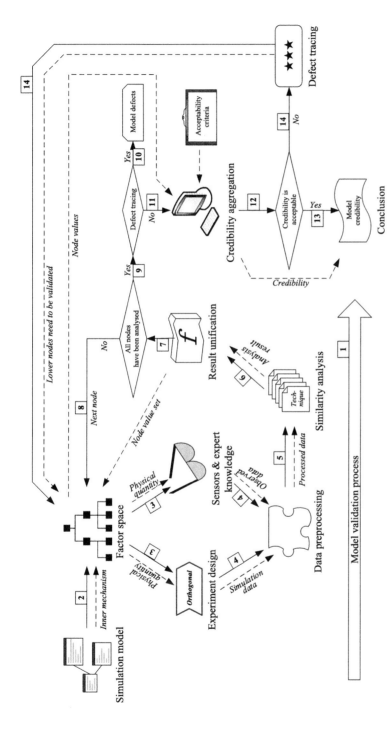

FIG. 3   A practical approach of model validation.

(3) Prepare validation data

Use the experiment design methods (orthogonal design (Su et al., 2016), Latin hypercube design (Shields and Zhang, 2016), etc.) to select the initial condition of the simulation model and run to obtain the simulation data, and use sensors or expert knowledge to obtain the observed data of the corresponding node in the factor space.

(4) Perform data preprocessing

Use appropriate methods (singular value elimination, time sequence consistency processing, moving average filtering, etc.) to perform data preprocessing on simulation and observed data to match the similarity analysis method in the next step.

(5) Perform similarity analysis

Use appropriate methods (statistics analysis, time domain analysis, frequency domain analysis, etc.) to "compare" the preprocessed simulation and observed data.

(6) Perform result unification

Use appropriate formula to convert the analysis results into a unified credibility description, according to the similarity analysis method used in the earlier step.

(7) Determine if all nodes are analyzed

If in the validation process, determine whether the final outputs of the model are all analyzed. If in the defect tracing process, determine whether the intermediate outputs and initial inputs of the model are all analyzed.

(8) Proceed to the next node

If the answer is "No" in Step 7, go back to the factor space, proceed to the next node which needs to be analyzed, and cycle to Step 3.

(9) Determine if in the defect tracing process

If the answer is "Yes" in Step 7, determine if it is in the defect tracing process.

(10) List model defects

If the answer is "Yes" in Step 9, list all detected nodes of model defect which induce the credibility deficiency.

(11) Perform credibility aggregation

If the answer is "No" in Step 9, use an appropriate method (taking the minimum, weighted average, neural network composition, etc.) to aggregate the partial credibility into the total one.

(12) Determine if the credibility is acceptable

According to the acceptability criteria of the model, determine if the total credibility is acceptable.

(13) Draw the conclusion

If the answer is "Yes" in Step 12, gather the partial credibility and the total credibility to present the conclusion of model validation.

(14) Perform defect tracing

If the answer is "No" in Step 12, go back to the factor space, perform defect tracing on lower nodes which are required to be analyzed further, and cycle to Step 3 until all nodes of the model defect are located.

2 FACTOR SPACE OF MODEL VALIDATION

Factor space is a systematically organized structure which contains all indicators and their relationship that influence model credibility (Fang et al., 2017a). Because of the complexity and intricate mechanism of the model, it is hard to achieve the credibility by a certain mathematical formula. In most cases, the total credibility of a model is determined by the decision-making way, such as:

$$O = \{a_1, a_2, ..., a_n\}$$
$$A = \{<n_1, n_2, ..., n_j>, <c_1, c_2, ..., c_k>\}$$
$$F_c = f(v_1, v_2, ..., v_j)$$

(1)

where O is a complex object which can be granted as the model. $a_1 \sim a_n$ are the subobjects or components after decomposition. A is a factor space which contains all influencing aspects of O. $n_1 \sim n_j$ and $c_1 \sim c_k$ are the indicators and their relationship. F_c is the final result. $v_1 \sim v_j$ are the partial results mapped with $n_1 \sim n_j$. f is the aggregation function.

Apparently, factor space A and aggregation function f are critical to the final result. Conventionally, A is often built by an AHP hierarchy (Saaty, 1997), and f is selected as the weighted average function. However, the relation between factors is not always linear, and the influence of the factor is not always transferred through layers one by one. Actually the similarity between the model and the "real" world is affected by the accumulation of error in the computational process of the model, but not decided by the weighted average value through adjacent layers in the AHP hierarchy, or something like that. Model validation needs a better method to reveal the factors and their influence on model credibility.

2.1 Network Definitions

Here, we introduce a network-based method (Fang et al., 2011; Fang et al., 2012) to build the factor space of model validation. Define the factor space as a directional graph of radial distributed nodes, which can be expressed as the quadruple:

$$F = \{<N, V>; <L, A>\}$$

(2)

where F is a factor space. N and V are the node set and value set; L and A are the link set and attribute set mapped with each other. Set the link direction as from the attribute-holding node to the attribute-receiving node. For example, a structural link of the traditional hierarchy has the direction from the child node to the parent node. According to the requirements of model validation, further develop element definitions of the network as in the following:

Def. 1 Define $N = \{n_1, n_2, ..., n_k\}$ as the "Node set" and $N = \hat{N} \cup \tilde{N}$, where $\hat{N}$ is the certain node set and $\tilde{N}$ is the uncertain node set. If $n_i \in \tilde{N}$, use $t(n_i)$ to indicate the node's type and $t(n_i) \in \{regular, sufficient, inherited\}$. If $n_i \in \tilde{N}$, use $c(n_i)$ as the transit condition of n_i, and $c(n_i) \in \{0, 1\}$.

Def. 2 Define $V = \{v_1, v_2, ..., v_k\}$ as the "Value set". If $t(N) \in \{regular, sufficient\}$, V and N are mapped with each other. Use $v_i = v(n_i)$ to indicate the value of node n_i which is mapped with v_i.

Def. 3 Define $L = \{l_1, l_2, ..., l_k\}$ as the "Link set" and $L = \hat{L} \cup \tilde{L}$, where $\hat{L}$ is the certain link set and $\tilde{L}$ is the uncertain link set. Use $t(l_i)$ to indicate the link's type, and $t(l_i) \in \{regular, sufficient, ultra, equivalent, contraditory, inherited, traced\}$. If $l_i \in \tilde{L}$, use $c(l_i)$ as the transit condition of l_i, and $c(l_i) \in \{0, 1\}$. Use $l = (n_o, n_d)$ to indicate a link from node n_o to node n_d and $n_o = b(l)$; $n_d = e(l)$.

Def. 4 Define $A = \{a_1, a_2, ..., a_k\}$ as the "Attribute set" and A is the one-one mapped with L. Use $a_i = a(l_i)$ to indicate the attribute of link l_i which is mapped with a_i. If $t(l_i) \in \{equivalent, inherited, traced\}$, then $a_i \in \varnothing$.

Make definitions below to indicate various types of network elements:

Def. 5 If $l \in L$, $t(l) = regular$ and $l = (n_o, n_d)$, define $a(l) \in [0, 1]$ as the weight distributed from n_d to n_o. If $N_c = \{n \mid (n, n_d) \in L_c\}$, $L_c \in \overline{L}$ and $t(L_c) = regular$, then $\sum_{i=1}^{k} a_k(l_k) = 1$, $l_k \in L_c$ and $k = d(L_c)$.

Def. 6 If $n \in \tilde{N} \wedge c(n) = 1$, which makes $n \in \hat{N}$, define n as the "Recovered node." If $n \in \tilde{N} \wedge c(n) = 0$, define n as the "Rubbish node," and it need to be deleted from the network.

Def. 7 If $l \in \tilde{L} \wedge c(l) = 1$, which makes $l \in \hat{L}$, define l as the "Recovered link." If $l \in \tilde{L} \wedge c(l) = 0$, define l as the "Rubbish link," and it need to be deleted from the network.

Def. 8 If $n \in \overline{N}$, $t(n) = sufficient$ and $v(n) = 1$, then the sufficient link from n breaks. If $v(n) = 0$, then n needs to supplement the additional brother node, which is defined as the "Shadow node" and of value 0.

Def. 9 If $l \in L$, $l = (n_o, n_d)$, and $t(l) = ultra$, define $a(l) \in [0, 1]$ as the "Acceptability threshold". If $v(n_o) < a(l)$, then $v(n_d) = 0$, and define n_o as the "Key node" of n_d, n_d as the "Super conduct node" of n_o. If $v(n_o) \geq a(l)$ then $l \in \tilde{L}$ and $c(l) = 0$.

Def. 10 If $l \in L$, $l = (n_o, n_d)$, and $t(l) = inherited \wedge t(n_o) = inherited$, then the subnetwork under n_d has to be replicated to n_o, and the new nodes are defined as inherited nodes, whose physical meaning will be given by n_o.

Def. 11 If $l \in L$, $l = (n_o, n_d)$, and $t(l) = contradictory$, define $.a(l) \in [0, 1].$ as the "Contradiction percentage" of n_o to n_d. $v(n_d) = \begin{cases} v(n_d) & v(n_d) \leq 1 - a(l) \cdot v(n_o) \\ 1 - a(l) \cdot v(n_o) & v(n_d) > 1 - a(l) \cdot v(n_o) \end{cases}$, where $v(n_o)$ is the source node value and $v(n_d)$ is the destination node value.

Def. 12 If $l \in L$, $l = (n_o, n_d)$, and $t(l) = equivalent$, define n_o and n_d as "Mirror node," and $v(n_d) = v(n_o)$.

Def. 13 If $l \in L, l = (n_o, n_d)$, and $t(l) = traced$, define n_d as the "Traced node." $v(n_d)$ is irrelevant to $v(n_o)$.

Make definitions below to express the network structure:

Def. 14 If $^\bullet n = \{x \mid x \in N \wedge (x, n) \in L\}$, define $^\bullet n$ as the "Preset" of n. If $n^\bullet = \{x \mid x \in N \wedge (n, x) \in L\}$, then define $n^\bullet$ as the "Postset" of n.

Def. 15 If $(n_o, n_d) \in L$, define n_o as the "Child node" of n_d, and n_d as the "Father node" of n_o. If $(n_o, n_a) \in L$ and $(n_d, n_a) \in L$, define n_o and n_d as the "Brother nodes." If $n^\bullet = \varnothing$, define n as the "Root node." If $n^\bullet \neq \varnothing$ and $^\bullet n \neq \varnothing$, define n as the "Branch node." If $^\bullet n = \varnothing$, define n as the "Leaf node."

Def. 16 Set power operator satisfies $(n^\bullet)^0 = n, (n^\bullet)^1 = n^\bullet, (n^\bullet)^2 = (n^\bullet)^\bullet \ldots$ If $n_d \in (n_o^\bullet)^{s_1}, n_d \in (n_o^\bullet)^{s_2}$, $\ldots, n_d \in (n_o^\bullet)^{s_k}$, define $S(n_o \to n_d) = \{s_1, s_2, \ldots, s_k\}$ as the "Distance set" from n_o to n_d. Nonnegative integers $s_1 \sim s_k$ are all distances from n_o to n_d. If n_0 is a root node, abbreviate $S(n \to n_0)$ as S_n.

Def. 17 If $N_L = \{n_1, n_2, \ldots, n_k\}$, $\forall n \in N_L$ makes $^\bullet n \cap N_L = \varnothing$ and $n^\bullet \cap N_L = \varnothing$, and $\forall n_o, n_d \in N_L$, $\forall s_k \in S_{n_o}, \forall s_m \in S_{n_d}$ makes $Max(s_k) = Max(s_m)$, define N_L as a "Layer" of the network, which is the $Max(s_k)$-th layer. Define $r(n_o)$ as the "Order" of node n_o and $r(n_i) = Max(s_k)$.

Def. 18 Define D as the "Depth" of the network, and $\forall n \in N, \forall s_k \in S_n, D = Max(s_k)$.

Def. 19 Define E as the "Span" of the network, while $N_{L(i)} = \{n_{1(i)}, n_{2(i)}, \ldots, n_{k(i)}\}$ is the node set of the i-*th* node layer, $i = 1, 2, \ldots, D$ and $E = Max[k(i)]$.

Def. 20 Define $N_a \cup L_a$ as a "Path" from node n_1 to n_d, while $N_a = \{n_1, n_2, \ldots, n_k\}, L_a = \{l_1, l_2, \ldots, l_k\}$, $l_i = (n_i, n_{i+1}) \in L_a, i = 1, 2, \ldots, k-1$, and $l_k = (n_k, n_d)$.

Def. 21 If $\exists s$ makes $n_d \in (n_o^\bullet)^s$. Define n_o to n_d as "Reachable." If $\forall s$ makes $n_d \notin (n_o^\bullet)^s$, define n_o to n_d as "Unreachable," and can be marked as $s(n_o \to n_d) = \infty$.

Def. 22 If $s(n_o \to n_d) > 1$ and $s(n_o \to n_d) \neq \infty$, define n_o as the "Offspring node" of n_d; n_d is the "Ancestor node" of n_o, and they are "Lineal relative nodes." If $s(n_o \to n_d) = \infty$ and n_o, n_d are not brother nodes, define n_o and n_d as "Collateral relative nodes."

2.2 Structural Rules

In order to use the network to perform model validation, we must define rules to regulate its structure and operation. The rules below should be followed when use the factor space network to validate simulation models.

Rule 1 If $n \in N$, then $^\bullet n \cup n^\bullet \neq \varnothing$. If $l \in L$ and $l = (n_1, n_2)$, then $n_1 \neq \varnothing$ and $n_2 \neq \varnothing$. (There is no isolated node or link in the factor space.)

Rule 2 If $N_0 = \{n \mid ^\bullet n = \varnothing\}$, then $d(N_0) = 1$ and $N_0 \subset N$. (There is only one root node in the factor space.)

Rule 3 If $n \in N, t(n) = ^{''}sufficient^{''}$ and $l = (n, n_d) \in L$, then $t(l) = ^{''}sufficient^{''}$. (The link which starts from a sufficient node is a sufficient link.)

Rule 4 If $n \in \widetilde{N}$, then $l = (n, n_d) \in \widetilde{L}$. (The link which has a source of uncertain node is an uncertain link.)

Rule 5 If $n_r \in \widetilde{N}$, $c(n_r)=0$, $N_c=\{n_1, n_2, \ldots, n_k\}$, $\forall n_c \in N_c$ makes n_c to n_r reachable, and $\forall n_s \in \hat{N}$ makes n_c to n_s unreachable, then $\forall n_c \in N_c$ makes $c(n_c)=0$. (If the offsprings of a rubbish node have no certain ancestors, they are all rubbish nodes.)

Rule 6 If $n_r \in \widetilde{N}$, $c(n_r)=0$, $l_{rb}=(n_r, n_i) \in \widetilde{L}$, and $l_{re}=(n_i, n_r) \in \widetilde{L}$, then $c(l_{rb})=c(l_{re})=0$. (The uncertain link which starts from or ends with a rubbish node is a rubbish link.)

2.3 Graphic Illustration

In order to express the factor space network visually, we define the necessary graph elements to provide a graphic illustration. The graph element set is mapped with all definitions and follows the rules. Fig. 4 shows an example of the graphic illustration of the factor space. The graphic illustration is explained below:

(1) Use single-lined figure (circle or rectangle) to present a regular node, double-lined figure to present sufficient node and a double-lined round-cornered rectangle to present the inherited node.
(2) Mark the node name or number in the node figure, and mark the node value and the transit condition outside nearby.
(3) Use a directional line segment to present the link, and use the line end to present the link type. The solid arrow end presents a regular link, circle end presents a sufficient link, double-arrow end presents an ultralink, equality-sign end presents an equivalent link,

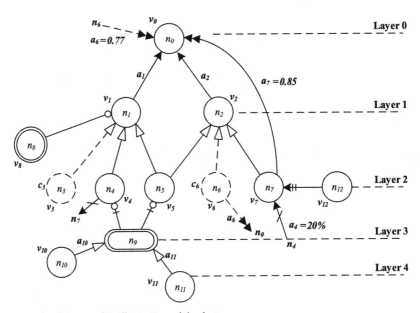

FIG. 4　An example of the graphic illustration of the factor space.

slash-sign end presents a contradictory link, hollow arrow end presents a traced link, and the slash-signed circle end presents an inherited link.

(4) Mark the attribute by the link. To avoid intersection of links, fold the link and mark the source and target node at the folded link.

(5) Use a solid-lined figure to present a certain element, and use a dot-lined figure to present an uncertain element.

2.4 Credibility Aggregation

Dynamic elements (uncertain and sufficient node/link) affect the structure of the factor space. The credibility aggregation cannot be performed unless the network is static, which means the dynamic elements have to be analyzed first.

Because the destination of a traced link has no effect on the value of its origin, the traced links do not contribute to credibility aggregation. The origin of a traced link usually represents the model input or output, where we can perform similarity analysis to get partial credibility directly. When going through the factor space downwards, the credibility aggregation stops when the tracing links appear.

When the dynamic analysis is done, the credibility aggregation can be achieved by the following procedure:

(1) Go through the factor space downwards, and stop at the destinations of traced links. Perform similarity analysis by comparing the simulation and real-world outputs, and get partial credibility on these nodes after result unification. Generally, the partial credibility can be achieved by

$$C(n) = 1 - \frac{\|O(n) - O(n')\|}{\|O(n')\|}\bigg|_{I(n)=I(n')} \tag{3}$$

where $I(n)$ and $O(n)$ are the simulation input and output of the node, $I(n')$ and $O(n')$ are the corresponding real-world input and output, and $\|\cdot\|$ represents the norm of the variant. According to the technique used, $C(n)$ can be achieved by statistics analysis, time domain analysis, or frequency domain analysis methods together with their result unification formulas.

(2) If the destination of the traced link is lack of real-world output, then go down along the path further to find a node whose real-world output is valid, and get the value of the node by similarity analysis. The partial credibility of the upper node which has no real-world output can be achieved by error analysis via the computational process of the model:

$$C(n_2) = 1 - \frac{\|f((2 - C(n_1)) \cdot O(n_1)) - f(O(n_1))\|}{\|O(n_2)\|} \tag{4}$$

where n_1 is the node which has real-world output, $C(n_1)$ is its partial credibility, and $O(n_1)$ is its simulation output. n_2 is the node which has no real-world output, $C(n_2)$ is its partial credibility, and $O(n_2)$ is its simulation output. f is the computational function of the model from n_1 to n_2.

(3) Use appropriate algorithm to aggregate the total credibility on the root node, by gathering the partial credibility on the destinations of the traced links along the decomposition paths of factor space:

$$C = f(v_i, v_{i+1}, \ldots, v_{i+k}) \tag{5}$$

where C is the total credibility; $v_i \sim v_{i+k}$ are the partial credibility on the destination nodes of the traced links; f is the aggregation function. f can be the method of taking the minimum, and weighted average, etc., and can be different across the layers in factor space. Take the weighted average method as an example. The partial credibility of a node in the path of links other than a traced link can be achieved by

$$v_i = \prod_{s=1}^{d} w_s \cdot \left[\sum_{k=1}^{p} (v_k \cdot w_k \cdot c_k) + (1 - v_i') \cdot \sum_{n=1}^{q} (v_n \cdot w_n) \right]$$
$$\sum_{k=1}^{p} w_k + \sum_{n=1}^{q} w_n = 1; w_s = \begin{cases} 0; & v_s < a(l_s) \\ 1; & v_s \geq a(l_s) \end{cases} \tag{6}$$

where v_i is the value of a node n_i in the path of links other than traced link; v_k is the value of certain and uncertain child nodes of n_i; c_k is the transit condition of an uncertain child node; w_k is the regular weight of the child node; p is the number of child nodes; v_i' is the value of sufficient child node of n_i; v_n is the value of shadow child nodes of n_i; w_n is the regular weight of the shadow child node; q is the number of shadow child nodes; w_s is the "ultraweight" of the ultralink which ends with n_i; v_s is the value of the ultralink origin; $a(l_s)$ is the acceptability threshold of the ultralink; and d is the number of ultralinks.

Because the value of a shadow node is 0, Formula (6) can be simplified as

$$v_i = \prod_{s=1}^{d} w_s \cdot \sum_{k=1}^{p} (v_k \cdot w_k \cdot c_k) \tag{7}$$

The total credibility of the root node can be aggregated by

$$v_0 = \sum_{i=1}^{q} \left[v_i \cdot \prod_{m=1}^{k_i} \left(w_m \cdot \prod_{m=2; r=1}^{k_i; p} w_{m,r} \cdot c_m \right) \right]$$
$$w_{m,r} = \begin{cases} 0; & v_{m,r} < a(l_{m,r}) \\ 1; & v_{m,r} \geq a(l_{m,r}) \end{cases} \tag{8}$$

where v_0 is the value of the root node n_0; v_i is the partial credibility on node n_i of traced link destination; q is the number of the traced link destination; w_m is the regular weight of the ancestor node of n_i; k_i is the ancestor node number of n_i; $w_{m,\,r}$ is the ultraweight of the ultralink on the path from n_i to n_0; p is the ultralink number; $v_{m,\,r}$ is the value of ultralink origin; $a(l_{m,\,r})$ is the acceptability threshold of the ultralink; and c_j is the transit condition of the uncertain ancestor node of n_i.

2.5 Priority Analysis

Because of the existence of ultralink, the destination of traced link has priority in the factor space network. Priority analysis can be realized by the fact that the bigger the possible discard of subnetwork is, the higher the priority of the node owns. Because the shadow node value is 0, it is no need to analyze the priority of shadow nodes.

Set $\widehat{L} \subseteq L, \forall \widehat{l} \in \widehat{L}, t\left(\widehat{l}\right) = {}^{"}ultra{}^{"}$; set $\widehat{L}_i \subseteq \widehat{L}, \widehat{L}_i \neq \varnothing$ as the ultralink set which is on the path from n_i to n_0; if it satisfies Condition 1: $\forall \widehat{l}_i \in \widehat{L}_i$ makes $v\left(b\left(\widehat{l}_i\right)\right) < a\left(\widehat{l}_i\right)$ and Condition 2: $\forall \widehat{l} \in \widehat{L} \wedge \widehat{l} \notin \widehat{L}_i$ makes $v\left(b\left(\widehat{l}\right)\right) \geq a\left(\widehat{l}\right)$, then the priority of node n_i is

$$p(n_i) = d(N_s) + d(N_r) \tag{9}$$

where $p(n_i)$ is the priority of n_i, N_s is the node set which needs to be validated if Condition 1 is true, and N_r is the node set which needs to be validated if both Conditions 1 and 2 are true. If $\widehat{L}_i = \varnothing$, then $d(N_s) = 0$, and Formula (9) can be simplified as

$$p(n_i) = d(N_r) \tag{10}$$

3 TECHNIQUES OF MODEL VALIDATION

We usually use appropriate techniques (similarity analysis methods) to achieve partial credibility on the destination of traced links in the factor space network. By the analysis mechanism, techniques can be categorized into "qualitative analysis" and "quantitative analysis." If the simulation and observed data are plenty, it is natural to perform quantitative analysis rather than qualitative one. This section discusses the quantitative analysis.

The destination of traced links is usually an output of the model, whatever it is a final one or an intermediate one. By the nature of the output, it can be categorized into "static performance" and "dynamic performance" of the model. Static performance is time invariant, and suitable to be analyzed by statistics analysis methods. Dynamic performance is time variant, and suitable to be analyzed by time domain and frequency domain analysis methods.

It has to be stated that any output of a model has more than one series of simulation data under different conditions. All these series of data contribute to reflect the model behavior. So it is more likely that a similarity analysis method is used more than once on the same node in the factor space, or even more than one method is used to get a comprehensive result.

3.1 Data Preprocessing

In most cases, the simulation and observed data cannot meet the requirements of the similarity analysis method, which makes that they have to be preprocessed before the comparison. Generally, there are three kinds of data preprocessing methods: singular value elimination, time series alignment, and moving average filtering. Moreover, singular value elimination and hypothesis test of statistics analysis need the data accords with normal distribution.

3.1.1 Normality Test

There are two typical normality test methods: W test (Shapiro-Wilk test) and D test (D′ Agostino test). The two methods have different statistical quantities.

(1) W test (Rahman and Govindarajulu, 1997)

Suppose $x_1, x_2, \ldots, x_n$ is the samples from population X and sequenced as $x_1 \leq x_2, \ldots, \leq x_n$. n is the volume of the samples and $n < 50$. For the hypothesis of $H_0: X$ is normally distributed by $N(\mu, \sigma^2)$ and $H_1: X$ is not normally distributed, use the statistical quantity below to perform a test:

$$W = \frac{\left[\sum_{i=1}^{m} a_{i,n}(x_{n+1-i} - x_i)\right]^2}{\sum_{i=1}^{n}(x_i - \overline{x})^2} \tag{11}$$

where m is the biggest positive integer which is less than or equal to $n/2$ and $\overline{x}$ is the average value of x_i series. $a_{i,n}$ is the calculation parameter, and it can be acquired in the table which provides data for each of the sample volume from 1 to 50. Given the significance level α, the rejection region of the hypothesis is $W \leq W_\alpha$. In other words, if $W > W_\alpha$ the population X is normally distributed, where W_α can be obtained by a W test lookup table.

(2) D test (D'Agostino et al., 1990)

Suppose $x_1, x_2, \ldots, x_n$ are the samples from population X, n is the volume of the samples. For the hypothesis of $H_0: X$ is normally distributed by $N(\mu, \sigma^2)$ and $H_1: X$ is not normally distributed, use the statistical quantity below to perform a test:

$$D = \frac{\sum_{i=1}^{n}\left(i - \frac{n+1}{2}\right)x_i}{(\sqrt{n})^3 \sqrt{\sum_{i=1}^{n}(x_i - \overline{x})^2}} \tag{12}$$

The approximate, normalized random variant of D is

$$Y = \frac{\sqrt{n}(D - 0.28209479)}{0.02998598} \tag{13}$$

Given the significance level α, the rejection region of the hypothesis is $Y \leq Y_{\alpha/2}$, where $Y_{\alpha/2}$ can be get by a table look-up.

3.1.2 Singular Value Elimination

There are two typical singular value elimination methods: Pauta method and Grubbs method. Both the methods require that the data population is normally distributed.

(1) Pauta method (Gao et al., 2014)

Suppose $x_1, x_2, ..., x_n$ is the samples from population X which is normally distributed by $N(\mu, \sigma^2)$, n is the volume of the samples. The average value of the samples is $\bar{x}$. The standard deviation can be calculated as

$$s = \sqrt{\frac{\sum\limits_{i=1}^{n}(x_i - \bar{x})^2}{n-1}} \tag{14}$$

Given the sample deviation $\Delta x_l = x_l - \bar{x}, l = 1, 2, ..., n$, if $|\Delta x_l| > 3s$ then treat x_l as a singular value. Because if $n \leq 10$ any x_i will satisfy $|\Delta x_i| \leq 3s$, the Pauta method is usually used when $n > 10$.

(2) Gubbs method (Grubbs, 1950)

Suppose $x_1, x_2, ..., x_n$ are the samples from population X, which is normally distributed by $N(\mu, \sigma^2)$, n is the volume of the samples. The average value of the samples is $\bar{x}$, and the sample deviation is $\Delta x_i = x_i - \bar{x}, i = 1, 2, ..., n$. The singular value can be determined by

$$|\Delta x_i| > k(n, \alpha)\sigma \tag{15}$$

α is the significance level, which shows that the probability of the error exceeds $\pm k\sigma$. α is usually selected as 0.01 or 0.05. k is determined by n and α, and can be get by a table look-up. σ is the variance of the sample.

3.1.3 Time Series Alignment

Usually the simulation and the observed data series are not mapped with each other by the time stamp or other common physical quantity on the x-axis. However, the typical time domain analysis methods require that the two series are aligned and each of the data is paired, such as TIC and GRA (gray relational analysis) (Deng, 1995). In this case, time series alignment has to be performed, which mainly uses the Lagrange interpolation (Golub and Ortega, 1993) to pair the two series.

Suppose the value of a real function $f(x)$ in the interval $[a, b]$ is $y_i = f(x_i), i = 0, 1, 2, ..., n$. To estimate the value of $f(x)$ on a point x in the interval $[a, b]$, search in the polynomial class M_n to find $y(x) \in M_n$ and let $y_i = f(x_i), i = 0, 1, 2, ..., n$. Use $y(x)$ as the estimation of $f(x)$. Call x is the interpolation point and $y(x)$ is the interpolation polynomial. It can be proved that $y(x) \in M_n$ is unique.

Use $L_i(x), i = 0, 1, 2, ..., n$ to present a polynomial of degree n, which satisfies

$$L_i(x_j) = \begin{cases} 0, & i \neq j \\ 1, & i = j \end{cases}, i, j = 0, 1, 2, ..., n \tag{16}$$

We can make an interpolation function as

$$L_i(x) = \frac{(x-x_0)(x-x_1)\cdots(x-x_{i-1})(x-x_{i+1})\cdots(x-x_n)}{(x_i-x_0)(x_i-x_1)\cdots(x_i-x_{i-1})(x_i-x_{i+1})\cdots(x_i-x_n)} \tag{17}$$

Then the Lagrange interpolation polynomial is

$$y(x) = \sum_{i=0}^{n} f(x_i)L_i(x) \tag{18}$$

3.1.4 Moving Average Filtering

Most of the frequency domain analysis methods require that the two datasets are smooth, such as the windowed spectrum estimation and maximum entropy spectrum estimation. However, the simulation and observed data are not smooth most of the time. Moving average filtering (Smith, 2013) is a low-pass filtering method, which picks up m sample data continuously before and after the current sample point, and calculates the moving average value of all points by sequence to eliminate the glitch interference.

Suppose the time series data $y(t)$ is formed by the certain part $f(t)$ and the uncertain part $e(t)$. Select m and let $m=2n+1$. Pick up m sample data before and after a sample point y_i, $i=n+1, n+2, ..., N-n$, and use the arithmetic average value of them as the filtered value f_i to eliminate the interference of e_i, as the formula shows:

$$f_i = y_i = \frac{1}{2n+1} \sum_{j=-n}^{n} y_{i+j}, i=n+1, n+2, ..., N-n \tag{19}$$

The two ends of y_i ($i=1, 2, ..., n, N-n+1, ..., N$) cannot be filtered by Formula (19), and it has to be added by some means. The moving average filtering of Formula (19) is actually an equal weighted average method. However, the nearer the sample data goes from y_i, it has more effect, and on the contrary it has less effect. So the sample points in the filtering interval $[i-n, i+n]$ have different weights on influencing the filtered value y_i:

$$f_i = y_i = \sum_{j=-p}^{q} w_j y_{i+j}, i=p+1, p+2, ..., N-q, \quad \sum_{j=p}^{q} w_j = 1 \tag{20}$$

where w_j is the weight coefficient. p, q are any of the positive integers less than m, and $m=p+q+1$. If $p=q=n$ and $w_j=1/(2n+1)$, Formula (20) turns into Formula (19). Practically, engineers usually use equal weighted center smoothing with $p=q=2\sim5$, or unequal weighted smoothing with $p=q=2\sim3$.

3.2 Statistics Analysis

Statistics analysis methods are mainly used to validate time-invariant outputs. They can get the primary characteristics of the time-invariant output which reflects the behavior of the physical quantity being validated. When the simulation is run multiple times, the statistical performance of the output approaches to the behavior of the model. Basically, there are two types of statistics analysis methods frequently used in model validation: parameter estimation and hypothesis test. Usually the methods require that the data population is normally distributed.

3.2.1 Parameter Estimation

Point estimation is mostly used to perform parameter estimation. Take the moment estimation (Linton, 2017) as an example. Mostly the population of time-invariant simulation and observed data are normally distributed. Suppose the mean value μ and the variance $\sigma^2 > 0$ of the population are unknown but exist. If $n \to \infty$, the order-k moment of the sample converges to the order-k moment of the population, which makes

$$\begin{cases} \mu_1 = E(X) = \mu \\ \mu_2 = E(X^2) = D(X) + [E(X)]^2 = \sigma^2 + \mu^2 \end{cases} \tag{21}$$

Then we can induce

$$\begin{cases} \mu = \mu_1 \\ \sigma^2 = \mu_2 - \mu_1^2 \end{cases}, \quad \begin{cases} \mu_1 = a_1 = \frac{1}{n}\sum_{i=1}^{n} X_i \\ \mu_2 = a_2 = \frac{1}{n}\sum_{i=1}^{n} X_i^2 \end{cases} \tag{22}$$

$$\begin{cases} \hat{\mu} = a_1 = \overline{X} \\ \hat{\sigma}^2 = a_2 - a_1^2 = \frac{1}{n}\sum_{i=1}^{n} X_i^2 - \overline{X}^2 = \frac{1}{n}\sum_{i=1}^{n}(X_i - \overline{X})^2 \end{cases} \tag{23}$$

The result shows that the mean value and variance of the sample are the estimation of mean value and variance of the population. Similarly, we can derive the same result for the maximum likelihood estimation. So in model validation, we can calculate the mean value and variance of the simulation and observed data, and compare them to determine the likelihood of whether they belong to one population. Moreover, we can also use interval estimation (DeGroot and Schervish, 2011a) to do the same thing.

3.2.2 Hypothesis Test

Usually we use parameter hypothesis test in model validation, which requires that the population is normally distributed. There are four types of parameter hypothesis test methods: u test, t test, χ^2 test, and F test. These four methods are used according to the different condition of the population parameter. Take u and t test (mean value tests) as examples.

(1) u test (DeGroot and Schervish, 2011b)

Suppose there are population $X \sim N(\mu_1, \sigma_1^2)$ and $Y \sim N(\mu_2, \sigma_2^2)$. $x_1, x_2, \cdots, x_{n_1}$ is the sample from X and its volume, mean value, and variance are $n_1, \overline{x}, s_1^2$. $y_1, y_2, \cdots, y_{n_2}$ is the sample from Y and its volume, mean value, and variance are $n_2, \overline{y}, s_2^2$.

If the variances of the two populations are known as σ_1, σ_2, we can test the hypothesis $H_0: \mu_1 = \mu_2$. If H_0 is valid, u is normally distributed by $N(0,1)$. Given a confidence level α, the border value $u_{\alpha/2}$ can be determined by table look-up, which makes

$$\begin{cases} u = \dfrac{\bar{x} - \bar{y}}{\sqrt{\dfrac{\sigma_1^2}{n_1} + \dfrac{\sigma_2^2}{n_2}}} \\[3ex] P\{|u| \geq u_{\alpha/2}\} = \alpha \end{cases} \tag{24}$$

If we know that a physical output of the model is normally distributed and the variance of the simulation and real-world output are σ_1, σ_2 (usually $\sigma_1 = \sigma_2$), with a given simulation and observed data samples, we can calculate the mean value of them. If $|u| \geq u_{\alpha/2}$, it means that the simulation output is not similar enough to the "real" world, and vice versa.

(2) t test (DeGroot and Schervish, 2011c)

If the variances of the two populations are unknown, we can test the hypothesis $H_0 : \mu_1 = \mu_2$. If H_0 is valid, t is distributed by $t(n_1 + n_2 - 2)$. Given a confidence level α, the border value $t_{\alpha/2}(n_1 + n_2 - 2)$ can be determined by a table lookup, which makes

$$\begin{cases} t = \dfrac{\bar{x} - \bar{y}}{S_w \sqrt{\dfrac{1}{n_1} + \dfrac{1}{n_2}}} \\[3ex] S_w = \dfrac{(n_1 - 1)s_1^2 + (n_2 - 1)s_2^2}{n_1 + n_2 - 2} \\[3ex] P\left\{ |t| \geq t_{\frac{\alpha}{2}}(n_1 + n_2 - 2) \right\} = \alpha \end{cases} \tag{25}$$

By a given simulation and observed data samples, we can calculate the mean value and variance of them. If $|t| \geq t_{\alpha/2}(n_1 + n_2 - 2)$, it means the simulation output is not similar enough to the "real" world, and vice versa.

3.3 Time Domain Analysis

Time domain analysis deals with time-variant physical output of the model which has a gradually changing curve in the X-Y plane. Usually the X-axis presents the time, and the Y-axis presents the quantity of the physical output. There are two major time domain analysis methods: TIC and GRA. They examine the distance or gradient between two data curves point by point, and use it as the error norm to calculate the similarity.

(1) TIC method (Kheir and Holmes, 1978)

The TIC method is an abbreviation of the Theil inequality coefficient, which uses an inequality coefficient called TIC to present the error norm between two data curves. Suppose there are two time series of $X = \{x_1, x_2, \ldots, x_n\}$ and $Y = \{y_1, y_2, \ldots, y_n\}$, we use the normalized mean square error as the error norm to present the difference between two time series, like the formula

$$\rho(X, Y) = \frac{\sqrt{\frac{1}{n}\sum_{i=1}^{n}(x_i - y_i)^2}}{\sqrt{\frac{1}{n}\sum_{i=1}^{n}x_i^2} + \sqrt{\frac{1}{n}\sum_{i=1}^{n}y_i^2}} \tag{26}$$

It presents a normalized distance error between two curves. If $\rho=0$ it shows that the two time series are totally equal. On the contrary, if $\rho=1$ it shows that the two time series are totally unequal. So the function of TIC is monotone decreasing. Because the distribution of ρ cannot be determined, engineers usually use the threshold of 0.3 to judge the similarity between the two time series. If $\rho \leq 0.3$, the two series are treated as "similar enough."

(2) GRA method (Deng, 1995)

GRA is abbreviated from gray relational analysis, which was proposed by Deng Junlong. It uses the gray relational degree to present the changing trend (distance, direction, speed, etc.) of the two series. Usually the GRA uses the distance and gradient to measure the relational degree.

Suppose $X_0 = \{x_0(1), x_0(2), ..., x_0(n)\}$ is an observed time series, and $X_i = \{x_i(1), x_i(2), ..., x_i(n)\}$ is the multiple simulation data series. There are several kinds of gray relational degree. The Deng's relational degree is the earliest proposed:

$$\gamma(X_0, X_i) = \frac{1}{n}\sum_{k=1}^{n}\gamma(x_0(k), x_i(k))$$

$$\gamma(x_0(k), x_i(k)) = \frac{\min_i \min_k |x_0(k) - x_i(k)| + \rho \max_i \max_k |x_0(k) - x_i(k)|}{|x_0(k) - x_i(k)| + \rho \max_i \max_k |x_0(k) - x_i(k)|} \tag{27}$$

where $|x_0(k) - x_i(k)|$ is the absolute difference of X_0 and X_i at moment k, $\min_i \min_k |x_0(k) - x_i(k)|$ is the minimum difference of poles, and $\max_i \max_k |x_0(k) - x_i(k)|$ is the maximum difference of poles. ρ is the discrimination coefficient. $\rho \in [0,1]$ and ρ is usually set to 0.5.

The modified relational degree uses a combined function to reduce the end error of Deng's relational degree:

$$\begin{cases} \gamma_m(X_0, X_i) = \gamma(X_0, X_i) - [\alpha R_m + (1-\alpha)R_e] \\ R_m = \frac{1}{N}\sum_{k=1}^{N}\frac{|x_0(k) - x_i(k)|}{|x_0(k)| + \varepsilon} \\ R_e = \frac{|x_0(N) - x_i(N)|}{|x_0(N)| + \varepsilon} \end{cases}, \varepsilon = \begin{cases} E[|X_0|] & x_0(k) < 0.01 \\ 0 & x_0(k) \geq 0.01 \end{cases} \tag{28}$$

where $\alpha \in [0,1]$ is a balance coefficient which is often selected as 0.5. R_m is the mathematical expectation of relative error. R_e is the relative error of end moment. N is the volume of data series. ε is the compensation factor.

3.4 Frequency Domain Analysis

Frequency domain analysis deals with time-variant physical output of the model which has a dramatically changing curve in the X-Y plane. Usually the X-axis presents frequency, and the Y-axis presents the power spectral density of the physical output. There are two major frequency domain analysis methods: windowed spectrum estimation and maximum entropy spectrum estimation. They examine the power spectral density between two data curves, and use the F test to determine whether the two series are similar enough.

(1) Windowed spectrum analysis (Montgomery and Conard, 1980)

The prerequisite of a spectrum analysis is that the two time series are stationary. If not, they have to be processed by some method first, such as the moving average filtering.

Suppose $\{X_t\} = \{X_t; t \in N\}$ is a general stationary series with the average value of $EX_t = \mu$, and its autocovariance function $r_x(t, t+k)$ is only relevant to the time interval k. Mark the average value and autocovariance function as μ_x and $r_x(k)$. The spectral density function can be expressed as

$$S_x(\omega) = \sum_{k=-\infty}^{+\infty} r_x(k)e^{-ik\omega}, \ \omega \in [-\pi, \pi], \ \sum_{k=0}^{\infty} |r_k| < \infty \tag{29}$$

where $\{r_k\}$ is the autocovariance function series of the stationary series. If the stationary series $\{X_t\} = \{X_t; t \in N\}$ is ergodic, the ensemble average of the sample can be substituted by the time average. Use the Fourier transform of $\widehat{r}_x(k)$, we can achieve the estimation of $S_x(\omega)$:

$$\widehat{S}_x(\omega) = \sum_{k=-N+1}^{N-1} \widehat{r}_x(k)e^{-ik\omega} = \frac{1}{N} \left| \sum_{l=1}^{N} x_l e^{-i\omega l} \right|^2 = I_N(\omega), (-\pi < \omega < \pi) \tag{30}$$

where $I_N(\omega)$ is the periodogram, which is not a consistent estimate of the spectral density. Although the average value is convergent to the real value, the variance does not tend to be zero. It can be proved that, if we use $2N-1 \ \widehat{r}_x(k)$ to estimate the spectral density, no matter how big the N is, the variance of the periodogram is always not less than the average square of the estimated value, which is $E^2[I(\omega)]$.

So the end influence of $\widehat{r}_k$ to the estimation of spectral density must be reduced. One of the easiest ways is omitting some end parts, which means to add a window to $\widehat{r}_x(k)$. Generally, define the window function $w(k)$ as

$$w(k) = \begin{cases} 0 & |k| \geq M \\ 1 & k=0 \end{cases}, |w(k)| \leq 1, w(k) = w(-k) \tag{31}$$

Then after adding the window function, the estimation of spectral density $\widehat{S}_x(\omega)$ is

$$\widehat{S}_x(\omega) = \sum_{k=-M}^{M} w(k)\widehat{r}_x(k)e^{-ik\omega}, \omega \in [-\pi, \pi] \tag{32}$$

where M is the cutoff point or maximum delay, which is often set as $\sqrt{N}$ or $\sqrt[3]{N}$. $w(k)$ is called as the time window or the delay window, whose width is $2M+1$.

(2) Maximum entropy spectrum estimation (Theodoridis and Cooper, 1981)

Burg adopted the concept of entropy and proposed the maximum entropy spectral estimation. Its main idea is that preserve the information out of the measuring interval as best as possible, and make prediction extrapolation to the area out of the sample data. The principle is when using the known autocorrelation function to make extrapolation to the unknown autocorrelation function, the entropy of the system probability space should be maximal.

If $\{x_t, t \in Z\}$ is normally stationary, the entropy can be calculated as

$$
\begin{cases}
H_X = \frac{1}{2}\log(2\pi e) + \frac{1}{4\pi}I(S) \\[2mm]
I(S) = \int\limits_{-\pi}^{\pi} \log S(\omega)d\omega
\end{cases}
\tag{33}
$$

where $I(S)$ is called as the "spectral entropy." Obviously the bigger the spectral entropy is, the more the randomness of the time series. This is consistent with the original definition of entropy. Then maximum entropy spectrum estimation can be expressed as: find an estimation of power spectral density $\widehat{S}(\omega)$ and make the spectral entropy $I(S)$ maximum. This problem can be solved by the Lagrange multiplier method.

It can be proved that for the normally distributed random process, the maximum entropy spectrum estimation is equivalent to the spectral estimation of the AR (auto-regressive) model, which makes

$$
\widehat{S}_n(\omega) = \frac{\sigma^2}{\left|1 + \sum\limits_{k=1}^{n} \varphi_k e^{-j\omega k}\right|^2}
\tag{34}
$$

where $(\varphi_1, \varphi_2, \cdots, \varphi_n, \sigma^2)$ satisfies the Y-W (Yule-Walker) formula:

$$
R_n \begin{bmatrix} 1 \\ \varphi_1 \\ \vdots \\ \varphi_n \end{bmatrix} = \begin{bmatrix} \sigma^2 \\ 0 \\ \vdots \\ 0 \end{bmatrix} \quad R_n = \begin{bmatrix} r(0) & r(1) & \cdots & r(n) \\ r(1) & r(2) & \cdots & r(n-1) \\ \vdots & \vdots & \ddots & \vdots \\ r(n) & r(n-1) & \cdots & r(0) \end{bmatrix} > 0
\tag{35}
$$

Maximum entropy spectral estimation needs to find the parameter $(\varphi_1, \varphi_2, \cdots, \varphi_n, \sigma^2)$. Levinson-Durbin proposed a fast recursion algorithm to solve the problem:

$$
\left(\varphi_1^{(1)}, \left(\sigma^{(1)}\right)^2\right), \left(\varphi_1^{(2)}, \varphi_2^{(2)}, \left(\sigma^{(2)}\right)^2\right), \cdots \left(\varphi_1^{(n)}, \cdots, \varphi_n^{(n)}, \left(\sigma^{(n)}\right)^2\right)
\tag{36}
$$

where the superscript represents the recursion times. The Levinson-Durbin recursion formula is

$$
\begin{cases}
\varphi_k^{(k)} = -\dfrac{r(k) + \displaystyle\sum_{i=1}^{k-1} \varphi_i^{(k-1)} r(k-i)}{\left(\sigma^{(k-1)}\right)^2} \\[4ex]
\varphi_i^{(k)} = \varphi_i^{(k-1)} + \varphi_k^{(k)} \varphi_{k-1}^{(k-1)}, i = 1, 2, \ldots, k-1 \\[2ex]
\left(\sigma^{(k)}\right)^2 = \left[1 - \left(\varphi_k^{(k)}\right)^2\right]\left(\sigma^{(k-1)}\right)^2 \\[2ex]
\left(\sigma^{(0)}\right)^2 = r(0)
\end{cases}
\tag{37}
$$

Levinson-Durbin method is fast and convenient. It can be proved that when the spectral density is smooth, the method can achieve an estimation of the spectral density with excellent precision. However, when the spectral density has sharp peaks, the Levinson-Durbin method usually cannot discriminate them or will have peak offsets in the estimation.

(3) The compatibility test of the spectral estimation

After using the spectral estimation to obtain the spectral density estimation $\widehat{S}(\omega)$, we need to perform the compatibility test to determine if the two samples of simulation and observed data belong to the similar population. If the spectral density of $\{x_t\}$ is $S(\omega)$, and its estimation is $\widehat{S}(\omega)$, for the windowed spectrum estimation Jenkins and Watts proved that when $\{x_t\}$ is normally distributed the formula below comes into existence, which shows that $r\widehat{S}(\omega)/S(\omega)$ is χ^2 distributed (DeGroot and Schervish, 2011d) by the degree of freedom r:

$$
r\widehat{S}(\omega)/S(\omega) \to \chi_r^2
\tag{38}
$$

Given the two data series of $\{x_t\}$ and $\{y_t\}, t = 1, 2, \ldots, N$; $S_x(\omega)$ and $S_y(\omega)$ are the spectral density of them, respectively, and $\widehat{S}_x(\omega)$ and $\widehat{S}_y(\omega)$ are the spectral density estimation achieved by windowed spectral estimation, with the hypothesis $H_0: S_x(\omega) = S_y(\omega)$, giving the formula

$$
\begin{cases}
F = \dfrac{\widehat{S}_x(\omega)}{\widehat{S}_y(\omega)} \to F(r, r) \\[3ex]
r = \dfrac{2N}{M \displaystyle\int_{-\infty}^{\infty} w(u)^2 du}
\end{cases}
\tag{39}
$$

where N is the series length, $w(u)$ is the window function, and M is the maximum delay. Further we can obtain the formula of F test (DeGroot and Schervish, 2011e):

$$
P\left\{\left[\frac{\widehat{S}_x(\omega)}{\widehat{S}_y(\omega)} < F_{\alpha/2, r, r}\right] \cup \left[\frac{\widehat{S}_x(\omega)}{\widehat{S}_y(\omega)} > F_{1-\alpha/2, r, r}\right]\right\} = \alpha
\tag{40}
$$

$F_{\alpha/2, r, r}$ and $F_{1-\alpha/2, r, r}$ can be obtained by table-lookup, and α is usually set as 0.01–0.05. When $\widehat{S}_x(\omega)/\widehat{S}_y(\omega) < F_{\alpha/2, r, r}$ or $\widehat{S}_x(\omega)/\widehat{S}_y(\omega) > F_{1-\alpha/2, r, r}$, the hypothesis should be denied, which means the power spectral density of the two data series are not consistent. Otherwise, the two data series have acceptable similarity. Engineers usually select the concerned frequency bands and make compatibility test on each of the frequency points in the bands.

4 RESULT UNIFICATION OF MODEL VALIDATION

Inevitably, different techniques adopt different metrics to present the similarity between simulation and observed data. For example, the hypothesis test achieves a Boolean result of acceptance or rejection, but the TIC analysis achieves a rational number in [0,1] interval with the monotone decreasing. The resultant forms are totally different. This problem causes difficulty in the credibility aggregation of the simulation model. The result has to be transformed into a unified description throughout the nodes in the factor space.

The result after unification needs to realize the following.

(1) Unified range: The result after unification must have the range of [0,1].
(2) Monotonically increasing: The result after unification must have a monotonically increasing characteristic, which means 0 represents totally incredible and 1 represents completely credible. The bigger the number is, the more credibility it has.
(3) Unified amplitude: The results after unification must have the same credibility extent according to the nature of the techniques used. This guarantees that the aggregation is rational.

In order to satisfy the three principles, the result unification formula must be derived by the nature of each technique. Here, we derive and propose some result unification formulas for statistics analysis, time domain analysis and frequency domain analysis methods.

4.1 Statistics Analysis Result Unification

We mainly talk about the hypothesis test of statistics analysis methods. If the alternative hypothesis is accepted, it means a small probability event happens. In this case, we can consider that the partial credibility is zero. If the original hypothesis is accepted, the level of the partial credibility should be discriminated. We can use the probability of false acceptance β to regulate the hypothesis test result into credibility description. Use $1-\beta$ as the partial credibility when the original hypothesis is accepted.

Use t test as an example to explain the solving process of $1-\beta$. Suppose that $\mu1$ and $\mu2$ are the mean value of simulation data population and reference data population, respectively. Build the hypothesis of $H0:\mu1=\mu2$, $H_1:\mu_1\neq\mu_2$, $\mu_1=\mu_2+\delta$. Regarding the simulation and observed sample, $\overline{X}$ and $\overline{Y}$ are the mean value, s_1^2 and s_2^2 are the variance, n and m are the volume, and δ is the maximum error that can be tolerated. The partial credibility can be derived as

$$\begin{cases} 1-\beta=2-T\left[T^{-1}\left(1-\frac{\alpha}{2}\right)-\frac{\delta}{S_w\sqrt{\frac{1}{m}+\frac{1}{n}}}\right]-T\left[T^{-1}\left(1-\frac{\alpha}{2}\right)+\frac{\delta}{S_w\sqrt{\frac{1}{m}+\frac{1}{n}}}\right] \\ S_\omega^2=\frac{(n-1)s_1^2+(m-1)s_2^2}{n+m-2} \end{cases} \tag{41}$$

4.2 Time Domain Analysis Result Unification

We mainly talk about TIC and GRA of the time domain analysis.

(1) TIC method

Practically there are two kinds of application of TIC in similarity analysis. One is the non-segmental calculation and the other is the segmental calculation. Regarding the non-segmental calculation, as the result of TIC has the range of $[0, 1]$ but with a monotonically decreasing characteristic, we can use $1 - \rho$ to make the unification from the TIC result to partial credibility, where ρ is the Theil inequality coefficient.

Regarding segmental calculation, if there are n segments, and m segments have the ρ exceeded threshold 0.3, it means the engineers can discriminate the difference of m partitions in total n partitions between simulation and observed data series. Consider the discrimination as an event A, and its probability is θ. Suppose the appearance times of A is X, then X belongs to the binomial distribution $b(n, \theta)$:

$$P(X = m \mid \theta) = C_n^m \theta^m (1 - \theta)^{n-m} \tag{42}$$

Consider θ is a random quantity, which can be described by a probability density distribution. Use the posterior Bayes estimation to calculate the distribution, which combines the prior Bayes estimation of the probability density and the information of the current sample. We can derive the mathematical expectation of θ:

$$E(\theta) = \int_0^1 \theta h(\theta \mid x) d\theta = \frac{m+1}{n+2} \tag{43}$$

This value is the correction of the discrimination in segmented TIC. The higher the value is, the lower the similarity achieved. So we can construct the result unification for segmented TIC:

$$C_{TIC} = \begin{cases} 1 - \dfrac{m+1}{n+2}, & \text{experienced} \\ 1 - \dfrac{m}{n}, & \text{nonexperienced} \end{cases} \tag{44}$$

(2) GRA method

The GRA method uses the gray relational degree γ to represent the similarity between two data series. γ has the range of $[0, 1]$ and a monotonically increasing characteristic, which is consistent with the definition of credibility. In order to regulate the acceptability threshold of GRA to the acceptability criteria of the model, we can use the result unification:

$$C_{GRA} = \begin{cases} \dfrac{1 - C_t}{1 - \gamma_t}(\gamma - \gamma_t) + C_t, & \gamma \in [\gamma_t, 1] \\ \dfrac{C_t \gamma}{\gamma_t}, & \gamma \in [0, \gamma_t] \end{cases} \tag{45}$$

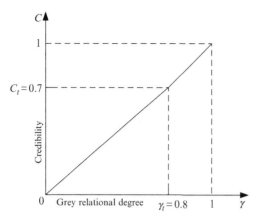

FIG. 5   The result unification of GRA.

where γ_t is the acceptability threshold of GRA, and C_t is the acceptability criteria of the model. Generally, we can set $\gamma_t = 0.8$, $C_t = 0.7$. The result unification can be illustrated as in Fig. 5.

4.3 Frequency Domain Analysis Result Unification

The result unification from frequency domain analysis to partial credibility is actually the work of F test result unification. Given the two data series of $\{x_t\}$ and $\{y_t\}$, $t = 1, 2, ..., N$, $\widehat{S}_x(\omega)$ and $\widehat{S}_y(\omega)$ are the spectral density estimation, and δ is the maximum tolerance of error, we can make the original hypothesis H_0 and the alternative hypothesis H_1 as

$$H_0 : \frac{S_y(\omega)}{S_x(\omega)} = 1, H_1 : \frac{S_y(\omega)}{S_x(\omega)} \neq 1, \frac{S_y(\omega)}{S_x(\omega)} = 1 + \delta \tag{46}$$

If β is the false acceptance and α is the significance level, we can derive $1 - \beta$ as

$$1 - \beta = 1 - F\left(F^{-1}\left(1 - \frac{\alpha}{2}\right)(1+\delta)\right) + F\left(\frac{1}{F^{-1}\left(1 - \frac{\alpha}{2}\right)}(1+\delta)\right) \tag{47}$$

If the final F test of the frequency domain analysis shows that the alternative hypothesis is accepted, the partial credibility of the similarity analysis is zero. Otherwise, if the original hypothesis is accepted, the partial credibility should be calculated by the above formula.

5 DEFECT TRACING OF MODEL VALIDATION

From the view of simulation engineers, validation should have the ability to locate model defects, if the conclusion of the credibility assessment is negative. In other words, the model credibility needs to have the traceability. Actually, the quantitative credibility of a rational number in [0,1] alone has no traceability. The number is helpful to determine the credibility level of the model, but cannot reveal the parts of the model that cause the credibility deficiency.

The factor space network gives the opportunity to locate model defects. A factor space possesses partial credibility on intermediate nodes, which can be used to trace the origin of the credibility deficiency. Here, we propose a method which combines orthogonal design and Sobol's method to find defect points out of traced links (Zhang et al., 2013), and use path tracing to find defect points on these links.

5.1 Orthogonal Design

Orthogonal design (Su et al., 2016) is a mathematical statistics method, which is used to solve the optimization of multifactor-multilevel experiments. According to the partial credibility in the factor space, categorize factors into two sets: negative factor set $P_1 = \{p_i \in P \mid 0 \leq v(s_i) < \delta\}$ and positive factor set $P_2 = \{p_i \in P \mid \delta \leq v(s_i) \leq 1\}$. δ represents the acceptability criteria. Without loss of generality, suppose $P_1 = \{p_1, p_2, ..., p_{n_1}\}$ and $P_2 = \{p_{n_1+1}, p_{n_1+2}, ..., p_n\}$, $1 \leq n_1 \leq n$.

According to the negative factors $p_2, p_3, ..., p_{n_1}$, select a suitable 2-level orthogonal table $L_a(2^c)$, which satisfies $c \geq n_1 - 1$. Draw $n_1 - 1$ columns from $L_a(2^c)$ optionally, and construct an extend table:

$$L_{2a}^T(2^{n_1}) = \begin{pmatrix} (1)_{a \times 1} & L_a(2^{n_1-1}) \\ (2)_{a \times 1} & L_a(2^{n_1-1}) \end{pmatrix} \tag{48}$$

Set the two levels of negative factor p_i as $v(s_i)$ and δ, $i = 1, 2, ..., n_1$, and we can obtain $2a$ experiments by the extended table $L_{2a}^T(2^{n_1})$. The concerned negative factor p_1 is located in the first column.

Use the credibility aggregation formula to calculate the total credibility by different experiments, and we can obtain the results of: $v_1(S), v_2(S), ..., v_{2a}(S)$. $|v_{i+a}(S) - v_i(S)|$, $i = 1, 2, ..., a$, reflects the credibility change when the negative factor p_1 is improved. We can use the normally distributed membership cloud (Fig. 6) to determine the deficiency level of p_1.

Use $|v_{i+a}(S) - v_i(S)|$ to calculate a five-dimensional vector $M_i = (m_{i1}, m_{i2}, m_{i3}, m_{i4}, m_{i5})$, $m_{ij} \geq 0$, $i = 1, 2, ..., a$, $j = 1, 2, 3, 4, 5$. Table 1 shows the parameters of each deficiency level in the membership cloud.

The deficiency level of the negative factor p_1 can be determined by the formula below and the maximum membership principle.

$$\overline{M} = \left(\sum_{i=1}^a m_{i1}/a, \sum_{i=1}^a m_{i2}/a, ..., \sum_{i=1}^a m_{i5}/a \right) \tag{49}$$

5.2 Sobol's Method

Sobol's method (Sobol, 1990) is a global sensitivity analysis method based on variance decomposition, which was proposed by I.M. Sobol in 1990. Its principle is to decompose the multivariant function into a constant, single-variant function and a combined-variant

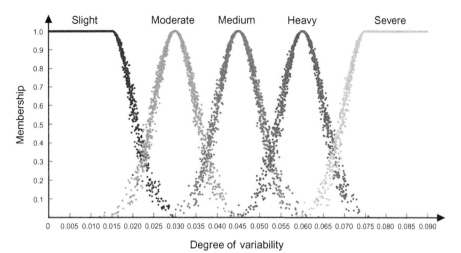

FIG. 6 The normally distributed membership cloud of the credibility deficiency level.

TABLE 1 The Parameters of Each Deficiency Level in the Membership Cloud

Defect Level	Parameter		
	Expectation	Entropy	Hyper Entropy
Slight	0.015	0.005	0.0005
Moderate	0.030	0.005	0.0005
Medium	0.045	0.005	0.0005
Heavy	0.060	0.005	0.0005
Severe	0.075	0.005	0.0005

function, and use the variance of multivariant function and each subitem to calculate the sensitivity of different variants.

If the model is modified to rectify the negative factor, it is possible to influence other factors related to the modified one. This requires sensitivity analysis on the negative factor. The more the sensitivity achieved, the more the possibility that it will have a negative influence on other related factors. Take the negative factor p_1 as an example.

First determine the factor set related to p_1: $M(p_1) = \{p_i \in P \,|\, s_i \cap s_1 \neq \varnothing\} \backslash \{p_1\}$, $i = 1, 2, \ldots, r$, and set the sampling interval of coupled factors p_i as: $[(1 - \tau) \times v(s_i), (1 + \tau) \times v(s_i)]$, $p_i \in M(p_1)$. τ represents the amplitude and usually $\tau = 10\%$.

In the sampling interval, use the Latin hypercube design (Shields and Zhang, 2016) to obtain the sample set. In order to guarantee the precision of sensitivity, the sample number should be no less than 200 times of the number of coupled factors. First make two separate Latin hypercube samplings, and get two experiment plans $A_{n \times r}$ and $B_{n \times r}$, where n represents the experiment times and r represents the number of coupled factors. Replacing the j-th column in plan A with the j-th column in plan B makes A_B^j, and replacing the j-th column in plan B with the j-th column in plan A makes B_A^j. Then the first-order sensitivity S_i and the total sensitivity S_{T_i} of p_i are

$$S_i = D_i/D$$
$$S_{T_i} = 1 - D_{-i}/D \tag{50}$$

In Formula (50), D is the total variance of the credibility aggregation function; D_i is the parameter variance of p_i; and D_{-i} is the irrelevant variance of p_i. They can be calculated by

$$\begin{cases} D = \dfrac{1}{n}\sum_{k=1}^{n} f^2(A)_k - \left[\dfrac{1}{n}\sum_{k=1}^{n} f(A)_k\right]^2 \\[2ex] D_i = \dfrac{1}{n}\sum_{k=1}^{n} f(A)_k f\left(B_A^j\right)_k - \left[\dfrac{1}{n}\sum_{k=1}^{n} f(A)_k\right]^2 \\[2ex] D_{-i} = \dfrac{1}{n}\sum_{k=1}^{n} f(A)_k f\left(A_B^j\right)_k - \left[\dfrac{1}{n}\sum_{k=1}^{n} f(A)_k\right]^2 \end{cases} \tag{51}$$

In Formula (51), $f(x)$ represents the total credibility aggregated by the experiment plan x, and the subscript k represents the k-th experiment in the plan.

Usually we take the total sensitivity as the criteria to determine whether the coupled factor p_i is sensitive to the total credibility, with a threshold of 0.6. If it is sensitive, when the origin factor (negative factor) p_1 is rectified, the engineers should pay attention to the chain reaction of the coupled factor p_i.

5.3 Path Tracing

Because the total credibility is achieved by the factor space network, which has traced links, we can use the partial credibility on the validation path of negative nodes to realize defect tracing. The path tracing can be performed by the following procedures:

(1) Use orthogonal design and Sobol's method to locate the negative factors other than the destination of traced links.
(2) Along the validation path in the factor space which contains the negative factors, make further validation to get partial credibility on the destination of traced links, and determine if it is a negative node by comparing with the acceptability criteria.
(3) If there is a destination of iteration paths, make validation of its initial input, related constants, and other variants which are irrelevant to the iteration variant (Fang et al., 2017b).
(4) When the validation path reaches the leaf node, the defect tracing is over.
(5) Collect all the negative nodes in the tracing, and take the nodes with the lowest order as the origin which induces the deficiency of the model credibility.

6 CASE STUDY

In this section, we use a 6-DOF (degree of freedom) flight vehicle model to present how the validation approach is applied, and explain the process of factor space building, similarity analysis, result unification, credibility aggregation, and defect tracing. The example shows a walk-through of the method in a practical way.

6.1 Introduction to the Model

The 6-DOF flight vehicle model is the fundamental part of aerodynamic simulation systems, whose credibility is critical to the correct application of the simulation. The model is mainly formed by the vehicle's dynamic and kinetic differential equations, which are used to achieve the position and attitude of the vehicle:

$$
\begin{cases}
\dfrac{dx}{dt} = V_x = V\cos\theta\cos\psi_v \\[2mm]
\dfrac{dy}{dt} = V_y = V\sin\theta \\[2mm]
\dfrac{dz}{dt} = V_z = -V\cos\theta\sin\psi_v
\end{cases}
\quad
\begin{cases}
m\left(\dfrac{dV_x}{dt} + g_x\right) = -C_x q S_M + P_x \\[2mm]
m\left(\dfrac{dV_y}{dt} + g_y\right) = C_y^\alpha(\alpha + \alpha_w)q S_M + P_y \\[2mm]
m\left(\dfrac{dV_z}{dt} + g_z\right) = -C_z^\beta(\beta + \beta_w)q S_M + P_z
\end{cases}
\tag{52}
$$

$$
[g_x, g_y, g_z] = f\left(x, y, z, R_{0x}, R_{0y}, R_{0z}, A_0, B_0, J, fM, R_a\right)
\tag{53}
$$

$$
\left[C_x, C_y^\alpha, C_z^\beta\right] = f\left(M_a, x, y, z, R_{0x}, R_{0y}, R_{0z}, A_0, B_0, fM, R_a\right)
\tag{54}
$$

$$
[\alpha + \alpha_w, \beta + \beta_w] = f\left(x, y, z, \phi, \psi, \gamma, R_{0x}, R_{0y}, R_{0z}, A_0, B_0, V_w, A_w\right)
\tag{55}
$$

$$
\begin{cases}
\dfrac{d\phi}{dt} = \omega_{y_1}\sin\gamma + \omega_{z_1}\cos\gamma \\[2mm]
\dfrac{d\psi}{dt} = \dfrac{1}{\cos\phi}\left(\omega_{y_1}\cos\gamma - \omega_{z_1}\sin\gamma\right) \\[2mm]
\dfrac{d\gamma}{dt} = \omega_{x_1} - \tan\phi\left(\omega_{y_1}\cos\gamma - \omega_{z_1}\sin\gamma\right)
\end{cases}
\tag{56}
$$

$$
\begin{cases}
J_{x_1}\dfrac{d\omega_{x_1}}{dt} + (J_{z_1} - J_{y_1})\omega_{y_1}\omega_{z_1} = M_{x_1} \\[2mm]
J_{y_1}\dfrac{d\omega_{y_1}}{dt} + (J_{x_1} - J_{z_1})\omega_{x_1}\omega_{z_1} = M_{y_1} \\[2mm]
J_{z_1}\dfrac{d\omega_{z_1}}{dt} + (J_{y_1} - J_{x_1})\omega_{x_1}\omega_{y_1} = M_{z_1}
\end{cases}
\quad
\begin{cases}
M_{x_1} = d''^\gamma_{3x}\ddot\delta_\gamma + d''^\gamma_{3z}\ddot\delta_\gamma \\[2mm]
M_{y_1} = b''^\psi_{3x}\ddot\delta_\psi + b''^\psi_{3z}\ddot\delta_\psi \\[2mm]
M_{z_1} = b''^\phi_{3x}\ddot\delta_\phi + b''^\phi_{3z}\ddot\delta_\phi
\end{cases}
\tag{57}
$$

$$
[\ddot\delta_\phi, \ddot\delta_\psi, \ddot\delta_\gamma] = f\left(x, y, z, \phi, \psi, \gamma, V_x, V_y, V_z, \omega_{x_1}, \omega_{y_1}, \omega_{z_1}, \ldots\right)
\tag{58}
$$

The model uses the dynamic and kinetic differential equations of the mass center motion (Eqs. 52–55) to calculate the flight vehicle's position. In the equations there are self-iteration variants of x, y, z, and cross iteration variants of ϕ, ψ, γ, which use the last moment value to solve the algebraic equations of the current moment.

TABLE 2   The Physical Meaning of the 6-DOF Flight Vehicle Model Variants

Variant	Meaning	Variant	Meaning	Variant	Meaning
x, y, z	Position	q	Dynamic pressure	M_a	Mach
ϕ, ψ, γ	Attitude angle	S_M	Sectional area	V_w	Wind velocity
V_x, V_y, V_z	Velocity	P_x, P_y, P_z	Thrust	A_w	Wind direction
g_x, g_y, g_z	Gravity acceleration	R_{0x}, R_{0y}, R_{0z}	Radius vector	$\omega_{x_1}, \omega_{y_1}, \omega_{z_1}$	Rotational angular velocity
m	Mass	A_0	Launch direction	$J_{x_1}, J_{y_1}, J_{z_1}$	Rotary inertia
θ	Trajectory tilt angle	B_0	Launch point latitude	$M_{x_1}, M_{y_1}, M_{z_1}$	Rotating moment
ψ_v	Trajectory deflection angle	J	Gravitational potential coefficient	$\delta_\phi, \delta_\psi, \delta_\gamma$	Rudder angle
$C_x, C_y^\alpha, C_z^\beta$	Aerodynamic coefficient	fM	Earth gravitational constant	$d_{3x}^\psi,\ d_{3z}^\psi$ $b_{3x}^\psi,\ b_{3z}^\psi$ $b_{3x}^\phi,\ b_{3z}^\phi$	Moment coefficient of inertia
$\alpha+\alpha_w$ $\beta+\beta_w$	Velocity transformation angle	R_a	Earth equator radius		

The model uses the dynamic and kinetic differential equations of the around mass center motion (Eqs. 56–58) to calculate the flight vehicle's attitude. In the equations there are self-iteration variants of ϕ, ψ, γ and cross iteration variants of x, y, z, which use the last moment value to solve the algebraic equations of the current moment.

The physical meaning of the variants in Eqs. (52)–(58) are illustrated in Table 2.

6.2 The Factor Space of Model Validation

Build the factor space network of model validation, as shown in Fig. 7.

The factor space shows that the model's credibility is aggregated by the partial credibility of mass center motion and around mass center motion, which are presented by the position x, y, z and attitude ϕ, ψ, γ. The aggregation function is taking the minimum when the partial credibility is not less than 0.8. If any of the partial credibility is less than 0.8, the total credibility of the model is forced to be zero.

Go down along the paths of x, y, z and ϕ, ψ, γ, we can see that the partial credibility of x, y, z is relevant to ϕ, ψ, γ and itself and the partial credibility of ϕ, ψ, γ is relevant to x, y, z and itself. This is a cross iteration, which makes it impossible to aggregate the total credibility upwards from the leaf nodes by conventional methods like AHP.

Meanwhile, because of the cross iteration between mass center motion and around mass center motion, when the total credibility is unacceptable, it is very likely to get an unacceptable partial credibility on all the nodes of x, y, z and ϕ, ψ, γ. This brings the difficulty of model

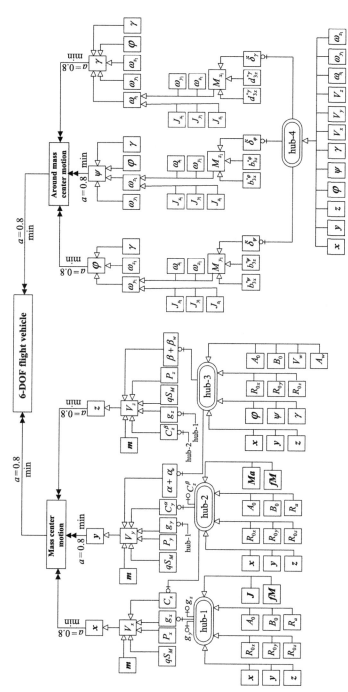

FIG. 7 The factor space of the 6-DOF flight vehicle model validation.

defect tracing. Because the comparison of a single-step simulation and observed data is meaningless to a continuous simulation, it is useless to locate the defect points by break-point analysis.

6.3 Similarity Analysis and Result Unification of Model Validation

First validate the final output of the model: x, y, z and ϕ, ψ, γ. Table 3 shows the simulation and observed data of the flight vehicle's position x, y, z.

The curves of the flight vehicle's position are shown in Fig. 8:

Use modified gray relational degree (Formula 28) to validate the nodes of x, y, z, and use result unification formula of GRA (Formula 45) to convert the results into partial credibility. We get the results of: $C(x)=0.662$, $C(y)=0.601$, $C(z)=0.636$. It is obvious that all the partial credibility is less than the acceptability threshold $a=0.8$ of the ultralinks. So the total credibility of the flight vehicle model is forced to zero.

Table 4 shows the simulation and observed data of the flight vehicle's attitude ϕ, ψ, γ.

Curves of the flight vehicle's attitude are shown in Fig. 9:

Also use modified gray relational degree to validate the nodes of ϕ, ψ, γ, and use the result unification formula of GRA to convert the results into partial credibility. We get the results of: $C(\phi)=0.694$, $C(\psi)=0.514$, $C(\gamma)=0.531$. The result is also unacceptable. Thus we need make defect tracing to locate what causes the model's credibility deficiency.

6.4 The Defect Tracing of Model Validation

According to the path tracing procedure (explained in Section 5.3), go downwards through the factor space and validate the nodes other than x, y, z and ϕ, ψ, γ, and meanwhile validate the initial input of x, y, z and ϕ, ψ, γ.

TABLE 3 The Simulation and Observed Data of the Flight Vehicle's Position x, y, z

Time (s)	Simulation x(m)	Observed x'(m)	Simulation y(m)	Observed y'(m)	Simulation z(m)	Observed z'(m)
0.100	150.757	150.755	20 799.960	20 799.960	0	0
0.200	294.326	294.315	20 799.850	20 799.850	0	0
0.300	437.884	437.861	20 799.670	20 799.670	0	0
......	......	......	......	......	......	......
25.100	34 685.460	34 323.240	16 460.620	16 592.120	118.865	114.550
25.200	34 814.140	34 448.610	16 422.750	16 555.970	119.754	115.399
......	......	......	......	......	......	......
49.800	62 228.430	60 814.420	1 368.693	2 639.631	403.060	392.137
49.900	62 315.850	60 897.090	1 287.806	2 567.734	403.434	392.968
50.000	62 402.990	60 979.460	1 206.855	2 495.826	403.761	393.779

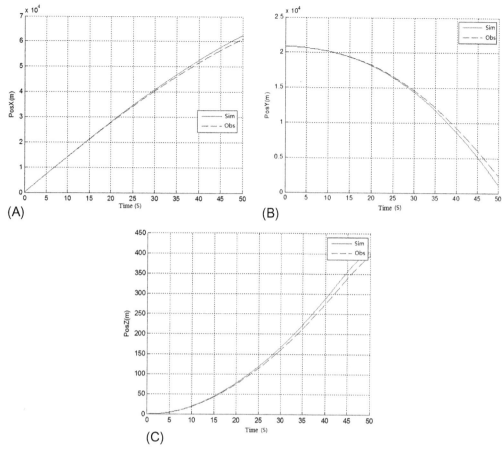

FIG. 8 Curves of the flight vehicle's position: (A) position x, (B) position y, and (C) position z.

TABLE 4 The Simulation and Observed Data of the Flight Vehicle's Position ϕ, ψ, γ

Time (s)	Simulation ϕ(°)	Observed ϕ' (°)	Simulation ψ(°)	Observed ψ' (°)	Simulation γ(°)	Observed γ' (°)
1.000	−3.125	−3.125	−0.363	−0.363	−6.201	−6.207
2.000	−5.449	−5.439	−0.567	−0.567	−5.072	−5.087
3.000	−6.219	−6.203	−0.555	−0.555	−4.543	−4.559
......	......	......	......	......	......	......
26.000	−20.101	−19.837	−0.455	−0.463	−1.568	−1.598
27.000	−20.536	−20.255	−0.46	−0.468	−1.543	−1.574
......	......	......	......	......	......	......
48.000	−33.716	−32.407	−0.189	−0.351	4.198	1.979
49.000	−34.601	−33.213	−0.009	−0.256	6.753	3.475
50.000	−35.459	−33.999	0.271	−0.127	10.814	5.431

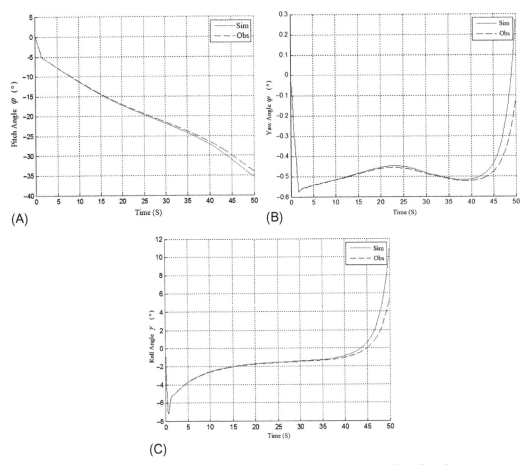

FIG. 9 Curves of the flight vehicle's attitude. (A) Pitch angle ϕ, (B) Yaw angle ψ, (C) Roll angle γ.

Given the initial values of the simulation and observed data, it is proved that $x(0) = x'(0) = 0$, $x(0) = x'(0) = 0$, $x(0) = x'(0) = 0$, $\phi(0) = \phi'(0) = 0$, $\psi(0) = \psi'(0) = 0$, $\gamma(0) = \gamma'(0) = 0$. Here we take the wind speed V_w and rudder angles δ_ϕ, δ_ψ, δ_γ as an example to explain the defect tracing. Other nodes have been proved to be acceptable.

Table 5 shows the simulation and observed data of the wind speed V_w.

The curve of the wind speed is shown in Fig. 10:

Use the modified gray relational degree to validate the node of V_w, and use the result unification formula of GRA to convert the results into partial credibility. We get the result of: $C(V_w) = 0.733$. Given $a = 0.8$ as the acceptability criteria, the partial credibility of the wind speed is unacceptable. This shows that it is a defect point that causes the credibility failure of x, y, z.

Further we analyze the node of rudder angles δ_ϕ, δ_ψ, δ_γ. Table 6 shows the simulation and observed data of the rudder angles.

TABLE 5 The Simulation and Observed Data of the Wind Speed V_w

Time (s)	Simulation V_w (m/s)	Observed V'_w (m/s)
0.100	13	13
0.200	13	13
0.300	13	13
......	......	......
25.100	16.37	15.564
25.200	16.601	15.785
......	......	......
49.800	18.386	22.279
49.900	18.187	22.014
50.000	17.987	21.748

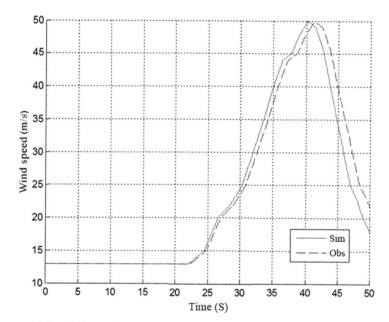

FIG. 10   The curve of the wind speed.

The curves of the rudder angles are shown in Fig. 11:

Also use the modified gray relational degree to validate the nodes of δ_ϕ, δ_ψ, δ_γ, and use the result unification formula of GRA to convert the results into partial credibility. We get the results of: $C(\delta_\phi)=0.659$, $C(\delta_\psi)=0.419$, $C(\delta_\gamma)=0.699$. The result is also unacceptable. This shows that they are defect points that cause the credibility failure of ϕ, ψ, γ.

TABLE 6   The Simulation and Observed Data of the Rudder Angles δ_ϕ, δ_ψ, δ_γ

Time (s)	Simulation δ_ϕ(°)	Observed δ_ϕ' (°)	Simulation δ_ψ(°)	Observed δ_ψ'(°)	Simulation δ_γ(°)	Observed δ_γ'(°)
1.000	−0.22	−0.216	1.824	1.823	−0.007	0.001
2.000	0.039	0.042	5.306	5.298	−0.017	−0.018
......	......	......	......	......	......	......
26.000	−0.169	−0.147	8.064	8.461	−0.046	−0.042
27.000	−0.15	−0.13	7.561	7.966	−0.041	−0.038
......	......	......	......	......	......	......
48.000	−0.009	−0.008	1.28	1.744	−0.009	−0.008
49.000	−0.011	−0.008	1.345	1.835	−0.012	−0.01
50.000	−0.014	−0.008	1.438	1.937	−0.017	−0.012

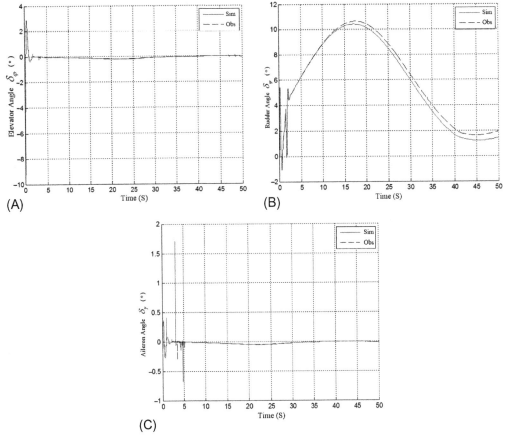

FIG. 11   The curves of the flight vehicle's rudder angles: (A) Elevator angle δ_ϕ, (B) Rudder angle δ_ψ, and (C) Aileron angle δ_γ.

6.5 Result Analysis of Model Validation

According to the results achieved by the approach for model validation, the partial credibility of the flight vehicle's position and attitude is less than the acceptability threshold $a = 0.8$. Thus, the total credibility of the model is zero, which means the model is not acceptable.

Go downwards through the factor space shown in Fig. 7, in the branches of the flight vehicle's position and attitude, we find the initial values of the cross iteration variants $x(0)$, $y(0)$, $z(0)$ and $\phi(0)$, $\psi(0)$, $\gamma(0)$ are all correct, but the partial credibility of wind speed V_w and δ_ϕ, δ_ψ, δ_γ are unacceptable.

Thus, we can make the conclusion that given the prerequisite of the model's computational functions are all correct, the deficiency of the total credibility is caused by the submodels of wind speed and rudder angles. This gives the model developers a direction to rectify the model and make it better to be applied in simulation. Moreover, because the rudder angles are outputs of the control model, the developers should also make an inspection of the flight vehicle's control model.

7 SUMMARY AND DISCUSSION

Model validation is the core of VV&A, which contributes to the simulation credibility assessment to a great extent. It involves various types of work, and needs appropriate techniques in different fields. Model validation is a system engineering but not a single job.

The major objective of model validation is to achieve the credibility and locate the defect factors which induce the possible deficiency of credibility. In order to accomplish a thorough model validation and get a comprehensive credibility, we have to build a factor space that reveals the model structure and computational mechanism. A network view is getting more and more adopted than an AHP hierarchy.

Similarity analysis is the most convincing way to measure how close the mode represents the real-world object in the application domain. Different techniques are selective on different nodes in the factor space, and the analysis result needs unification to a partial credibility description, which allows credibility aggregation to achieve the total credibility of the model.

The traceability of the model credibility is an issue which is often neglected but actually one of the most concerned thing by the simulation developer and the user. The orthogonal design and Sobol's method can provide a way to locate the negative factor and assess its defect level. Meanwhile, we can perform the path tracing in the factor space to find model defect origins.

However, there are some more specific problems remained. How to use and select multiple model execution data to perform the similarity analysis? How to realize a white-box validation when some intermediate outputs of the model are unobservable in the real world? How to make a similarity analysis when the output is in some special form (e.g., PWM; pulse width modulation)? How to develop a synthetic environment which is open to provide services to model validation instead of separated tools? The questions need future work to have better and practical answers.

Model validation is the essence of VV&A, and its story never ends.

References

D'Agostino, R.B., Albert, B., D'Agostino, J.R.B., 1990. A suggestion for using powerful and informative tests of normality. Am. Stat. 44 (4), 316–321.

DeGroot, M.H., Schervish, M.J., 2011a. Probability and Statistics, fourth ed. Pearson, Boston, MA, pp. 485–493.

DeGroot, M.H., Schervish, M.J., 2011b. Probability and Statistics, fourth ed. Pearson, Boston, MA, pp. 587–596.

DeGroot, M.H., Schervish, M.J., 2011c. Probability and Statistics, fourth ed. Pearson, Boston, MA, pp. 576–586.

DeGroot, M.H., Schervish, M.J., 2011d. Probability and Statistics, fourth ed. Pearson, Boston, MA, pp. 626–630.

DeGroot, M.H., Schervish, M.J., 2011e. Probability and Statistics, fourth ed. Pearson, Boston, MA, pp. 599–603.

Deng, J.L., 1995. Grey relational analysis: a new method for multivariate statistical analysis. J. Statist. Res. 3, 43–48.

Fang, K., Ma, P., Yang, M., 2012. The MAD network for virtual protocol systems credibility evaluation. Comput. Integr. Manuf. Syst. 18 (5), 1054–1060.

Fang, K., Yang, M., Zhang, Z., 2011. In: The MAD network for credibility evaluation of computer simulation.Proceedings of 13th IEEE Joint International Computer Science and Information Technology Conference, Chongqing. vol. 03. pp. 636–642.

Fang, K., Zhou, Y., Zhao, E., 2017a. Discussion for the factor space of simulation model validation. J. Syst. Eng. Electron. 39 (11), 2592–2602.

Fang, K., Zhou, Y., Zhao, K., 2017b. Validation method for simulation models with iteration operation. J. Syst. Eng. Electron. 39 (2), 445–450.

Gao, Z.Q., Liu, B., Gao, H., Meng, X.W., et al., 2014. The correlation between the cylinder pressure and the ion current fitted with a Gaussian algorithm for a spark ignition engine fuelled with natural-gas-hydrogen blends. Proc. Inst. Mech. Eng. D 228 (12), 1480–1490.

Golub, G., Ortega, J.M., 1993. Scientific Computing: An Introduction with Parallel Computing. Academic Press, Boston, MA, pp. 91–135.

Grubbs, F.E., 1950. Sample criteria for testing outlying observations. Ann. Math. Stat. 21 (1), 27–58.

IEEE, 1997. Recommended Practice Guide for Distributed Interactive Simulation – Verification, Validation and Accreditation. IEEE, USA, pp. 2–8.

Ke, F., Yang, M., Wang, Z., 2005. In: The HITVICE VV&A environment.Proceedings of 2005 Winter Simulation Conference, Orlando, FL, USA. vol. V1-4. pp. 1220–1227.

Kheir, N.A., Holmes, W.M., 1978. On validating simulation models of missile systems. Simulation 30 (4), 117–128.

Li, H., Philip Chen, C.L., Yen, V.C., et al., 2000. Factor spaces theory and its applications to fuzzy information processing: two kinds of factor space canes. Comput. Math. Appl. 40, 835–843.

Linton, O., 2017. Probability, Statistics and Econometrics. Academic Press, London, UK, pp. 151–174.

Montgomery, D.C., Conard, R.G., 1980. Comparison of simulation and flight-test data for missile systems. Simulation 34 (2), 63–72.

Rahman, M.M., Govindarajulu, Z., 1997. A modification of the test of Shapiro and Wilk for normality. J. Appl. Stat. 24 (2), 219–236.

Saaty, T.L., 1997. In: Modeling unstructured decision problems: a theory of analytical hierarchy.Proceedings of the First International Conference on Mathematical Modeling. vol. V1. pp. 59–77.

Shields, M.D., Zhang, J.X., 2016. The generalization of latin hypercube sampling. Reliab. Eng. Syst. Saf. 148, 96–108.

Smith, S.W., 2013. Digital Signal Processing: A Practical Guide for Engineers and Scientists. Newnes, Amsterdam/Boston, pp. 277–284.

Sobol, I.M., 1990. On sensitivity estimates for nonlinear mathematical models. Matem. Mod. 2 (1), 112–118.

Su, L.S., Zhang, J.B., Wang, C.J., 2016. Identifying main factors of capacity fading in lithium-ion cells using orthogonal design of experiments. Appl. Energy 163, 201–210.

Theodoridis, S., Cooper, D.C., 1981. Application of the maximum entropy spectrum analysis technique to signals with spectral peaks of finite width. Signal Process. 3 (2), 109–122.

Wang, P., 1992. Factor space and descriptions of concepts. J. Softw. (1), 30–40.

WfMC, 1995. The Workflow Reference Model. Available from: http://www.wfmc.org.

WfMC, 1996. Workflow Management Coalition Audit Data Specification. Available from: http://www.wfmc.org.

Zhang, L., 2011. In: Model engineering for complex system simulation. Keynote speech.58th CAST (Chinese Association for Science and Technology) Forum on New Academic Views, October 15, Lijiang, Yunnan, China.

Zhang, L., Shen, Y., Zhang, X., et al., 2014. In: The model engineering for complex system simulation.Proceedings of the I3M Multiconference, September 10–12, Bordeaux, France.

Zhang, Z., Fang, K., Wu, F., et al., 2013. In: Detection method for credibility defect of simulation based on Sobol' method and orthogonal design.Proceedings of the 2013 Asia Simulation Conference, Singapore.

Further Reading

Ke, F., Zhao, K., Zhou, Y., 2017. In: A method for obtaining the credibility of a simulation model.Proceedings of EMSS (European Modeling & Simulation Symposium) 2017, Barcelona, Spain.

Quantitative Measurements of Model Credibility

Megan Olsen*, Mohammad Raunak*

*Department of Computer Science, Loyola University Maryland, Baltimore, MD, United States

1 INTRODUCTION

The field of modeling and simulation (M&S) has grown rapidly in usage and influence, with computational simulations created to study many fields by computer scientists, engineers, and experts in the studied domain such as natural science (physics, chemistry, biology, etc.), computing, engineering, operations research, ecology, and social science. The 2016 Winter Simulation Conference, for example, had 20 tracks including 7 tracks focused on different fields to which simulation is applied. Government agencies, especially defense and the defense-related industry, rely heavily on simulation of systems of all scales (FDA, 2010; Zeltyn et al., 2011), and billions of dollars and human lives are dependent on the changes made based on these studies.

A variety of M&S approaches are used in each domain, including system dynamics models, models that focus on event-based changes to system state (discrete-event simulation [DES]), and models that focus on autonomous entities interacting in a spatial environment (agent-based model [ABM] or individual-based model). The choice of approach depends on the type of system being studied and the purpose of the study. For instance, a hospital could be studied using any of these approaches, such as an ABM to study crowd control in hallways, or a DES to study flow of patients through an ED. Increasingly, models are defined that are a hybrid of these techniques.

For a model to be used and trusted, it must be credible. There are many steps that contribute to building a credible model, including technical steps such as verification and validation (V&V). There are also steps involving people such as ensuring that those in management or decision-making positions understand the assumptions of the model, their involvement, and the interactions between them and the model developers (Law, 2015). This chapter focuses on the technical aspect of building a credible model, specifically validation.

1.1 The Role of Verification and Validation in Modeling and Simulation

Models in general are developed to facilitate understanding and analysis of often large and complex systems. Simulation models, in particular, are developed and utilized to study natural systems such as population mobility or cancer cell growth (Abbott, 2002); systems that are yet to be built such as missile defense or nuclear reactors (Ender et al., 2010); or systems for which making changes in the real-life/runtime environment to study their impact is very expensive or dangerous, such as the process of patient care in a hospital ED (Raunak et al., 2009) or automotive control (Ray et al., 2009). Whenever a simulation model is developed, it is constructed with some purpose in mind. Often the objectives include better understanding the dynamics of the system being modeled, or for performing "what-if" analyses to identify optimum policy or resource allocation to be implemented on the real system.

A crucial step of any simulation-based study, whether the system under study (SUS) is mechanical/electrical, natural, or theoretical, is to ensure that the simulation model is credible, as established by the process and results of verifying and validating the model. Establishing model credibility thus includes that the model (a) is internally consistent with no known errors (verification) and (b) mimics the SUS's behavior to a level of confidence necessary for making the model useful for its intended application (validation).

Without proper V&V, predictions with confidence cannot be made about the SUS based on simulation results. Validation in particular is a crucially important part of building confidence in the results derived from a simulation model. In essence, verification is testing *how* the model was built, and validation is assessing *what* has been modeled and whether the model can sufficiently reproduce the behavior of the real system for its simulation purpose. There are many steps within the M&S process where each must be applied to achieve high confidence in the M&S (Balci, 2010).

1.2 The Simulation Validation Process

Researchers have studied the development of valid and credible simulation models for decades, with a large part of this research effort focusing on the process and principles of V&V (Sargent, 2011; Balci, 1998, 2010; Banks, 1998; Kleijnen, 1995; Law, 2015). The high-level activities of developing a simulation model (Fig. 1) begin with the study of the system in the real world, which we refer to as the SUS. From the analysis of the SUS, a conceptual model is developed. There is much debate about what constitutes a conceptual model; to some it is a set of planned abstractions, to others it is a model written using a formalism, and to many it is

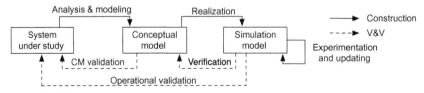

FIG. 1   The high-level steps of simulation development. The system under study is used as a reference for validation of both the conceptual model and the implemented simulation model. Verification ensures that the simulation model correctly represents the conceptual model. Data validity is not shown.

something in the middle and may include a number of documents (Bair and Tolk, 2013). For this chapter, we assume that the conceptual model is a concrete artifact that includes the assumptions and abstractions being used to design a model, as well as a working definition of the structure and expected behaviors, as possible. Beyond that the format is irrelevant.

Although not shown in Fig. 1, ensuring the validity of the data used to run the model is also necessary. Sargent describes three processes for data validity: "(1) collecting and maintaining data, (2) testing the collected data using techniques such as internal consistency checks, and (3) screening the data for outliers and determining if the outliers are correct" (Sargent, 2010). These steps must be performed before the data can be used.

The first set of model validation tests are performed to ensure that the conceptual model, whatever form it is in, is consistent with the SUS it is intended to represent and the purpose with which it is being created. Once a validated conceptual model exists, a software implementation of the conceptual model is constructed as the simulation model.

Before experiments can be performed on the simulation model with the expectation of deriving insights about the SUS, it must be both verified and validated as faults may occur during the translation of the conceptual model to the computerized model. Such faults are discovered and fixed using the usual software testing and verification techniques, benefiting from many years of software testing and verification research (Bertolino, 2007). Operational validation is performed to give credence to the computerized model as a valid model of the SUS for the model's purpose. This last set of validation is the most expansive (Sargent, 2010; Banks, 1998; Balci, 1998).

More than 75 verification, validation, and testing (VV&T) techniques are presented in the Handbook of Simulation for discrete-event-based simulation models, categorized into four groups: informal, static, dynamic, and formal (Balci, 1998). Informal techniques, such as reviews, walkthroughs, and inspection, are commonly used approaches for validating simulation models. The dynamic techniques are useful primarily for verifying the simulation code and are usually not directly applicable for model validation. The state of the art in validation primarily includes such techniques as manual review and inspection, degenerative tests, comparing with other models simulating the same SUS, and visualization. Additional dynamic techniques specifically for validation often rely on data from the SUS, and are preferred when data are available (Sokolowski and Banks, 2010; Sargent, 2007). Standard operational validation techniques include animation, face validation, results validation, Turing tests, and comparisons to other models (Sargent, 2010).

Many standard validation techniques are not as easily applied to complex system simulations, which are particularly difficult to validate (Taylor et al., 2013). Often expert knowledge is the safest approach for validating these systems, and recent work has examined how to best organize and utilize that knowledge (Reynolds, 2010; Reynolds and Wimberly, 2011). Stochasticity is one of the primary sources of difficulty in complex system validation. One recent approach to this problem is Bayesian Statistical Model Checking, which builds on previous statistical model checking algorithms and scales to larger problems (Jha, 2010; Jha et al., 2009). Robust generative validation has also been suggested, specifically for ABMs (Yilmaz et al., 2011), which are generally both stochastic and complex. Additionally, metamorphic validation can provide a pseudo-oracle when data of the real system doe not exist for comparison (Raunak and Olsen, 2015). As Nicol points out, properly performing V&V is a real challenge in M&S, but must be solved for the field to move forward (Taylor et al., 2013).

1.3 The Need for Quantifying Model Credibility

Model credibility is acutely connected with the validation aspect of V&V activities performed throughout the development and usage life cycle of a simulation model. However, model validation-related activities have generally been characterized as a necessarily qualitative approach in the sense that there is no standard way to quantify the level of confidence gained in the model through validation. In recent years, simulation experts have asked the crucial question of whether the field of M&S, as it is now practiced, is sufficiently scientific. One concern is the lack of reproducibility and reuse of simulation models, caused by issues such as unavailable code, unclear validation, or poorly defined conceptual models. It is not clear that all simulation practitioners believe that these goals are necessary for the field, but it is generally agreed that a simulation's results cannot be credible if the simulation is not validated (Uhrmacher et al., 2016).

Another concern is to have unifying theories and frameworks throughout the M&S cycle. One of the crucial areas of this cycle, validation of simulation models, has suffered from lack of a rigorous standard. It is well accepted that validation of simulation models is difficult to ascertain, especially in published papers where validation details are often not included (Raunak and Olsen, 2014b; Fone et al., 2003). Moreover, validity is not usually independently verified due to the lack of replication of simulation models (Taylor et al., 2013; Bair and Tolk, 2013; Uhrmacher et al., 2016). To trust the recommendations based on any simulation study, a well-documented and communicated V&V effort is necessary. It is also a crucial part of enabling the results of a simulation experiment to be replicated. If a simulation model cannot be properly validated, then it is unlikely that its results will be easily reproduced if the model is implemented in a different way.

To properly communicate and document the validation process, it is necessary to ascertain how much validation has been performed. This assessment is crucial in answering a set of related research questions: how much validation is sufficient, and when should one stop validation activity on a simulation model? There is a parallel set of research questions in the software testing community: how much testing is sufficient, and when to stop testing software? That research has been successfully guiding the software testing community over the last few decades, leading to the development of useful metrics such as statement coverage, branch coverage, mutation coverage, and other structural coverage-based testing criteria (Zhu et al., 1997). Useful metrics such as combinatorial coverage have also been developed to measure the coverage of the input space of an application. All of these metrics and methods lead to quantifying how much testing on a software system has been performed, and thus determining how confident one can be about the software being fault-free.

In the area of M&S, there is a gap between the validation recommendations of researchers and the actions of the rapidly growing population of simulation practitioners (Raunak and Olsen, 2014b). Although there are simulations that are adequately validated, this is not true for all. Without a metric to capture and communicate the level of validation performed on a simulation model, and thus a quantified measure of the model credibility, it is difficult to communicate if a model has achieved an appropriate level of confidence for decisions made about the original system based on its findings. The rest of this chapter describes a process for quantifying validation coverage to enable communication and documentation of the validation process and confidence level in the model's correctness for increased model credibility, reproducibility, and reuse. After the process is described, two ABM examples and one DES example are used to demonstrate its application.

2 A GENERIC FRAMEWORK FOR QUANTIFIED VALIDATION COVERAGE

Many steps are required to obtain model credibility. Model validation is the most important part of these steps requiring rigorous technical work, but it is often not well documented, and therefore is the focus of this chapter. For any simulation model of reasonable complexity, it is usually infeasible to perform complete (100%) validation. The scenario is no different in the case of verifying software systems. It is well known and accepted that exhaustive testing, testing with all possible input values and operating condition, is infeasible for any reasonably sized software. Nevertheless, research in testing has led to quantitative measurements for increasing confidence in the software systems being fault free. Similarly, in simulation modeling we must validate our models as best as possible with the challenges of resource constraints in terms of time, money, and data. One crucial aspect of this effort is to properly document the details of the validation performed in some standardized fashion.

To properly validate a simulation model, and to articulate as well as quantify the validation work, we must be able to determine the aspects of the SUS that can be validated within the simulation model. This step is required regardless of the type of simulation model being validated, or the way in which it was developed. The framework outlined in this chapter is based on the four main properties shown in the experimental frame (Fig. 2): purpose, behavior, data, and structure. The purpose guides us, defining why the model was built and what type of questions may be asked of it. The behavior and structure of the model are based on that purpose. Behaviors are essentially the observable external manifestation, which can be considered as pairs of input and its related output; given a certain state/input of the model, how should it act? The structure is the internal design of the system, such as the steps of the system or the components of the system. Data are viewed in two pieces: the data used to create the

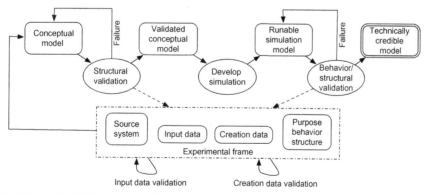

FIG. 2   The high-level validation process. We purposefully leave out the verification steps, and other steps toward building a credible model that do not focus on validation, to avoid overwhelming the reader.

model are validated through structural validation, and the data used to execute the model are validated using data validation. When we discuss data in this chapter, we will generally be referring to the latter instance.

The difference between behavior and structure for validation may initially cause trouble for some readers. Behaviors include any observable emergent information, such as the rate in which patients are able to move through an ED in a discrete-event-based ED simulation, or the way scent diffuses in the environment of an ABM of predator and prey. Structure in an ABM refers to the underlying agent properties, environment, and agent interactions, whereas structure in a DES includes the logical flow and details of the events, resources, their constraints, and interactions. The structure should ideally match the underlying structure of the SUS, albeit some details are abstracted out in the model. Some structural elements can be explicitly and independently validated through the conceptual model. The rest are validated indirectly through behavior validation. This difference is because structure invariably affects the behavior, and thus behavior validation indirectly validates the structure. However, to have full confidence in a model, performing indirect structure validation is not sufficient; we need to also perform direct structural validation, where applicable, to ensure that the model is correctly representing the processes and environment.

In Fig. 2, we show the high-level validation aspect of the simulation process, particularly to highlight the relationship between structural, behavioral, and data validation. A more detailed process of when to apply behavioral validation can be seen in Klügl (2008), the high-level simulation modeling process can be seen in Law (2015), and a detailed process of V&V can be seen in Balci (2012). The experimental frame contains all information related to the source system, including its data, purpose, behavior, and structure. Validation requires referring back to this frame. The conceptual model is validated via structural validation, the runnable simulation model is validated using operational validation that validates both behavior as well as its underlying structure, and then the model is calibrated and further validated to ensure that calibration does not break the model. At the end of this process, we get a model that is adequately validated for its purpose. Note that validation does not occur only at the end, but also throughout the process. Whenever verification or validation fails on the system, we return to the model to make improvements or modifications and then the subsequent steps are repeated. We do not show verification steps in Fig. 2 as they are not part of the confidence calculation, which is the focus of this chapter. Nevertheless, verification is crucial to ensure that the runnable model is credible in terms of its technical elements.

To calculate the level of confidence in the model, one has to consider all three levels in which the model should be validated: behavior, structure, and data. Each level has a different set of relevant validation techniques that are already in the literature, and will be discussed in the following sections.

2.1 Structural Validation

The structure of a model is indirectly validated through behavioral validation, but for the highest possible confidence in a model it should be directly validated as well. Forrester and Senge (1980) recommend five of the tests listed in Table 1 for explicitly validating structure. Although many techniques have the word "verification" in the name, they are validation tests as they compare the model to the real system.

TABLE 1 Proposed Ranking of the Most Common Structure Validation Techniques With Their Maximum Possible Level of Confidence

Technique (st)	Acronym	Potential Confidence ($c(st)$)	Creation Data	Other
Parameter verification	PV	2	Yes	No
Dimensional consistency	DC	2	Yes	No
Structure verification	SV	3	No	Yes
Extreme condition	EC	3	No	Yes
Boundary adequacy	BA	4	Yes	Yes

Notes: The first two tests are relevant for model creation data, the middle tests are relevant for other structural elements, and the last test is relevant in both cases. The abbreviations are used in the later examples.

We can consider two types of structure: data used to design the model ("creation data," not to be confused with input data for running the simulation), and structural elements that are not data such as process, interactions, environment, and abilities. Likewise, we can consider the validation techniques to each apply to one or both types of elements. Parameter verification and dimensional consistency tests validate the model creation data used for the design of the model. The parameter verification test evaluates whether the values adequately represent the real system. The dimensional consistency test can reveal incorrect structure assumptions when run with the parameter verification test, and is another simple way to increase model confidence.

Structural verification and extreme condition tests analyze the environment, interactions, process, and abilities in the structure. The structure verification test verifies that any structure found in the model is also found in the source system, by comparison to the system's organization, decision making, and assumptions. Generally the structure verification test is initially performed by the modeler, but for full confidence it should also be performed by an expert on the real system (often known as a domain expert or subject matter expert). Extreme condition tests at the structural level determine that the model can accept and likely function with extreme values and scenarios; if it cannot, then a change in structure is needed.

The boundary adequacy test determines if the correct level of abstraction has been chosen by analyzing both model purpose and model boundaries, which correlates to both types of structural elements. Detailed discussion of these techniques can be found in Forrester and Senge (1980). Most of these tests are focused on the conceptual model, as they check the abstractions, assumptions, and data used to design the model as opposed to testing a running simulation model as would be done via operational validation. The exception is extreme condition tests, which in modern systems need to be run on an executable simulation model, depending on the type of conceptual model and the problem being solved. This framework supports extreme condition tests being run at different stages as appropriate for the model, at the modeler's discretion.

To have full confidence in the structural integrity and validity of a simulation model, all structural validation tests should be successfully administered. However, like other stages of the model development and experiments, there may be constraints that prevent a particular structural validation test from being applied, or a scenario in which the test only validates some aspects of the structure. In those cases, the confidence in the model must be calculated.

To calculate confidence in the structural integrity, we must determine the level of confidence we have gained via our validation activities. Each structural validation technique may be applied with varying levels of success. In this case, each structural validation technique st_i has the potential confidence level $c(st_i)$ attainable via successful application as shown in Table 1. This confidence level is based on Forrester and Senge (1980) and Zengin and Ozturk (2012); the scale of the numbers is not important, only their values relative to each other.

Each structural validation test may validate a specific percentage of the model. For instance, we may test all extreme conditions in our model, but perhaps only 80% of the elements were validated by this test. We have thus gained confidence in our model, but not full confidence as we are aware of extreme condition test failures that must be fixed before trusting the model's results in all cases. To determine our overall confidence in the model structure, let $S = \{s_0, s_1, ..., s_n\}$ be the set of all structural elements to be validated, where each s_i is either a model creation data element ($s \in S_M$) or not ($s \in S_S$); thus $S = S_M \cup S_S$. We determine to what extent each structural element has been validated via each relevant validation test (st_k) as a value $v(s_i, st_k) \in [0, 1]$. The calculation of the level of confidence we have achieved via applying structural validation technique st_k is $p(st_k)$ as shown in Eq. (1), where $S_K = S_M$ if st_i is a model creation data technique, $S_K = S_S$ if it is a procedural structure technique, or $S_K = S$ if the technique is applicable to both structure types.

$$p(st_k) = \begin{cases} \dfrac{\displaystyle\sum_{s \in S_K} v(s, st_k)}{|s \in S_K|} & : |s \in S_K| > 0 \\ -1 & : o.w. \end{cases} \tag{1}$$

If there are five structural elements that can be tested via extreme condition tests (ec), and three of the elements are fully validated with extreme condition tests but two are only 50% validated, then $p(ec) = 0.8$. We propose that overall confidence in the structure of the model may be quantified as in Eq. (2), where R represents the relevant structural validation techniques, that is, $R = \{st_k | p(st_k) \neq -1\}$, and the potential confidence $c(r)$, where $r \in R$, is defined as in Table 1.

$$sc = 100 * \frac{\displaystyle\sum_{r \in R} c(r) * p(r)}{\displaystyle\sum_{r \in R} c(r)} \tag{2}$$

If all relevant tests are applied successfully to each structural element, we gain the highest possible confidence in the model's structure (100%). If, as in our simple example, all structural validation tests are fully passed except extreme condition tests which have a P value of .8, then our overall confidence in the structure of this model is 95.7%. Recall that this confidence value is in the correctness of the structure matching the source system, not the percentage of the model that has been validated.

2.2 Behavioral Validation

To calculate our confidence in the behavioral validity of a model, we first determine the behaviors of the model that need to be validated. We consider behaviors to encompass an observable result from the model, paired with the inputs or parameters that define the system that creates that output. In general, behaviors are the aspects of a simulation that most people first consider when discussing validation, and are generally validated using operational validation techniques.

To aid in determining the behaviors of a model, we propose utilizing the *aspects* of the relevant modeling type as described in the next section for finding validatable elements. How to define behaviors will vary depending on the type of simulation model being validated. Once we have identified all behaviors that must be validated, we must determine the applicable validation techniques for each of them. A single operational validation technique may apply to many different behaviors, and each behavior will often be validated with more than one applicable technique.

Not all validation techniques will contribute equally toward building confidence in the simulation model. By applying any validation technique successfully, we increase our confidence in the validity of the model. However, which validation technique(s) should be applied for each behavior? We propose utilizing rankings such as those in Table 2. Each commonly accepted technique is listed with a suggested relative level of confidence increase from its successful application.

The proposed ranking is based on a survey of the most commonly used validation techniques (Raunak and Olsen, 2014a), a discussion of their uses (Sargent, 2010), and the authors' experience. We rate animation the lowest as it can be misleading and hide underlying issues, but rank trace data and results validation the highest as they can be powerful techniques when the required data exists. We rate expert validation (face validation, Turing test) slightly below the validation involving data due to the potential for human error. We have ranked metamorphic testing similar to the level of model comparison as it can be highly effective in finding anomalies in the model. Like some other techniques listed in the table, metamorphic validation may not be easily applicable for all types of models. Additional techniques exist and can be ranked as necessary given the purpose of the model. This ranking is

TABLE 2 Ranking of the Most Common Validation Techniques With the Maximum Level of Confidence They Can Provide for a Given Behavior

Technique	Abbreviation	Maximum Confidence
Animation	A	3
Degenerate tests	DT	4
Internal validity	IV	5
Turing tests	TT	5
Face validation (expert)	FV	7
Sensitivity analysis	SA	7
Metamorphic validation	MV	8
Model comparison	MC	8
Trace data	TD	10
Results validation	RV	10

Notes: The abbreviations are used in the later examples.

subjective and used only for the purpose of illustration. More research and case studies from the community will be needed to establish a more generally accepted ranking of validation techniques.

Given a simulation model to validate, techniques are chosen based on their usefulness in creating the highest possible confidence in the model. There may be more validation techniques that are applicable than are necessary to achieve a reasonable level of high confidence in the model. For the highest model confidence, however, we must apply every applicable validation technique successfully. If it is not possible to apply all applicable techniques for any reason, we can use the following process to calculate our level of confidence.

The calculation of behavior confidence takes into account a simulation model's behaviors, the applicable validation techniques by behavior, and the successfully applied validation techniques. Let $B = \{b_0, b_1, ..., b_n\}$ be the set of behaviors to be validated for a particular simulation model, and $T = \{t_0, t_1, ..., t_m\}$ be the list of available validation techniques for operational validation such as those in Table 2. Each of these elements has an associated confidence weighting $w(t_i)$, defined as in Table 2. For b_i, the weight $w(b_i)$ is determined based on the impact of this behavior for model validation, and $w(b_i) > 0$. For instance, a behavior may have a very strong effect on overall validation, or may be almost trivial. In the simplest case all behavior weights are equal.

When performing validation it is possible that a single validation technique is applied successfully and validates multiple behaviors simultaneously. Therefore, a validation technique must be paired with some set $D \subseteq P(B)$, where $|D| \geq 1$ and $P(B)$ denotes the power set of B. A validation technique t_i may be used multiple times when validating a single simulation, each time used to validate a different set of behaviors by focusing on a different aspect of the simulation model and/or using different sets of available data or experts. We must consider a single validation approach $v_i \in V$ as a triplet $v_i = \langle t_i, d_i, c_i \rangle$, where $t_i \in T, d_i \in D$ is a set of b_js, and c_i is the confidence gain from applying t_i to d_i. There can only be one $t_i \in v_i$, as each validation technique is considered independently, although it may validate more than one b_i with one application. Each b_j and t_i may be duplicated as many times as necessary in V. Fig. 3 demonstrates the relationship between v, t, and b.

The $c_i \in v_i$ represents the confidence gain from t_i's successful application. It is based on the weight of each behavior in d_i represented as $w(b_j)$, and the weight of the validation technique $w(t_i)$ as seen in Eq. (3). Therefore, higher importance behaviors contribute more to our confidence in the system. Additionally, more powerful validation techniques contribute more to our confidence than less powerful techniques.

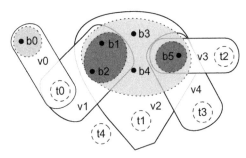

FIG. 3   A visualization of each element of a v_i. Each t_i is a validation technique, each b_j is a behavior, each *dotted circle* around a set of b_j is a d_i, and each *full shape* is a v_i. Note that each b_j can be in multiple v_i. The size of the shape represents c_i.

$$c_i = w(t_i) * \sum_{j=0}^{|d_i|} w(b_j), \quad \text{where } b_j \in d_i \tag{3}$$

Recall that confidence is based on what validation techniques are applied successfully. Each validation technique successfully applied raises our confidence in the validity of the model. To quantify this level we must define the highest possible confidence. We base the maximum confidence on the potential validation approaches $P = \{v_0, v_1, ..., v_p\}$, where $P \subseteq V$; and which validation approaches succeeded, $A = \{v_0, v_1, ..., v_a\}$, where $A \subseteq P$. We assume that any validation approach v_i that succeeded, did so for all $b_j \in d_i$. Finally, we may calculate our confidence in the behaviors across the entire simulation model bc as in Eq. (4).

$$bc = 100 * \frac{\sum c_i | c_i \in v_i \wedge v_i \in A}{\sum c_i | c_i \in v_i \wedge v_i \in P} \tag{4}$$

Our level of confidence in the behavioral aspect of the simulation model is therefore a real number, $0 \le bc \le 100$, where 100 represents the highest possible confidence. Note that, just like passing of all test cases in a test suite does not guarantee that a software is fault-free, confidence in a simulation model does not equal correctness. Thus the confidence value does not imply that the simulation model is correct. A confidence level of 100 implies that the simulation model has been validated to the fullest extent possible by currently known techniques.

2.3 Data Validation

During the process of developing and experimenting with simulation models, we deal with two types of data: the data used to build the model and the data used to run the model. By the term data validity in this case, we refer to the validation of data used to run the model. Data used for building the model are part of the model structure and thus validated through structural validity.

We propose that "Goodness of Fit" and "Face Validity" are the minimum validation techniques necessary for data validation (Sargent, 2005). Either technique could be applied multiple times, as for instance the goodness-of-fit test should be used to validate each distribution within the model. Additional techniques can also be added as appropriate.

All applicable validation tests should be run on all input data to achieve maximum confidence that we are using the correct data for our simulation. To calculate confidence in the data, we need to enumerate the list of data $I = \{i_0, i_1, ..., i_r\}$. This list is part of the model's experimental frame, and we assume that the modelers will collect it as part of their modeling and validation work.

Let $DP = \{dp_0, dp_1, ..., dp_p\}$ denote the set of applicable data validation techniques. For each dataset i_j we have the set of successfully applied data validation techniques $ADP_j \subseteq DP$. If we assume that all data validation techniques are equally important, then $c(i_j)$, the confidence in each dataset i_j, can be computed by Eq. (5). By dataset we refer to any set of data used to run the model, such as a distribution or input value range.

$$c(i_j) = 100 * \frac{|ADP_j|}{|DP|} \tag{5}$$

A simulation modeler may decide that some input data are more important to validate than others. In that case, we propose incorporating the relative importance of the input data by attaching weights to each i_j denoted as $w(i_j)$. The overall confidence in data validity is then defined as in Eq. (6).

$$dc = \frac{\sum_{j=0}^{|I|} c(i_j) * w(i_j)}{\sum w(i_j)} \tag{6}$$

2.4 Overall Confidence

Once we have confidence in the behavior, structure, and input data of the simulation model, we can define confidence in the model as a whole. We propose that having proper behaviors and structures are equally important. However, as behavioral validation indirectly validates some aspects of structure, and direct structural validation is slightly limited in whether the final simulation model truly represents the SUS correctly due to its focus on face and data validation on the conceptual model, their validation confidence levels should not equally influence our overall confidence in the simulation model. Additionally, data validation is crucial for correct behavioral validation, but very hard to perform completely across all data for the system. In reality, the data used for running a simulation are often considered to be valid based on its source and collection method. In many scenarios, such level of data validation may be sufficient. We therefore propose the following calculation for overall model confidence mc:

$$mc = 0.5bc + 0.3sc + 0.2dc \tag{7}$$

where bc, sc, and dc are as defined previously. If no structural validation is performed, the highest possible confidence in the model's correctness is 70%; no data validation means a max of 80%; and no behavioral validation means a max of 50%, which we would consider inadequate. However, generally all three will be performed, often achieving less than 100% confidence level in each area. The previous calculation will provide a high-level view of overall confidence.

2.5 Publication of Model Confidence

We propose that model confidence should not only be calculated for each model, but that high-level details of those calculations should be included in published papers. In the examples in the next section we provide a table format that includes a listing of validatable elements, the level of validation achieved via validation techniques, and the resulting model confidence calculations sc, bc, dc, and mc. For suitably complex models it will not be possible to include a full table in a publication with current chapter page limits; however, at the bare minimum these four values should be included in the chapter with a summary of the validation performed, with the full table available on the authors' website, or an appendix. For simulation models that are made publicly available, this information should accompany the files as well.

2.6 Module Validation

Complex simulations are generally made of many modules that may be validated separately, and then the overall system validated as a whole. Each module should be validated as described earlier, with an overall mc score calculated. These values should be reported. The structure, behavior, and data of the integrated model must also be validated, and calculated as earlier. The final confidence level in the overall model will be a weighted average of the confidence scores of the independent modules and the integration of the modules. The weights for each module should be relative to their size and impact on the overall system.

3 SPECIFIC FRAMEWORK: ABMS

To apply the coverage calculation on an ABM, the first step is to define the structural and behavioral elements of the model and the input data that would ideally be validated. Structure relates to the structure of agents in their attributes and abilities, as well as the environment in terms of its design, topology, methods of interaction by agents, and the relationships of agents. Observable behaviors in an ABM may relate to resources, the environment, agent actions or state changes, and agent interactions (Fig. 4). Structure is generally defined in the conceptual model documents, and behaviors are expected to be observed when the simulation model is run. Recall that behaviors that are not predictable, such as unexpected emergent

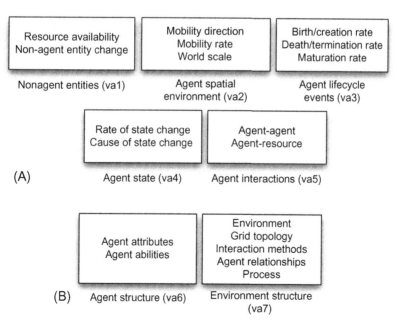

FIG. 4 The aspects of an ABM that should be validated, to be used as a guide for determining validatable elements. For any given model not all aspects may be relevant. Data may relate to any aspect. (A) Behavior aspects; (B) structure aspects.

behavior or the results of a what-if scenario, are not validatable. Instead, the modeler should focus on validating the aspects of the model that can be shown to match the SUS's known behavior, such that the results of such what-if scenarios can be trusted.

After determining what elements of the model must be validated, and validation is performed, the calculation of coverage is the same as previously discussed. The role of structure, behavior, and data will of course vary based on the type of system modeled. To demonstrate its general applicability, the following examples display the application of this approach to two different types of agent-based simulations: a predator-prey model on a two-dimensional grid and a networked ABM.

3.1 Example: Tasmanian Devils

We examine an agent-based predator-prey model of Tasmanian Devils, a carnivorous marsupial found only in the Tasmanian island of Australia that suffers from a deadly transmittable cancer known as Devil Facial Tumor Disease. The purpose of this simulation is to study the rate of interactions between devils, and whether or not a newly proposed road through their habitat will endanger their survival as the disease is transmitted by biting, which primarily occur during eating. A road could theoretically increase biting as devils eat road kill.

In this simulation, devils move in a two-dimensional grid either randomly or toward prey if the prey's scent is within their neighborhood. As devils are primarily scavengers, prey appears in random locations at a fixed rate to mimic carcasses. Prey emits a scent gradient that decays over time, which devils follow when hungry. Devil hunger decreases as they eat, and is of a fixed level each day. The simulation logs how frequently devils meet while eating, to help determine the overall rate of disease transmission through biting. The simulation supports adding a road, which provides another source of food via roadkill, to ask how this concentrated food source affects devil interactions (Fay et al., 2011).

3.1.1 Defining Validatable Elements

The simulation must be validated to show that the nonroad scenario adequately represents current devil interactions, so that we can add a road into the simulation and ask how it affects devil interactions. The way in which the road is added must also be validated.

A large portion of this model that must be validated is structure as opposed to behavior (Table 3). The structure includes information on how the environment has been abstracted, how the devils' behaviors have been abstracted, and how the road will be implemented. The first six structural elements are related to the environmental or agent representation, and the rest are data used to design the model.

The primary behavioral element to validate is the devils' social network, which is not explicitly modeled but emerges and should represent known social structures of the Tasmanian Devils for the model to be valid. Additionally, the rates at which devils eat should mimic their eating habits in the wild, denoting that they way they find prey adequately represents the real system. The way scent is diffused should appear reasonable in level, rate, and distance. If the structure has been properly validated, and the simulation model has been verified to match that conceptual model, then the results can be trusted for studying the impact of the road on devil interactions.

TABLE 3 Structural, Behavioral, and Data Elements of the Tasmanian Devils Model That Must Be Validated

Structural Validation					
	Applied Techniques				
Element	**PV**	**DC**	**SV**	**EC**	**BA**
Foraging area			1	1	1
Road representation			0.5	0	1
Devil prey finding strategy			1	0	1
Roadkill representation			1	0	1
Devil movement strategy			1	0	1
Time representation			1	0	1
Prey creation data (appearance, size, decay)	1	1			1
Devil population density	1	1			1
Devil movement rate	1	1			1
Roadkill appearance rate	1	1			1
World size	1	1			1
Structural confidence by technique ($p(st_i)$)	1	1	0.916	0.167	1
Overall structural confidence (sc)			80.34%		

BEHAVIORAL VALIDATION			
Element	$w(b_j)$	Applicable Techniques	Applied Techniques
Devil social network	1	A, FV, SA, MV, RV	A, FV, SA, RV
Devil hunger over time	0.75	A, FV, SA, MV, RV	A, FV, SA, MV
Scent diffusion	0.75	A, FV, SA	A, FV, SA
Devil attraction to prey	0.75	A, FV, SA, MV	A, FV, SA
Overall behavioral confidence (bc)		76.82%	

DATA VALIDATION			
Element	w_i	Applicable Techniques	Applied Techniques
Scent diffusion rate	1	GF, FV	GF, FV
Overall data confidence (dc)		100%	
Model confidence (mc)		82.51%	

Notes: For structure we note the success of applied techniques, and for behavior and data we list the applicable techniques and their success. Technique acronyms are defined in Tables 1 and 2. For each set we show the calculation, and then the final validation coverage.

3.1.2 Validation

To track the level of confidence gain via validation of the system, we must note the applicable validation techniques and which ones have been applied. Table 3 shows the validation techniques for each structure, behavior, and data element. For structural validation, each element is listed with a value between 0 (not applied or failed) and 1 (fully successful) denoting the level of successful application of each relevant validation technique. The data used in creating the conceptual model, including prey appearance/size/decay, devil density, devil movement levels, roadkill appearance rates, and world size were validated using parameter verification and dimensional consistency. Each of these rates is based on data on the real system, and verified by a subject matter expert. These rates combined with the structural aspects of the foraging area, road representation, how devils find prey, how roadkill and prey are represented, and how time is represented were examined together by a subject matter expert to determine that adequate boundaries were chosen for the abstraction and that the structure was correct. In this case, we see that the model creation data are fully validated using all three relevant techniques, extreme condition testing was only performed for one element, and structural verification was fully successful on all structure elements except the road representation. This validation leads to a 80.34% confidence level in the structure. For each element that we further validate with extreme condition testing, we can increase our confidence in the structure by 3.57%.

For behavioral validation the applicable techniques are listed by acronym, followed by a list of applied techniques. Sensitivity analysis was used to determine how all behaviors are affected by changing the prey size, prey appearance rate, prey decay rate, scent diffusion rate, and devil movement rate. Although most of these rates are part of the data used to create the model, they can be tweaked in calibration to ensure that the abstraction leads to the expected behaviors. Metamorphic validation was used to validate devil hunger over time, and results validation was used to compare the devil social network to known studies on interaction networks between Tasmanian Devils in the wild. Animation was used to validate all behaviors.

The values of each of these techniques are used in combination with the behavior's weight to calculate confidence gained (c_i) by applying validation techniques to specific behaviors, which are not denoted in the table. In this example, based on Table 3 we have the following validation approaches for our model ($v_i = \langle t_i, d_i, c_i \rangle$), which are used to calculate our confidence bc:

1. $v_0 = \langle A, (b_0), 3 \rangle$, applied
2. $v_1 = \langle A, (b_1, b_2, b_3), 6.75 \rangle$, applied
3. $v_2 = \langle FV, (b_0, b_1, b_2, b_3), 22.75 \rangle$, applied
4. $v_3 = \langle SA, (b_0, b_1, b_2, b_3), 22.75 \rangle$, applied
5. $v_4 = \langle MV, (b_0), 8 \rangle$, not applied
6. $v_5 = \langle MV, (b_1), 6 \rangle$, applied
7. $v_6 = \langle MV, (b_3), 6 \rangle$, not applied
8. $v_7 = \langle RV, (b_0), 10 \rangle$, applied
9. $v_8 = \langle RV, (b_1), 7.5 \rangle$, not applied

Our final model confidence in the Tasmanian Devils model is 82.51%. Applying results validation to devil hunger changes over time would increase model confidence to 86.55%.

Applying all missing behavioral validation techniques increases the model confidence to 94.102%, as behavioral validation has the strongest weight in our overall confidence calculation. This confidence level provides a summary of the model confidence, but Table 3 provides important details on what aspects of the model are best supported by the applied validation techniques.

3.2 Example: Gossip Propagation

The second example is an ABM of gossip propagation in a social network, which is naturally modeled as a graph. Each node in the graph represents a person capable of spreading gossip about some target node. Some nodes are liars, which mutate the bit string that is passed as gossip. The gossip propagates from node to node based on the strength of connection between those nodes and how long the gossip has been propagated so far. Gossip that has been propagating for some time has a weaker strength than fresh gossip, which will cause nodes to no longer wish to spread the information. Each node must determine what to believe about a particular piece of gossip, and the simulation compares the overall belief of the network based on various decision strategies. After a piece of gossip is no longer spreading through the network, nodes are assigned a fitness score denoting how close their beliefs were to the truth. The purpose of this simulation is to study human gossip decision rules and gossip propagation (Laidre et al., 2013).

3.2.1 Defining Validatable Elements

The elements of the model that could be validated are shown in Table 4. The environmental structure in this model includes the agent network, the definition of who can message an agent, and the process of gossip spread. The agent structure in this model includes the agents' property of an agent's memory. It also includes an agents' abilities: strategy for choosing gossip to believe, an agent's decision to share gossip, and the approach to distorting gossip before sharing. The behaviors that must be validated in this model include the gossip fidelity decrease, the rate of sharing with other agents, and the changes to agent beliefs as gossip is spread. The input data to validate are the number of liars and the number of observers, as well as a message length and loss rate. The final input data are the heterogeneity of agents in terms of their gossiping strategies.

3.2.2 Validation

Table 4 shows the confidence in the correctness of the model based on which of the possible validation techniques were applied. The format of this table was explained in Section 3.1. We have high confidence in the structure of the model as boundary adequacy succeeded on all structural elements, extreme condition tests were applied and succeeded on the majority of elements, and structural verification succeeded fully on all but three elements. The creation data on the number of agents were also successfully validated against data on social network sizes in human populations. Most of the applicable behavior validation techniques were also successfully applied, using the following validation approaches:

TABLE 4 Structural, Behavioral, and Data Elements of the Networked ABM Gossip Model That Must Be Validated

Structural Validation					
	Applied Techniques				
Element	**PV**	**DC**	**SV**	**EC**	**BA**
Agent network			1	1	1
Messaging structure			0.75	0	1
Heterogeneity of agents			0.75	1	1
Gossip spread process			1	0	1
Agent memory			1	1	1
Choice strategy			1	1	1
Sharing strategy			0.75	1	1
Distortion approach			1	0	1
Number of agents	1	1			1
Structural confidence by technique	1	1	0.906	0.625	1
Overall structural confidence (*sc*)			89.95%		

BEHAVIORAL VALIDATION			
Element	$w(b_j)$	**Applicable Techniques**	**Applied Techniques**
Gossip fidelity decrease	1	A, SA, FV	A, SA, FV
Agent belief change	1	A, SA, FV, RV	A, SA, FV
Neighbor sharing	1	A, SA, FV, MV	A, SA, FV, MV
Liar impact propagation	1	A, SA, FV, MV	A, SA, FV, MV
Gossip propagation ends	1	A, SA	A, SA
Observer placement impact	1	A, SA, MV	A, SA, MV
Overall behavioral confidence (*bc*)		91.8%	

DATA VALIDATION			
Element	w_i	**Applicable Techniques**	**Applied Techniques**
Percentage of liars	0.5	FV	FV
Percentage of observers	0.5	FV	FV
Message length	0.5	FV	FV
Message loss rate	1	GF, FV	FV
Heterogeneity of agents	1	GF, FV	FV
Overall data confidence (*dc*)		71.43%	
Model coverage (*mc*)		87.231%	

Notes: For structure we note the success of applied techniques, and for behavior and data we list the applicable techniques and their success. Technique acronyms are defined in Tables 1 and 2. For each set we show the calculation, and then the final validation coverage.

1. $v_0 = \langle A, (b_0, b_1, b_2, b_3, b_4, b_5), 18 \rangle$, applied
2. $v_1 = \langle SA, (b_0, b_1, b_2, b_3, b_4, b_5), 42 \rangle$, applied
3. $v_2 = \langle FV, (b_0, b_1, b_2, b_3), 28 \rangle$, applied
4. $v_3 = \langle RV, b_1, 10 \rangle$, not applied
5. $v_4 = \langle MV, b_2, 8 \rangle$, applied
6. $v_5 = \langle MV, b_3, 8 \rangle$, applied
7. $v_6 = \langle MV, b_5, 8 \rangle$, applied

This model has been very well validated for behaviors, with the lowest validation confidence in the data. For this type of model it can be difficult to ascertain the correct input to exactly mimic a human system; however, given the strong structural and behavioral validation, the model can still be trusted to give interesting insights on how gossip strategies and communication networks may impact the spread of gossip. The way the results are interpreted are only dampened by the lower input data validation score. Thus, denoting the confidence in structure, behavior, and data separately can increase our understanding of a model's validity and the types of questions that may be asked of it.

4 SPECIFIC FRAMEWORK: DISCRETE-EVENT MODELS (DES)

DESs differ from ABMs in their focus on discrete sequences of events, as opposed to autonomous agents making decisions. As with ABMs, the first step for computing model confidence requires listing all structural and behavioral elements of the model as well as input data that should be validated. Structure includes how the logical flow and resources mimic the real system, including resource attributes, the queuing model, and how events are designed and ordered. Observable behaviors are often related to the rates at which resources are used, or characteristics or rates of requests for resources as the simulation runs (Fig. 5). Input data can be related to either structural or behavioral element.

4.1 Example: Hospital Emergency Department

We illustrate the process of quantifying model credibility for DES by applying it on a DES of patient flow in the ED of a hospital (Raunak et al., 2009; Raunak and Osterweil, 2012). This simulation model was developed to model the Emergency Department of BayState Medical Center in West Springfield, MA. The model focuses on the patient care path of walk-in patients in an ED. The purpose of the model is to investigate the impact of process changes in ED patient care, such as immediate placement of an incoming patient inside the ED when a bed is available, followed by bedside registration in parallel with patient treatment. To understand how to compute the credibility score on the simulation model, we need to be acquainted with the patient care process in an ED.

Patient care at a typical ED encompasses a series of sequential activities starting with the new patient's arrival. A patient may arrive as either a walk-in or through an ambulance. The walk-in patients are first seen by a triage nurse for determination of a triage acuity level. The patient is then sent to the registration clerk. The registration clerk enters the patient's

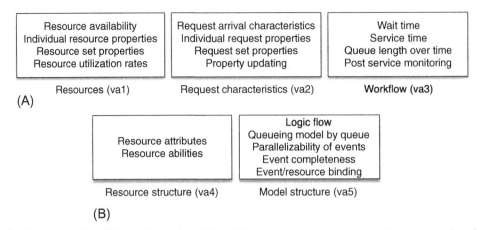

(A)

(B)

FIG. 5   The aspects of an DES model that should be validated. For any given model not all aspects may be relevant. Data may relate to any aspect. (A) Behavior aspects; (B) structure aspects.

insurance and other information into the system and generates an ID band. The patient then waits in the waiting room if a bed is not available. Once a bed becomes available, the patient is placed in the bed and is assessed by a nurse, followed by the attending doctor. The doctor's assessment may result in additional tests or procedures. Afterward the doctor makes a final assessment and decides whether to admit or discharge the patient. The simulation cycle ends with the patient either discharged or admitted into the inpatient unit of the hospital. The simulation model defines makeup and availability of a number of resources such as doctors, nurses, clerks, beds, X-ray machines, etc. Patient arrival rates and different service time distributions are examples of input data needed to run the executable simulation of this model.

4.1.1 Defining Validatable Elements in the ED Model

The ED model needs to represent a reasonable patient arrival distribution, resource combinations with specific attributes and abilities, the treatment process, and potential congestion. All elements must be validated to achieve a high confidence level for the credibility of the model.

The structural elements of the ED model include the process of treatment for a patient as defined by the ED workflow. The structural elements also include attributes and abilities of resources such as doctors, nurses, registration clerks, and beds. The abilities define what tasks can be performed by a particular agent resource such as a doctor or nurse. For example, a nurse can perform an initial assessment of a patient, and assign an initial acuity level. A doctor is capable of ordering tests and performing procedures on the patient. Similarly, attributes of a bed resource define if it has facilities particularly suited for certain types of patients. Table 5 shows the structural elements that need to be validated: ED workflow, treatment bed attributes, and abilities of a number of resources such as doctors, nurses, and registration clerks. There is also one model creation data in this model: the rate at which patients are arriving, which must match the distribution in a real hospital.

This particular simulation model, based on its purpose, needs to mimic the real-world process and the experience of walk-in patients in an ED. Thus the behavior validation needs to ensure that such emergent information as the average waiting time experienced by patients and the average length of stay (LOS) for patients are consistent with the defined workflow and allocated resource mix. Additionally, other observable information such as the utilization rate of the resources (doctors, nurses, beds, clerks, etc.) should also be validated.

The input data elements, which are required for running an ED simulation, include the mix of resources used for a particular simulation run. This mix includes the number of doctors, nurses, registration clerks, and beds. Moreover, the distribution of tests and procedures performed on the patients are important input data for running the ED simulation model.

4.1.2 Computing Model Credibility Score for ED Model

To compute the confidence gained by validating the ED simulation, we identify all applicable validation techniques for structural, behavioral, and input data validation and the techniques that have been successfully applied in each case. Table 5 shows a summary of the validation techniques that were applied and how they contribute toward computing a measurement of the ED model's credibility. The format of the table was described in Section 3.1.

We have relatively high confidence in the structural elements of the ED model. All structural elements were fully verified by a subject matter expert (SV), and validated by checking the boundary adequacy (BA). Extreme condition tests were successfully applied on some of the structural elements, but not all. The applied validation techniques result in an 87.14% confidence level in the structure.

The behavior validation section of Table 5 lists the abbreviated names of both the applicable validation techniques and the ones that were applied to validate each of the behavior elements. Degenerate tests, face validation, model comparison, metamorphic validation, and results validation were applicable to all behavior elements and they were all successfully applied. Even though sensitivity analysis was applicable, no behavioral element of this particular ED model was validated using this technique. To compute the confidence level gained by applying these validation techniques on the behaviors in Table 5, we define the following validation approaches for our model ($v_i = \langle t_i, d_i, c_i \rangle$):

1. $v_0 = \langle DT, (b_0, b_1, b_2, b_3, b_4, b_5), 16 \rangle$, applied
2. $v_1 = \langle FV, (b_0, b_1, b_2, b_3, b_4, b_5), 28 \rangle$, applied
3. $v_2 = \langle SA, (b_0, b_1, b_2, b_3, b_4, b_5), 28 \rangle$, not applied
4. $v_3 = \langle MC, (b_0, b_1, b_2, b_3, b_4, b_5), 32 \rangle$, applied
5. $v_4 = \langle MV, (b_0, b_1, b_2, b_3, b_4, b_5), 32 \rangle$, applied
6. $v_5 = \langle RV, (b_0, b_1, b_2, b_3, b_4, b_5), 40 \rangle$, applied

The behavior computation results in a confidence level of 84.09% in the behavior of our ED model. The input data are all primarily validated using face validation. In this example, we

TABLE 5 Structural, Behavioral, and Data Elements of the Emergency Department Model

		Structural Validation			
	Applied Techniques				
Element	**PV**	**DC**	**SV**	**EC**	**BA**
ED workflow			1.0	1.0	1
Treatment bed attributes			1.0	0	1
Doctor abilities			1.0	0.5	1
Nurse abilities			1.0	0.5	1
Registration clerk abilities			1.0	0	1
Patient arrival rate	1.0	1.0			1
Struc. conf. by technique ($p(st_i)$)	1.0	1.0	1.0	0.4	1.0
Overall structural confidence (sc)			87.14%		

		BEHAVIORAL VALIDATION		
Element	$w(b_j)$	Applicable Techniques		Applied Techniques
Avg. wait time for bed	1	DT, FV, SA, MC, MV, RV		DT, FV, MC, MV, RV
Avg. LOS in ED	1	DT, FV, SA, MC, MV, RV		DT, FV, MC, MV, RV
Doctor util. rate	0.5	DT, FV, SA, MC, MV, RV		DT, FV, MC, MV, RV
Nurse util. rate	0.5	DT, FV, SA, MC, MV, RV		DT, FV, MC, MV, RV
Reg. clerk util. rate	0.5	DT, FV, SA, MC, MV, RV		DT, FV, MC, MV, RV
Bed util. rate	0.5	DT, FV, SA, MC, MV, RV		DT, FV, MC, MV, RV
Overall behavioral confidence (bc)		84.09%		

		DATA VALIDATION		
Element	w_i	Applicable Techniques		Applied Techniques
No. of doctors	0.5	FV		FV
No. of nurses	0.5	FV		FV
No. of reg. clerks	0.5	FV		FV
No. of beds	0.5	FV		FV
Distr. of tests performed	1	GF, FV		GF, FV
Distr. of procedures performed	1	GF, FV		GF, FV
Overall data confidence (dc)		100%		
Model confidence (mc)		88.19%		

Notes: For structure we note the success of applied techniques, and for behavior and data we list the applicable techniques and their success. Technique acronyms are defined in Tables 1 and 2. For each set we show the calculation, and then the final model credibility score.

achieve 100% input data validation. With all three parts of our model confidence at hand, we can now compute the overall model credibility score for our ED example as in Eq. (8).

$$mc = 0.3*87.14 + 0.5*84.09 + 0.2*1.0 = 88.19\% \tag{8}$$

In this example, we have a relatively high model credibility score since most of the validatable elements have validation techniques that were applicable for validating different validatable elements of the model.

5 SUMMARY AND DISCUSSION

There are many activities necessary to produce a credible model, including demonstrating V&V, useful animations, and buy-in by managers or higher-level decision makers (Law, 2015). A key step is validation of a model, increasing confidence that the model matches the SUS. Although validation is done to some extent on most models, there is no standard on how to quantify the amount of confidence gained through validation. As can be seen from the survey of 192 health care simulation papers in Raunak and Olsen (2014b), less than half of papers provide a detailed validation section, and almost 24% of papers make no mention of validation at all.

This chapter described a model confidence metric for quantifying the amount of validation performed, and thus the level of confidence gained, on a simulation model. Structure is validated on the conceptual model, behavior and structure are validated in the simulation model, and input data are validated to ensure that they adequately match the real-world SUS. Validatable elements of structure and behavior are determined, potential validation techniques are listed, and as validation is successfully performed our confidence increases. The calculation of model confidence represents how many of these relevant techniques were successful, weighted by importance.

The usage of this model confidence calculation was demonstrated with an agent-based predator-prey model, an agent-based networked model, and a discrete-event hospital simulation model. These examples show a variety of applications, the suggested table format for sharing information on the confidence calculation, and how to interpret the results of the model confidence score. In each case a standard template was used to summarize the validation that was performed and the level of confidence gained in the model's correctness.

Although this chapter focuses on ABMs and DES models, the model confidence calculation is applicable to any simulation paradigm. For instance, hybrid models and DEVS models are easily analyzed using this approach. Hybrid models are becoming more common, where aspects of different paradigms are used in the same model. A hybrid ABM and DES model, for example, would be validated by validatable elements being found for each approach, likely following the modular system process described in this chapter. DEVS can be used to describe systems that are event-based in both inputs and outputs (Zeigler and Muzy, 2017), and thus naturally fit with the DES approach described in this chapter as outlined in Olsen and Raunak (2015).

We strongly encourage the use of this model confidence score to communicate the credibility of the model as related to its validation. This model confidence score is of a similar structure to coverage calculations in software testing. Just as in software testing, a coverage criterion is only one element necessary for quantifying validation of a simulation model. We urge our fellow researchers to consider additional model confidence calculations to increase the trust in model results, and the ability to reuse models.

References

Abbott, R., 2002. Cancersim: A Computer-Based Simulation of Hanahan and Weinberg's Hallmarks of Cancer (Ph.D. thesis), University of New Mexico.

Bair, L.J., Tolk, A., 2013. Towards a unified theory of validation. In: WSC '13. Proceedings of the 2013 Winter Simulation Conference: Simulation: Making Decisions in a Complex World13.IEEE Press, Piscataway, NJ, USA, pp. 1245–1256.

Balci, O., 1998. Verification, Validation, and Testing. John Wiley & Sons, New York, NY, pp. 335–393.

Balci, O., 2010. Golden rules of verification, validation, testing, and certification of modeling and simulation applications. SCS Model. Simul. Mag. 4.

Balci, O., 2012. A life cycle for modeling and simulation. Simulation 88 (7), 870–883.

Banks, J., 1998. Handbook of Simulation: Principles, Methodology, Advances, Applications, and Practice. John Wiley & Sons, New York, NY.

Bertolino, A., 2007. Software testing research: achievements, challenges, dreams. In: Proc. of ICSE Future of Software Engineering (FOSE), pp. 85–103.

Ender, T., Leurck, R.F., Weaver, B., Miceli, P., Blair, W D., West, P., Mavris, D., 2010. Systems-of-systems analysis of ballistic missile defense architecture effectiveness through surrogate modeling and simulation. IEEE Syst. J. 4, 156–166.

Fay, G., Olsen, M., Gran, J., Johnson, A., Weinberger, V., Carja, O., 2011. Agent-based model of Tasmanian Devils examines spread of devil facial tumor disease due to road construction. In: Proceedings of the International Conference on Complex Systems.

FDA, 2010. Infusion pump software research at FDA. http://www.fda.gov/MedicalDevices/ProductsandMedicalProcedures/GeneralHospitalDevicesandSupplies/InfusionPumps/ucm202511.htm. (Accessed 8/1/2017).

Fone, D., Hollinghurst, S., Temple, M., Round, A., Lester, N., Weightman, A., Roberts, K., Coyle, E., Bevan, G., Palmer, S., 2003. Systematic review of the use and value of computer simulation modelling in population health and health care delivery. J. Public Health 25 (4), 325–335.

Forrester, J.W., Senge, P.M., 1980. Tests for building confidence in system dynamics models. In: System Dynamics, TIMS Studies in Management Sciences, vol.14, pp. 209–228.

Jha, S., 2010. Model Validation and Discovery for Complex Stochastic Systems. (Ph.D. thesis), Carnegie Mellon University.

Jha, S.K., Clarke, E., Langmead, C., Legay, A., Platzer, A., Zuliani, P., 2009. A Bayesian approach to model checking biological systems. In: Computational Methods in Systems BiologySpringer, Berlin, Heidelberg.

Kleijnen, J., 1995. Verification and validation of simulation models. Eur. J. Oper. Res. 82, 145–162.

Klügl, F., 2008. A validation methodology for agent-based simulations. In: Proceedings of SACpp. 39–43.

Laidre, M.E., Lamb, A., Shultz, S., Olsen, M., 2013. Making sense of information in noisy networks: human communication, gossip, and distortion. J. Theor. Biol. 317, 152–160.

Law, A.M., 2015. Simulation Modeling and Analysis, fifth ed. McGraw-Hill, New York, NY.

Olsen, M., Raunak, M., 2015. A method for quantified confidence of DEVS validation. In: Proceedings of SpringSim TMS/DEVS.

Raunak, M., Olsen, M., 2014a. Quantifying validation of discrete event models. In: Proceedings of the Winter Simulation Conference.

Raunak, M., Olsen, M., 2014b. A survey of validation in health care simulation studies. In: Proceedings of the 2014 Winter Simulation Conference, pp. 4089–4090.

Raunak, M., Olsen, M., 2015. Simulation validation using metamorphic testing (WIP). In: Proceedings of the Summer Simulation Conference.

Raunak, M., Osterweil, L., 2012. Resource management for complex, dynamic environments. Trans. Softw. Eng. 39 (3), 384–402.

Raunak, M., Osterweil, L.J., Wise, A., Clarke, L.A., Henneman, P.L., 2009. Simulating patient flow through an emergency department using process-driven discrete event simulation. In: Proceedings of the Software Engineering in Health Care (ICSE 2009), Vancouver, Canada.

Ray, A., Morschhaeuser, I., Ackermann, C., Cleaveland, R., Shelton, C., Martin, C., 2009. Validating automotive control software using instrumentation-based verification. In: Proceedings of the IEEE/ACM International Conference on Automated Software Engineering.

Reynolds, W.N., 2010. Breadth-depth triangulation for validation of modeling and simulation of complex systems. In: IEEE Intelligence and Security Informatics Conference, Workshop on Current Issues in Predictive Approaches to Intelligence and Security Analytics (PAISA), pp. 190–195.

Reynolds, W.N., Wimberly, F., 2011. Simulation validation using causal inference theory with morphological constraints. In: Proceedings of the 2011 Winter Simulation Conference.

Sargent, R.G., 2005. Verifying and validating simulation models. In: Proceedings of the 2005 Winter Simulation Conference.

Sargent, R.G., 2007. Verifying and validating simulation models. In: Proceedings of the 2007 Winter Simulation Conference, pp. 124–137.

Sargent, R.G., 2010. Verifying and validating simulation models. In: Proceedings of the 2010 Winter Simulation Conference, pp. 166–183.

Sargent, R.G., 2011. Verifying and validating simulation models. In: Winter Simulation Conference, Colorodo.

Sokolowski, J.A., Banks, C.M., 2010. Modeling & Simulation Fundamentals. Wiley, New York, NY.

Taylor, S.J., Khan, A., Morse, K.L., Tolk, A., Yilmaz, L., Zander, J., 2013. Grand challenges on the theory of modeling and simulation. In: Proceedings of the Symposium on Theory of Modeling & Simulation-DEVS Integrative M&S Symposium, p. 34.

Uhrmacher, A.M., Brailsford, S., Liu, J., Rabe, M., Tolk, A., 2016. Panel—reproducible research in discrete event simulation—a must or rather a maybe? In: Proceedings of the Winter Simulation Conferencepp. 1301–1315.

Yilmaz, L., Zou, G., Balci, O., 2011. A robust evolutionary strategy for generative validation of agent-based models using adaptive simulation ensembles. In: Proceedings of the 2011 Winter Simulation Conference.

Zeigler, B., Muzy, A., 2017. From discrete event simulation to discrete event specified systems (DEVS). In: International Federation of Automatic Control (IFAC), pp. 9–14.

Zeltyn, S., Marmor, Y.N., Mandelbaum, A., Carmeli, B., Greenshpan, O., Mesika, Y., Wasserkrug, S., Vortman, P., Shtub, A., Lauterman, T., Schwartz, D., Moskovitch, K., Tzafrir, S., Basis, F., 2011. Simulation-based models of emergency departments: operational, tactical, and strategic staffing. ACM Trans. Model. Comput. Simul. 21 (4), 24:1–24:25.

Zengin, A., Ozturk, M.M., 2012. Formal verification and validation with DEVS-suite: {OSPF} case study. Simul. Model. Pract. Theory 29, 193–206.

Zhu, H., Hall, P.A., May, J.H., 1997. Software unit test coverage and adequacy. ACM Comput. Surv. 29 (4), 366–427.

A Comprehensive Method for Model Credibility Measurement

Yuanjun Laili, Lin Zhang, Gengjiao Yang

School of Automation Science and Electrical Engineering, Beihang University, Beijing, China

1 INTRODUCTION

Simulation has been increasingly used to study complex systems in real world, which provides a faster, cheaper, and more flexible alternative than physical experiments (Zeigler et al., 2000). With the expansion of simulation objects, it is more and more imperative to compose a number of domain models to implement a comprehensive and dynamic system analysis (Zeigler and Zhang, 2015).

In a cross-domain system, multidisciplinary models are required to interact with each other as modular elements. Their activities, temporal information, function transformation, and the related knowledge base are changed autonomously with a series of deterministic or nondeterministic logics. The interaction process among these models introduces more complex time constraints and evolutionary characteristics. On one hand, they should be sufficiently flexible with add-on uncertainty features to adapt themselves to a dynamic synthetics along with changing environments (Zeigler, 1990). On the other hand, they ought to be well organized throughout its life cycle to ensure proper support for system execution.

As a simulation system that is not credible has no practical significance and application value (Law and Kelton, 2000), the credibility evaluation of the system models is of the most importance.

The credibility of a simulation model can be roughly defined as the degree of its adaptability to the simulation purpose, which is usually determined by a similarity between the model and the target system (Jiao et al., 2007). This is no longer suitable for a complex composite system for two reasons. First, the dynamic essence of a complex system determines that the similarity of a model within a limited number of testing is not convincible since both the model and the prototype system are changing over time. Second, the input data and management issues on a model other than its similarity are also decisive factors on its credibility for a simulation purpose. Therefore, the credibility of a simulation model is redefined in NASA-STD-7009 (2013) as the quality to elicit belief or trust in its long-term performance.

Generally, a simulation model is evaluated by verification, validation, and accreditation (VV&A) (Balci, 1997) activities to decide whether a model is able to work correctly, whether it is consistent with the simulation purpose and whether it is adaptable to the practical use. As the verification can be accomplished by some logical checking methods introduced from software engineering (Berard et al., 2001; Sargent, 2013) and the accreditation is carried by users or experts' documentation (Gass, 1993; Sargent et al., 2015), most research attention in modeling and simulation (M&S) has been paid on model validation (Sargent, 2015; Schmidt, 2006; Teknomo, 2016).

At present, there are three main methods to evaluate the credibility of simulation models: quantitative analysis, qualitative analysis, and comprehensive analysis. The existing qualitative methods for validating complex simulation systems are based on questionnaire design and expert scoring (Bai et al., 2017; Ma et al., 2017; Teferra et al., 2014). On one hand, a questionnaire is designed for a specific model, which is subjective and hard to extend. On the other hand, experts should be certificated with high professional knowledge to perform scoring.

Quantitative analysis refers mainly to the consistency validation of simulation results with the real-world objects (Feinstein and Cannon, 2001; Huang et al., 2013; Pater et al., 2014). The measurement of accuracy between model outputs and real data is made according to different sorts of performance criteria. It has high accuracy and strong objectiveness. However, quantitative analysis usually requires a large number of reference data from real world, which is rare or even unavailable for most simulation objects. Moreover, many important internal and external features other than the simulation results have yet been considered.

In comprehensive analysis, some intelligent algorithms, such as Bayesian algorithms (Mahadevan and Rebba, 2005), metaheuristics (Chiappone et al., 2016) and neural networks (NNs) (Zhu et al., 2015), and so on, are introduced with a large amount of historical simulation data to mapping the nonlinear relationships between model validation metrics (Liu et al., 2008) and subject matter expert scores. However, researches on mining model internal factors account for its overall credibility are still limited.

According to the above-mentioned problems, this chapter presents a broader framework to unify the complete process of model credibility measurement and provides a guideline for the user to evaluate different sorts of underlayer models in a composite complex system. First of all, the NASA Standard on M&S (2013) is introduced to guide the establishment of an indicator system for a model by its internal and external available features. Afterward, the existing qualification, quantification, and intelligent methods are classified and incorporated to map the model features, the historical data, and the simulation results to different indicators (i.e., credibility metrics). Last but not the least, we elaborate how to integrate these indicators to finally generate an objective credibility score for the user to understand whether the model is credible enough and how to improve it.

2 RELATED WORKS

Quantitative analysis, qualitative analysis, and comprehensive analysis which combine both quantitative study and qualitative study are widely used to validate the credibility of simulation models. Quantitative analysis is to establish a mathematical model to represent the relationships between the simulation results and the evaluation indicators. Qualitative analysis is a kind of method which depends on rigorous techniques for gathering high-quality data and the credibility of experts. Comprehensive analysis research is set up by using different learning algorithms. There is a great debate between those evaluation methods in early literature (Beck, 1993; Williams et al., 2004). In practice, these three methods have different advantages and disadvantages (Dubois and Prade, 1998). A consensus has been gradually reached that the important challenge is how to match appropriate methods to empirical questions and issues.

Most of the research on qualitative analysis is based on questionnaire design and expert scoring. For example, Schruben (1980) proposed the Turing test method that is based on relevant experts to process output data from a simulation model. Beydoun et al. (2013) developed a set of evaluation methods for models based on experts scoring, simulation requirements, and simulation environment. Buchmann et al. (2016) analyzed the relationship between agent heterogeneity, model structure, and the detailed data used to represent model performance. Regarding the mutual trust of agents, Schreiber designed a multiagent network model for the coevolution of agents (Schreiber and Carley, 2013). Schmidt et al. (2007) established a method of calculating the credibility of a certain model based on fuzzy theory. However, Velayas and Levary (1987) indicated that this method has a strong subjectivity; qualitative aspects of the evaluation and the actual use of emptied weight may lack the expert scoring. In addition, most of the qualitative methods are time consuming and inextensible. Therefore, qualitative analysis is not widely used in isolation.

Currently, quantitative analysis research on evaluating the credibility of a simulation model is based on the traditional evaluation method and the comparison between the simulation data and the practical data (Sarin et al., 2008). For example, Acar (2015) indicates that the prediction capability of metamodeling can be improved by combining various types of models in the form of a weighted average ensemble in order to minimize the root mean square cross-validation error (RMSE-CV) and the root mean square error (RMSE). To ensure the performance, stability, and security of simulation models, researchers provide a set of general indices used for credibility evaluation of simulation model by using mathematical error, information theory, parameter estimation, nonparametric test, and distance judgments (Hora and Campos, 2015; Sterling and Taveter, 2009). These indices can be used not only for the multiagent simulation model evaluation in the fields of transportation, manufacturing, economy, and so on (Lee et al., 2015; Papalambros et al., 2010; Schoenharl and Madey, 2008), but also for the evaluation of machine learning algorithms, such as fitting NNs and clustering algorithms (Liu et al., 2013; Vinh et al., 2010; Wang and Chen, 2014). However, in these studies, the opinions of experts in the relevant scientific research are ignored. So it can only be used for models with a high level of data integrity and consistency. In some models, such as the simulation workflow, quantifiable indices are on a small scale (Louzao et al., 2011). It is hard for quantitative analysis to evaluate the simulation credibility, especially in the strategic analysis of a composed model with insufficient simulation data.

Comprehensive analysis is a newly emerging method for the validation of simulation models. It combines the subjective expert scoring and the objective calculation of model performances by using historical training data. Typical examples include the credibility evaluation theory based on probability and evidence (Ferson, 2009; Zeigler and Nutaro, 2016), fuzzy set theory (Liu and Yang, 2009; Martens et al., 2007), multiple attribute decision-making theory (Papalambros et al., 2010), knowledge-based system (Min et al., 2010), and so on. Considering the frequent changes in simulation requirements and the complex mechanism of real objects, a number of methods on evaluating the credibility of simulation model based on stochastic probability distributions have been proposed in recent years, such as validation based on cumulative density function comparison, and so forth (Bogojevi and Anin, 2016; Velayas and Levary, 1987). Ferson (2009) and Ferson et al. (2008) designed the u-pooling region index. Li et al. (2014) proposed multivariate probability integral transformation (PIT). Chan (2011) and Chan et al. (2010) investigated the use of interaction statistics as a metric for detecting emergent behaviors from the agent-based simulation. Dornheim et al. proposed a hybrid linear expectation model to calculate the reliability of complex system automatically and efficiently (Dornheim and Brazauskas, 2011). Liang et al. (2014) proposed a new method of reliability measurement which is based on a dynamic Bayesian network. However, research on the impact of internal factors of using comprehensive analysis is still limited.

In general, the research on validation technology of simulation system has matured. Many standards and documents are formed while the study of model credibility evaluation is also in progress. The biggest challenge has become how to efficiently and effectively select specific indicators and the most suitable methods to form a comprehensive credibility evaluation. Currently, none of the existing methods has considered the relationship construction between the model credibility and its internal factors other than its simulation results. Few studies have considered the management issue, evolutionary dynamics, and uncertain characteristics of a simulation model.

Hence, a guideline for model credibility measurement under a dynamic requirement is important and necessary to a complex system.

3 A GENERAL FRAMEWORK FOR MODEL CREDIBILITY MEASUREMENT

In this section, we briefly describe the life cycle of a reusable simulation model in an ever-changing environment. Based on the NASA-STD-7009 standard, an indicator system is designed to demonstrate a full-dimensional view on the issues which are able to influence model credibility directly. By selecting available indicators, the existing methods are incorporated into five steps to generate a comprehensive evaluation process on model credibility measurement.

3.1 The Life Cycle of a Reusable Model

Generally, the life cycle of a model is drawn by its requirement, design, construction, VV&A, application, and maintenance states. Due to the reuse scheme in model engineering, the life cycle of a reusable model for a complex system can be specialized as the whole process since it is initiated from a prototype and until it is expired from the current simulation environment, as shown in the top layer of Fig. 1.

A reusable model defines the model which is capable of being applied to another process after the current simulation is finished. To overcome the domain barrier, the reusable model should embrace the most common features across multiple domains and be able to assemble

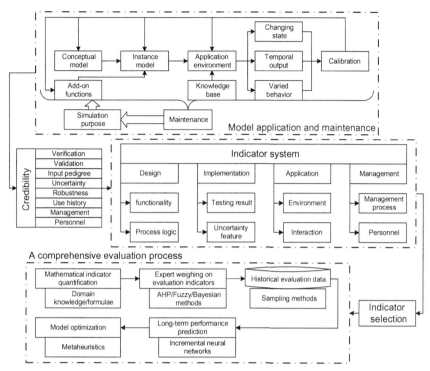

FIG. 1   A comprehensive evaluation process on model credibility measurement.

with different domain functions. Therefore, we define its basic prototype with common features as the conceptual model and the domain-dependent feature as the add-on function. For example, the pedestrian model in either a railway station or a street belongs to the same object and can be crossly applied for different simulation purposes. By combining the conceptual model and some add-on functions, an instance model is established to represent an application version of the reusable model.

In a particular simulation system, a model should act in line with its knowledge rules and perform a systematic evolutionary process with temporal outputs, changing states, and varied behaviors. A calibration strategy is also introduced as a selective stage to adjust the application environment, the instance model, and the conceptual model for the new requirement. After the simulation is finished, a maintenance mechanism is carried out to update the instance model, and store some critical information for further reuse.

Because the instance model is constructed based on the conceptual model and the add-on functions, operated in a dynamic interactive environment and driven by its historical data, the instantiation, application, and maintenance steps together determine whether the reusable model is credible for the current simulation purpose from different angles.

3.2 An Indicator System for Model Credibility Measurement

Focusing on credibility evaluation, the NASA-STD-7009 provides eight factors to describe the fidelity of a model, which are verification, validation, input pedigree, result uncertainty, result robustness, use history, M&S management, and people qualification. They cover fully the main issues during the model life cycle. Thus the standard can be seen as a reference on establishing multidimensional metrics for different models.

As demonstrated in the middle layer of Fig. 1, an indicator system for the credibility measurement should be able to cover the design, implementation, application, and maintenance features of a model.

In the first place, the verification is executed in the design stage in which the processing logic and its related functionalities are tested. It is normally represented by a Boolean variable to decide whether the model is right logically and functionally. The validation is placed in the implementation stage to make sure the long-term performance of a model satisfies the simulation requirement in a steady environment. It involves the consistency of some testing results corresponding to the practical demands, the capability maturity, the matching degree to the target object, the fitness of different objectives, the violation of constraints, and so on.

In the second place, the input pedigree is measured in the application stage to find whether some knowledge base or environmental data is introduced to drive model execution. The uncertainty factor is contained in both the implementation stage and the application stage to embody the static and dynamic features of a model. In the implementation stage, the instantiation of a model involves a lot of probabilistic parameters and several response rules, which enforces the model output change over time. In the application stage, more complex environmental influences and interactions produce much more varying states and behaviors to the model. Besides, the robustness factor which is reflected in the application stage should be evaluated if there exists distinct disturbance in the simulation environment.

In the third place, the use history of a model can be introduced to assess its success rate, quality, and degree of adaptation on the current simulation purpose, if available. This factor further includes historical user scoring, historical application domain, use of frequency, etc. Last but not the least, the M&S management factor and the personnel factor reveals the overall quality and professionality of a model from another side. They can be measured by the degree of standardization, the qualification level, the completeness of the related documentation, the update period, and so forth.

The mapping between the eight standard factors and the evaluation scopes is illustrated in Fig. 2. On account of various domain features and simulation environments, the exact metrics for different credibility factors should be set according to the available data extracted from the model. It is noted that not all of these eight factors are necessary when no available data can be obtained from its corresponding scope.

3.3 The Comprehensive Evaluation Process

As is known, the existing VV&A methods are established to evaluate only one aspect of a simulation model or a composite simulation system. To cover all factors mentioned in the NASA-STD-7009 fully, we propose a unified and extensive process to evaluate the credibility of a simulation model, as illuminated in the bottom layer of Fig. 1. Based on an indicator system, it consists mainly of five steps, that is, (1) indicator quantification, (2) indicator weighting, (3) critical data sampling, (4) performance prediction, and (5) model optimization. Steps (1), (2), and (5) are necessary for every model while the other two are selective.

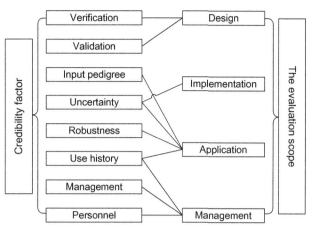

FIG. 2   The evaluation scopes of the eight credibility factors.

(1) Indicator quantification

Indicator quantification means to calculate the value of each metric listed in the indicator system. It includes the estimation of simulation result consistency, the assessment of model capability maturity, the deduction of its uncertainty and sensitivity, the calculation of some other intermediate variables, and so on. This step covers what exactly the existing validation methods do. For instance, statistical testing can be introduced to calculate the consistency of model output with its practical target. Fuzzy set theory can be applied to rate the capability maturity. Bayesian methods can be used to calculate the metrics related to model uncertainty and sensitivity. If a metric is unable to be quantified, the classical qualification methods such as the analytic hierarchy process (AHP) and gray relational analysis (GRA) can then be adopted to incorporate expert experiences and generate a synthetic score for it.

(2) Indicator weighting

To integrate multiple metrics and generate an objective credibility score, the most commonly used way is the weighting scheme with expert scoring, as shown in Eq. (1):

$$E = 100 \sum_{i=1}^{n} \omega_i X_i \tag{1}$$

where E represents the credibility score of a specific model, n refers to the number of metrics in the indicator system, w_i indicates the weights of the ith metric, and it satisfies $\sum_{i=1}^{n} \omega_i = 1$. For more intuitive understanding, the credibility score is usually scaled to the range $[0, 100]$. In this scheme, experts are required to give a rate between every two metrics to determine which of them is more important. The state-of-the-art qualification methods are applied again to provide the value of the weights.

(3) Critical data sampling

To understand the long-term quality of a model throughout its life cycle, the critical data related to its application configuration, its metric values, and the corresponding credibility score should be stored. On one hand, this data is the basis to assess the use history factor. On the other hand, it can be applied as a group of training samples for predicting further model features and states. However, we cannot store all critical data of any single model because of the limit memory space. Then we need the sampling strategy to help us determine which data to be stored. Currently, most typical sampling strategies, such as the Monte Carlo sampling, Gaussian sampling, and Gibbs sampling, are adaptable to select critical samples from hundred times of simulations, so the details would not be covered here.

(4) Performance prediction

To simulate a complex system in a dynamic environment, the model should be able to frequently update its structure and behaviors along with its neighborhood components. Clearly, a model is credible at present does not fully represent that it is credible in its whole life cycle. Therefore, performance prediction is a very important way to understand a model's long-term quality. Especially with the development of multiagent-based simulation (MAS) and system of systems (SoS), it becomes more and more imperative. Although few studies have focused on this topic, existing classification and regression algorithms, such as NNs, support vector machine (SVM), and Gaussian mixture model (GMM), are able to be directly used to construct the relations between the model features, their metric values, and the credibility score.

Once the relations have been established, steps (1) and (2) can be replaced by this training model to directly estimate the credibility score automatically. The evaluation process will then be largely accelerated without cumbersome expert scoring and mathematical quantification.

(5) Model optimization

The credibility score of a model is calculated not only to evaluate whether it is suitable for the current simulation purpose but also to improve it. Model optimization refers to the reverse adjustment and update process to guarantee the high credibility of a model.

If the credibility score is obtained by steps (1) and (2), we can figure out which metric got low value and which got high weight. According to the mapping relations between the metrics and their dependent variables, the drawbacks of the model are easy to be inferred. In this case, a deterministic calibration method can be introduced to adjust the model. If the score is deduced by a prediction method, some critical data of the model and similar ones should be extracted to analyze the reason. Irrespective of the reason for deduction or for the calibration, it is very common and efficient to use metaheuristics, such as the genetic algorithm (GA), particle swarm optimization (PSO), and memetic algorithm (MA), to adjust the model blindly and search for an optimal configuration with a high credibility score.

4 THE CREDIBILITY EVALUATION OF SIMULATION WORKFLOW MODEL

To verify the feasibility of the above comprehensive process, an experimental analysis based on a simulation workflow model is carried out.

4.1 The Simulation Workflow Instance

Simulation workflow is a top-level model for the design and analysis of the complex system. It can be constructed in XML scheme. In this chapter, it is represented as a directed graph $S = (\mathbf{N}, \mathbf{E})$. The node set $\mathbf{N}$ of the directed graph consists of three types, active node N_a, logical node N_l, and event node N_e.

The active node is a detailed description of a short simulation subprocess with a specific environment. It specifies the simulation parameters, interfaces, and prerequisites of the related underlayer model with a group of events.

The logical node is defined as the AND/OR/NOR conditions among different active nodes. It is designed as a compensation of the edges to describe clearly the execution conditions of each active node and make them cooperated in a strict order.

Besides, the event node represents the start event (when the simulation prerequisites are satisfied and the simulation parameters are well configured), the stimulate event (which can be seen as an outside precondition for an active node) and the end event (when all of the active nodes are finished) of a specific process.

With a group of directed edges, these active nodes can be designated to guide the related underlayer models separately in a distributed manner. Typically, it has no input and output data during simulation. Fig. 3 shows an example of the simulation workflow.

4.2 Indicator Quantification and Weighting for the Simulation Workflow Model

To analyze the simulation workflow from a quantitative perspective, we consider mainly 16 internal features, as listed in Table 1, which can be quantitatively calculated from the specific simulation workflow. We directly use the eight factors as evaluation metrics and provide eight simple equations to represent the relation between the features and the metrics.

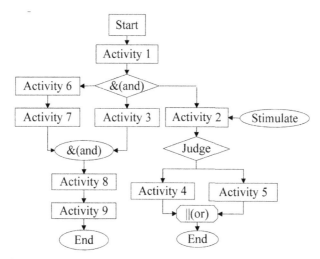

FIG. 3   An example of the simulation workflow.

TABLE 1 Notations of the Internal Features in Simulation Workflow

Symbol	Name	Range	Symbol	Name	Range
P_{match}	Interface matching degree between nodes	[0, 1]	$N_{history}$	Number of historical data	[0, 200]
$P_{integrity}$	Parameter configuration integrity	[0, 1]	P_{hist_cons}	The consistency of the historical configuration	[0, 1]
$\hat{T}$	Estimated execution time(s)	[30, 150]	$N_{stimulate}$	Number of external stimulate events	[0, 10]
$\overline{T}$	Average practical execution time(s)	[30, 150]	N_{para}	Number of incentive parameters	[0, 100]
V_t	Variance of execution time	[0, 3]	N_{ex_para}	Number of external incentive parameters	[0, 20]
N_o	Average number of overtime activities	[0, 100]	P_f	Average failure rate of active node	[0, 1]
N_{active}	Number of active nodes	[0, 100]	P_s	Success rate of historical usage	[0, 1]
N_{logic}	Number of logical nodes	[0, 100]	N_{model}	Number of models linked to the workflow	[0, 100]

Assume the eight evaluation indices to be $\{Xi \in [0,1] | i \in [1,8]\}$. Then the above-mentioned equations can be expressed as follows:

$$X_1 = P_{integrity} \cdot P_{match} \tag{2}$$

$$X_2 = P_{match} \left(1 - \frac{|\overline{T} - \hat{T}|}{\hat{T}} V_t \right) \tag{3}$$

$$X_3 = 1 - \frac{N_{logic} + N_{stimulate}}{N_{logic} + N_{active} + N_{stimulate}} \cdot \frac{N_{ex\_para}}{N_{para}} \tag{4}$$

$$X_4 = P_{integrity} \cdot P_s \cdot e^{-\left(\frac{N_{stimulate} + N_{ex\_para}}{N_{active} + N_{logic} + N_{para}} \right)} \tag{5}$$

$$X_5 = e^{-\frac{N_{model}}{N_{active}}} \tag{6}$$

$$X_6 = P_{hist\_cons} \cdot P_s \cdot e^{-\frac{1}{N_{history}}} \tag{7}$$

$$X_7 = (1 - P_f) \cdot e^{-\frac{N_{model} \cdot N_o}{N_{active}^2}} \tag{8}$$

$$X_8 = P_{hist\_cons} P_{integrity} P_s (1 - P_f) \tag{9}$$

It should be noticed that all of these equations can be replaced or modified into any other forms in accordance with the simulation objects and the working environment of the model.

To further obtain the relationship between these evaluation indices and the final credibility value, we adopt the classical AHP algorithm to incorporate expert scoring on the weighting process. According to the mechanism of AHP, experts need to judge the importance of the eight indices and finish the judgment matrix as shown in Table 2. The mathematics method is used to test the consistency of each matrix and obtain the eigenvector, so that the weight relationship of each factor can be obtained. Due to the limited space, the process of AHP will not be repeated here.

Clearly, it requires two steps to structure the two-layer relations between the workflow features, the evaluation metrics, and the credibility value. Both quantitative deductions of the evaluation metrics by some empirical equations and qualitative calculation of their weights based on expert scoring should be carried out for a model. How to establish the direct influence of these features on the final model credibility is still challenging. To solve this problem, we apply two offline learning algorithms and two incremental learning algorithms for efficient validation of simulation workflow in the next section.

4.3 Online Establishment of Empirical Evaluation Model

In order to make full use of the historical data and implement more efficient validation, we adopt two offline algorithms, that is, single hidden layer back-propagation (BP) NN and extreme learning machine (ELM) (Huang et al., 2006), and two online algorithms, that is, evolving neofuzzy neuron (eNFN) (Silva et al., 2014) and fast incremental Gaussian mixture model (FIGMM) (Pinto and Engel, 2015), to train the empirical evaluation model.

In this section, a landing simulation workflow for the aircraft is adopted to verify the performance of the proposed procedure and compare the four selected learning algorithms in generating the empirical evaluation model. There are a total of 2000 historical data for aircraft landing with different environment and different workflow structures. All the data come from a real simulation system, which includes hundreds of simulation workflows for different aircraft with changing flight missions.

Table 3 provides an evaluation sample for a specific simulation workflow. According to Eq. (2)–(9), the eight evaluation indices can be scored as shown in Table 4.

In qualitative analysis, the experts need to judge the importance of the eight metrics according to the basic AHP. Take the scoring case shown in Table 5 as an instance. The eigenvectors of the eight subfactors are 0.1175, 0.1107, 0.1412, 0.0989, 0.1248, 0.0831, 0.1507, and 0.1731, which are the final weights of them. By Eq. (9), the final credibility is 90.25. Then the data from Table 3 can be used as the input and the final credibility as the output of the training and testing samples.

Based on these historical evaluation data, the prediction results of BP and ELM are shown in Table 6.

It can be seen that BP and ELM have the great ability of prediction when the amount of training data are large. It is clear that the prediction error of the ELM is smaller than that of BP as the training data source is a significant amount, yet BP can show a better-fitting effect

TABLE 2 Judgment Matrix of Eight Subfactors

Score	Completeness	Accuracy	Independence	Uncertainty	Robustness	Historical Use	Reliability	Reproducibility
Completeness	1	2						
Accuracy	0.5	1						
Independence			1					
Uncertainty				1				
Robustness					1			
Historical use						1		
Reliability							1	
Reproducibility								1

TABLE 3   An Evaluation Sample for a Specific Workflow

Index	Value
P_{match}	1
$P_{integrity}$	0.9039
$\hat{T}$	142.85
$\overline{T}$	143.71
V_t	3.5152
N_o	3
N_{active}	27
N_{logic}	6
$N_{history}$	24
P_{hist_cons}	0.9632
$N_{stimulate}$	5
N_{para}	21
N_{ex_para}	7
P_f	0.0392
P_s	1
N_{model}	3

TABLE 4   Quantitative Scores of the Eight Evaluation Indices

	Quantitative Value	On 100 Scale
Completeness	0.9039	90
Accuracy	0.8847	88
Independence	0.9035	90
Uncertainty	0.8007	80
Robustness	0.8948	89
Historical use	0.9239	92
Reliability	0.9490	94
Reproducibility	0.9254	93

TABLE 5 Judgment Matrix of the Eight Evaluation Indices

Score	Verification	Accuracy	Input	Uncertainty	Robustness	Historical Data	People	Management
Verification	1	1.277	0.783	1.63	0.783	1.277	0.783	0.613
Accuracy	0.783	1	0.783	1.63	0.783	1.277	0.783	0.613
Input	1.277	1.277	1	1.277	1.63	1.63	0.783	0.783
Uncertainty	0.613	0.613	0.783	1	0.783	1.63	0.783	0.613
Robustness	1.277	1.277	0.613	1.277	1	1.63	0.783	0.783
Historical data	0.783	0.783	0.613	0.783	0.613	1	0.613	0.481
People	1.277	1.277	1.63	1.277	1.277	1.63	1	0.783
Management	1.63	1.63	1.277	1.63	1.277	2.08	1.277	1

TABLE 6 Experiment Results of BP and ELM

Name	Amount of Training Data	Amount of Testing Data	Average Prediction-Error	Average Prediction-Error in Percentage	Percentage of Prediction-Error > 2	Percentage of Prediction-Error > 5
BP	1900	100	0.6175	0.7313	0.03	0
ELM	1900	100	0.4511	0.5430	0.01	0
BP	1500	500	0.9434	1.131	0.110	0.002
ELM	1500	500	0.4434	0.547	0.014	0
BP	50	1950	1.9404	2.6678	0.3508	0.04221
ELM	50	1950	2.7028	3.694292	0.605	0.1292

as the data source is limited. The average percentage of prediction error is maintained at about 1%, which is certainly in the great ability of prediction. BP can stabilize the prediction-error value by 3% though the data source is limited. In summary, BP and ELM can be a good simplification way of calculating the credibility of simulation workflow instead of experts.

Similarly, the experimental results of eNFN are shown in Figs. 4–8, while the results of FIGMM are shown in Fig. 9.

As shown in Fig. 4, the prediction error decreases as the amount of input data increases at the beginning, the prediction error will be stable in a certain area when it is decreased to a certain extent, Due to the limited number of samples, the prediction error are not as small as the offline learning algorithms after the stability. Fig. 5 shows the prediction error of 500 data sets based on this algorithm. Fig. 6 shows the prediction error of last 1500 datasets based on this algorithm. The absolute error of the average error is about 3.9125.

The above experimental results are obtained using the triangular membership function. We can also use the Gaussian curve membership function to replace the triangular member-ship function as a comparison task. Fig. 7 shows the prediction error of 500 data sets based on the Gaussian curve membership function. Fig. 8 shows the prediction error of the last 1500

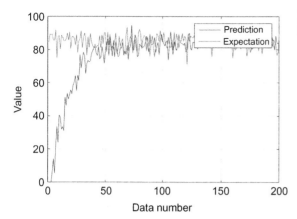

FIG. 4 Output value of eNFN with triangle membership functions.

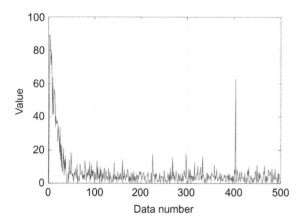

FIG. 5 Prediction error (absolute value) of eNFN with triangle membership functions.

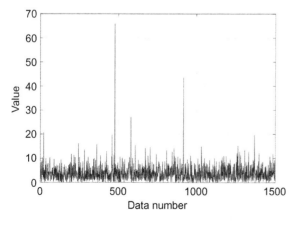

FIG. 6 Prediction error (absolute value) of eNFN with triangular membership functions.

FIG. 7   Prediction error (absolute value) of eNFN with Gaussian curve membership functions.

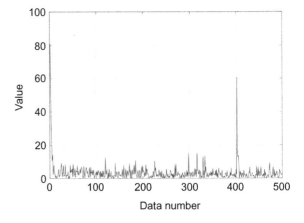

FIG. 8   Prediction error (absolute value) based on the evolving neoneuron algorithm after stabalized (Gaussian curve membership functions).

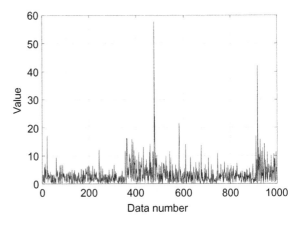

FIG. 9   Prediction value of FIGMM.

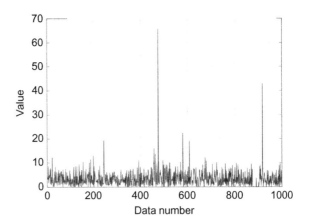

data sets based on this algorithm. The mean prediction error based on the membership function of the Gaussian curve is slightly reduced to 3.221 compared with the triangle membership function.

Compared with the above eNFN, the mean prediction error of FIGMM is 3.5574 as shown in Fig. 9, which is slightly smaller than the triangle membership function-based eNFN and larger than the Gaussian membership function-based one.

Compared with these four learning algorithms, the results of offline learning algorithms are in high precision, but it cannot improve their learning ability as the database increased or the evaluation environment changed. On the contrary, incremental learning algorithms allow new data to update the empirical model and make the prediction shift with time. Although the prediction error is increased to some extent, it is much more efficient in such evaluation tasks and can be extended to more dynamic circumstances.

To be more specific, the results show that BPNN and ELM can be a good way of simplification for calculating the credibility of simulation workflow without expert scoring. ELM shows the highest accuracy while the average prediction error is about 0.45. BPNN shows a high accuracy with an average prediction error of 1.94. Although FIGMN and ENFN have bigger prediction errors, they are still in great learning ability.

5 CONCLUSION

This chapter focused on incorporating multiple VV&A methods and the state-of-the-art intelligent method to evaluate the credibility of a simulation model. Considering mainly the simulation requirement of a complex system and its dynamic environment, the critical issues that influence model credibility during its configuration, application, and maintenance procedures are elaborated. Based on a NASA standard, we illustrated how to establish a complete indicator system to cover different features of a simulation model and discussed the different situations. More importantly, a comprehensive process for model credibility measurement was established and detailed. It is able to unify the evaluation process of different simulation models, largely accelerate it, and extend it to make a further model improvement. A case study based on a simulation workflow model was also carried out to verify the feasibility of such a process.

References

Acar, E., 2015. Effect of error metrics on optimum weight factor selection for ensemble of metamodels. Expert Syst. Appl. 42 (5), 2703–2709.

Bai, J., Wang, L., Zhang, G., Zhang, R., 2017. In: Improved feature selective validation in evaluating uncertainty analysis results in EMC simulation. Applied Computational Electromagnetics Society Symposium (ACES), 2017 International. IEEE.

Balci, O., 1997. In: Verification validation and accreditation of simulation models. Proceedings of the Simulation Conference, 1997.

Beck, C.T., 1993. Qualitative research: the evaluation of its credibility, fittingness, and auditability. West. J. Nurs. Res. 15 (2), 263–266.

Berard, B., Bidoit, M., Finkel, A., Laroussinie, F., Petit, A., Petrucci, L., Schnoebelen, P., 2001. Systems and Software Verification: Model-Checking Techniques and Tools. Springer, Berlin.

Beydoun, G., Low, G., Bogg, P., 2013. Suitability assessment framework of agent-based software architectures. Inf. Softw. Technol. 55 (4), 673–689.

Bogojevi, N., Anin, V.L., 2016. The proposal of validation metrics for the assessment of the quality of simulations of the dynamic behaviour of railway vehicles. Proc. Inst. Mech. Eng. F J. Rail Rapid Transit 230 (2), 585–597.

Buchmann, C.M., Grossmann, K., Schwarz, N., 2016. How agent heterogeneity, model structure and input data determine the performance of an empirical ABM—A real-world case study on residential mobility. Environ. Model. Softw. 75, 77–93.

Chan, W.K.V., 2011. In: Interaction metric of emergent behaviors in agent-based simulation.Winter Simulation Conference.

Chan, W.K.V., Son, Y.J., Macal, C.M., 2010. In: Agent-based simulation tutorial-simulation of emergent behavior and differences between agent-based simulation and discrete-event simulation.Simulation Conference.

Chiappone, S., Mauro, R., Sferlazza, A., 2016. Traffic simulation models calibration using speed-density relationship. Expert Syst. Appl. 44 (C), 147–155.

Dornheim, H., Brazauskas, V., 2011. Robust-efficient credibility models with heavy-tailed claims: a mixed linear models perspective. Insur. Math. Econ. 48 (1), 72–84.

Dubois, D., Prade, H., 1998. Possibility theory: qualitative and quantitative aspects. Quantified Representation of Uncertainty & Imprecision, vol. 1, pp. 169–226.

Feinstein, A.H., Cannon, H.M., 2001. In: Fidelity, verifiability, and validity of simulation: constructs for evaluation. Developments in Business Simulation and Experiential Learning.

Ferson, S., 2009. Validation of imprecise probability models. Int. J. Reliab. Saf. 3 (1), 1–22.

Ferson, S., Oberkampf, W.L., Ginzburg, L., 2008. Model validation and predictive capability for the thermal challenge problem. Comput. Methods Appl. Mech. Eng. 197 (29–32), 2408–2430.

Gass, S.I., 1993. Model accreditation: a rationale and process for determining a numerical rating. Eur. J. Oper. Res. 66 (2), 250–258.

Hora, J., Campos, P., 2015. A review of performance criteria to validate simulation models. Expert. Syst. 32 (5), 578–595.

Huang, G.B., Zhu, Q.Y., Siew, C.K., 2006. Extreme learning machine: theory and applications. Neurocomputing 70 (1), 489–501.

Huang, Z., Du, P., Kosterev, D., Yang, S., 2013. Generator dynamic model validation and parameter calibration using phasor measurements at the point of connection. IEEE Trans. Power Syst. 28 (2), 1939–1949.

Jiao, P., Tang, J.B., Zha, Y.B., 2007. Amelioration and application of similar degree method for simulation credibility evaluation. J. Syst. Simul. 19 (12), 2658–2660.

Law, A.M., Kelton, W.D., 2000. Simulation modeling and analysis. J. Am. Stat. Assoc. 275 (3), 248–293.

Lee, J.S., Filatova, T., Ligmann-Zielinska, A., Hassani-Mahmooei, B., Stonedahl, F., Lorscheid, I., Voinov, A., Polhill, G., Sun, Z., Parker, D.C., 2015. The complexities of agent-based modeling output analysis. J. Artif. Soc. Soc. Simul. 18 (4), 1–4.

Li, W., Chen, W., Jiang, Z., Lu, Z., Liu, Y., 2014. New validation metrics for models with multiple correlated responses. Reliab. Eng. Syst. Saf. 127 (6), 1–11.

Liang, J., Bai, Y., Bi, C., Sun, Z., Yan, C., Liang, H., 2014. In: Adaptive routing based on Bayesian network and fuzzy decision algorithm in delay-tolerant network.IEEE International Conference on High PERFORMANCE Computing and Communications & 2013 IEEE International Conference on Embedded and Ubiquitous Computing.

Liu, F., Yang, M., 2009. An optimal design method for simulation verification, validation and accreditation schemes. Society for Computer Simulation International.

Liu, F., Yang, M., Wang, Z., 2008. VV&A solution for complex simulation systems. Int. J. Simul. Syst. Sci. Technol. 9 (1), 21–29.

Liu, M.Y., Tuzel, O., Ramalingam, S., Chellappa, R., 2013. Entropy-rate clustering: cluster analysis via maximizing a submodular function subject to a matroid constraint. IEEE Trans. Pattern Anal. Mach. Intell. 36 (1), 99–112.

Louzao, M., Pinaud, D., Péron, C., Delord, K., Wiegand, T., Weimerskirch, H., 2011. Conserving pelagic habitats: seascape modelling of an oceanic top predator. J. Appl. Ecol. 48 (1), 121–132.

Ma, Y., Lu, L., Lu, J.J., 2017. Safety evaluation model of urban cross-river tunnel based on driving simulation. Int. J. Inj. Control Saf. Promot. 24 (3), 1–10.

Mahadevan, S., Rebba, R., 2005. Validation of reliability computational models using Bayes networks. Reliab. Eng. Syst. Saf. 87 (2), 223–232.

Martens, J., Put, F., Kerre, E., 2007. A fuzzy-neural resemblance approach to validate simulation models. Soft. Comput. 11 (3), 299–307.

Min, F.Y., Yang, M., Wang, Z.C., 2010. Knowledge-based method for the validation of complex simulation models. Simul. Model. Pract. Theory 18 (5), 500–515.

NASA-STD-7009, 2013. NASA Technical Standards System (NTSS). National Aeronautics and Space Administration, Washington, DC.

Papalambros, P., Barbat, S., Yang, R.J., 2010. Comparing time histories for validation of simulation models: error measures and metrics. J. Dyn. Syst. Meas. Control. 132 (6), 768–778.

Pater, P., Seuntjens, J., Naqa, I.E., Bernal, M.A., 2014. On the consistency of Monte Carlo track structure DNA damage simulations. Med. Phys. 41(12):121708.

Pinto, R.C., Engel, P.M., 2015. Correction: a fast incremental Gaussian mixture model. PLoS One 10(10):e0141942.

Sargent, R.G., 2013. Verification and validation of simulation models. Eur. J. Oper. Res. 7 (1), 12–24.

Sargent, R.G., 2015. An interval statistical procedure for use in validation of simulation models. J. Simul. 9 (3), 232–237.

Sargent, R.G., Goldsman, D.M., Yaacoub, T., 2015. In: Use of the interval statistical procedure for simulation model validation.Winter Simulation Conference.

Sarin, H., Kokkolaras, M., Hulbert, G., Papalambros, P., Barbat, S., Yang, R.J., 2008. A Comprehensive Metric for Comparing Time Histories in Validation of Simulation Models With Emphasis on Vehicle Safety Applications. Lapland University of Applied Sciences, Kemi, pp. 1275–1286.

Schmidt, D.C., 2006. Model-driven engineering. IEEE Comput. Soc. 39 (2), 25.

Schmidt, S., Steele, R., Dillon, T.S., Chang, E., 2007. Fuzzy trust evaluation and credibility development in multi-agent systems. Appl. Soft Comput. 7 (2), 492–505.

Schoenharl, T.W., Madey, G., 2008. In: Evaluation of measurement techniques for the validation of agent-based simulations against streaming data.International Conference on Computational Science.

Schreiber, C., Carley, K.M., 2013. Validating agent interactions in construct against empirical communication networks using the calibrated grounding technique. IEEE Trans. Syst. Man Cybern. B 43 (1), 208–214.

Schruben, L.W., 1980. Establishing the credibility of simulations. Simulation 34 (34), 101–105.

Silva, A.M., Caminhas, W., Lemos, A., Gomide, F., 2014. A fast learning algorithm for evolving neo-fuzzy neuron. Appl. Soft Comput. 14 (1), 194–209.

Standards.nasa.gov, 2013. NASA-STD-7009 | NASA Technical Standards System (NTSS). Available from: https://standards.nasa.gov/standard/nasa/nasa-std-7009. Accessed 14 March 2018.

Sterling, L., Taveter, K., 2009. The Art of Agent-Oriented Modeling. The MIT Press, Cambridge, MA.

Teferra, K., Shields, M.D., Hapij, A., Daddazio, R.P., 2014. Mapping model validation metrics to subject matter expert scores for model adequacy assessment. Reliab. Eng. Syst. Saf. 132 (132), 9–19.

Teknomo, K., 2016. Microscopic Pedestrian Flow Characteristics: Development of an Image Processing Data Collection and Simulation Model. Tohoku University, Sendai, Japan.

Velayas, J.M., Levary, R.R., 1987. Validation of simulation models using decision theory. Simulation 48 (3), 87–92.

Vinh, N.X., Epps, J., Bailey, J., 2010. Information theoretic measures for clusterings comparison: variants, properties, normalization and correction for chance. J. Mach. Learn. Res. 11, 2837–2854.

Wang, Y., Chen, Y., 2014. A comparison of Mamdani and Sugeno fuzzy inference systems for traffic flow prediction. J. Comput. 9 (1), 12–21.

Williams, J., Ryan, J., Hadjidemetriou, C., Misailidou, C., Afantiti-Lamprianou, T., 2004. Credible Tools for Formative Assessment: Measurement AND Qualitative Research Needed for Practice. Warrencountyschools Org.

Zeigler, B.P., 1990. Object-Oriented Simulation With Hierarchical, Modular Models: Intelligent Agents and Endomorphic Systems. Academic Press Professional, Inc., San Diego.

Zeigler, B.P., Nutaro, J.J., 2016. Towards a framework for more robust validation and verification of simulation models for systems of systems. J. Def. Model. Simul. 13 (1), 3–16.

Zeigler, B.P., Praehofer, H., Kim, T.G., 2000. Theory of Modeling and Simulation: Integrating Discrete Event and Continuous Complex Dynamic Systems. Academic Press, USA (Colección Libros Y Materiales Educativos).

Zeigler, B.P., Zhang, L., 2015. Service-Oriented Model Engineering and Simulation for System of Systems Engineering. Springer International Publishing, Switzerland.

Zhu, M., Chen, X., Luo, X., Liu, F., 2015. Software credibility assessment model based on back propagation network. Int. J. Comput. Syst. Eng. 2 (2), 66.

10

Quality Assessment and Quality Improvement in Model Engineering

U. Durak*, I. Stürmer[†], T. Pawletta[‡], S. Mahmoodi[§]

*German Aerospace Center (DLR), Braunschweig, Germany †Model Engineering Solutions UK Ltd, London, UK ‡Wismar University of Applied Sciences, Wismar, Germany §Clausthal University of Technology, Clausthal-Zellerfeld, Germany

1 INTRODUCTION

Model engineering is an emerging discipline addressing the whole modeling life cycle, aiming at the same time for low system and software development costs, and high product quality. This can be achieved with a systematic, standardized, and quantifiable methodology that consists of theories, processes, technologies, standards, and tools (Zeigler and Zhang, 2015; Zhang et al., 2014). With modeling- and simulation-based approaches to systems and software engineering on the rise, model quality has become an integral part of system quality (D'Ambrogio and Durak, 2016). While system quality assessment and improvement has been well addressed, the model quality issues still lack the attention they deserve.

A system architecture is defined as the blueprint of a system or systems, which enables systems engineers to visualize the proposed systems, analyze the problem, and specify the solution architecture (Wang and Dagli, 2008). Executable system architecture models are becoming increasingly popular (Tolk and Hughes, 2014). From the early steps of systems engineering, simulation of architecture models are used to conduct analysis through computational experimentation. The importance of model-centric engineering can be seen in the fact that model-based development has turned into a standard software development approach for various industrial domains, such as the automotive (Broy et al., 2013) and aerospace (Amundson et al., 2015) sectors. ED-218/DO-331 (RTCA/EUROCAE, 2012a), the model-based development and verification supplement to ED-12C/DO-178C Software Considerations in Airborne Systems and Equipment Certification (RTCA/EUROCAE, 2012b), describes two kinds of models: *specification models* and *design models. Specification models* are used for abstract representation of functional requirements, performance characteristics, interface descriptions, or safety properties of software systems, whereas *design models* refer to an abstract specification of the software to be developed on component level. These design models[1] specify, for example, internal data structures, data and control flows, or the software architecture itself. DO-331 promotes simulation for verification of *specification models* and *design models*. Simulation is becoming a crucial part of model-based development, allowing early validation of system and software properties before software implementation has even started.

In this chapter, we presented an overview of model quality assessment and improvement methods and techniques based on the standards and practices from automotive and aeronautics domain. The chapter first introduced the quality aspects in model development where the indicators of model quality as well as the sources of error in model engineering were discussed. We also highlighted the reference workflows for model-based development and the integrated quality assurance approaches for model engineering. Afterward, various model quality assurance procedures were discussed one by one. Types of constructive procedures, such as process improvement, modeling guidelines, model checking and repair, model refactoring, and tool qualification were introduced using references from recent literature and examples from our own research where appropriate. In terms of analytical procedures, the verification in model-based development was examined, and then complemented with model-based testing.

2 QUALITY ASPECTS OF MODEL DEVELOPMENT

Although certain communities still use textual modeling, graphical modeling is today an industry standard. In graphical model development, model size, and accordingly model complexity are described using a graphical combination of the number of blocks, their interconnection, subsystems/super blocks and hierarchical levels (Stürmer et al., 2010). The calculation of the resulting *model volume* is based on the Halstead metrics for code, which has been adapted to software models. Stürmer and Pohlheim (2012) further noted that the large models in the automotive domain may reach up to 15,000 blocks, 700 subsystems, and 16 hierarchical levels. The flight dynamics model of the German Aerospace Center (DLR) Advanced Technology Research Aircraft (ATRA) Airbus A320 developed in MATLAB/Simulink is about 25,000 blocks, and 2000 subsystems. Such a scale in graphical modeling makes quality analysis, assessment, and improvement a challenging task. It requires an integrated approach for quality assurance of models as a part of the entire systems development process supported by a proper combination of tools, methodologies, and techniques.

The aim of model quality assurance is to identify model problems as early as possible. The common quality assurance methods that are applied can be classified under three topics: use

---

[1]Design models are also often called *implementation models*, since they define implementation details and are often used as a basis for automatic controller code generation.

of modeling guidelines, manual review of models against requirement specifications, and testing the models with simulation. Besides the model being the core artifact of quality assurance, other artifacts of model engineering, such as textual requirements, test specifications, test reports, and review reports are also addressed.

Fig. 1 presents a graphical representation of the indicators of model quality. It starts with the requirements management where the realization levels of requirements indicate the model quality. In model analysis, model complexity measurements, checking of guideline compliancy, model reviews, and static testing can be listed as key activities for reaching a high model quality. Testing from its specification to its reporting is vital on the quality of the model. Finally, issue tracking is an important quality management task.

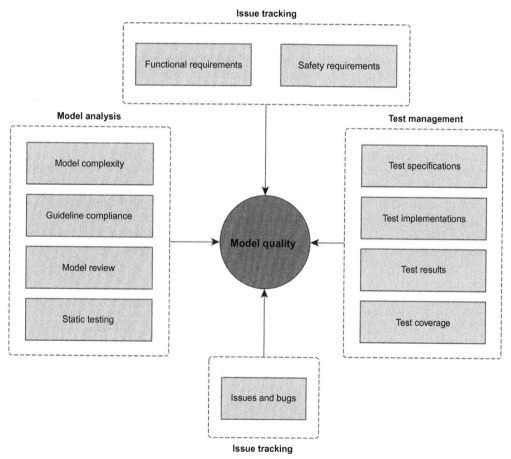

FIG. 1 Indicators of model quality. *Redrawn from Stürmer, I., Pohlheim, H., 2012. Model quality assessment in practice: how to measure and assess the quality of software models during the embedded software development process. In: Embedded Real Time Software and Systems, Toulouse, France.*

2.1 Model-Based Development and Reference Workflow

Today, model-based development is widely applied in various safety-critical domains such as the automotive and aerospace sector. Safety-critical domains are characterized by the strict requirements they impose on the engineering processes. Accordingly, ISO 26262 is the standard for functional safety of electrical, electronical, and software components of road vehicles (ISO, 2011). DO 178C and its European equivalent ED-12C are about software consideration in certification of airborne systems (RCTA/EUROCAE, 2012b). Both ISO 26262 and ED-12C/DO-178C address the model-based development aspects with special annexes that are devoted to this approach. Keeping these standards in mind, a model-based development process is depicted in Fig. 2. The seamless use of executable models is characteristic of systems design in model-based development.

First, the *specification model* is developed based on the textual requirements. It describes the behavior of the system to be developed. It includes algorithm specifications for transformation of input signals, events, and states which are almost always described using floating point arithmetic. The *design model* is developed through revision of the specification model by the implementation experts with regard to the requirements of the production code, such as the realization of fixed point arithmetic or substituting model elements that are not supported by the particular code generator.

Code generation is a model-to-text transformation which results in targeted source code. Eventually, the code is compiled and an object code is reached for a particular system element. The object code is ultimately deployed on a target platform, such as microcontroller (MCU) or digital signal processor (DSP).

Many publications present model-based development workflows, each of which utilizes a particular set of tools. Similar to what has been presented in Fig. 2, Estrada et al. (2013) introduced best practices in the MathWorks ecosystem for DO-178 compliant model-based development. A further example would be from Eisemann (2016) who discusses how dSPACE Target Link (dSPACE, 2017) and tools by BTC Embedded Systems (2017), namely BTC EmbeddedSpecifier and BTC EmbeddedTester, can be integrated into a model-based

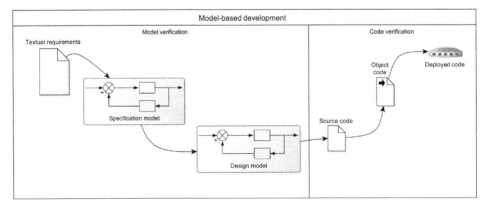

FIG. 2   Model-based development workflow.

development workflow based on MathWorks products to achieve up to highest ED-12C/DO-178C Design Assurance Level.

2.2 Sources of Error in Model Development

The sources of error in model development were listed by Stürmer et al. (2005) as design errors, arithmetic errors, tool errors, hardware errors, runtime errors, and interface errors. Design errors are caused due to inappropriate construction of a model. It can be a failure of the specification model not conforming to its requirements, or a design model that does not correspond to its specification model. Arithmetic errors usually arise in design models due to imprecise representation or improper conversions. Tool errors correspond to issues that come up due to bugs in the toolchain or an inappropriate configuration of the tool itself. Any problem in the target environment may lead to hardware errors. Problems with scheduling and/or resource mismatches are categorized under runtime errors. Lastly, interface errors occur due to problems between the generated code and the wrapper software or custom code, such as drivers or the operating system API. All these error types need to be addressed by the activities relating to model quality assurance.

2.3 Integrated Quality Assurance

The assurance of quality in model engineering can be achieved through constructive procedures, such as adaptation of standards and guidelines, and analytical procedures, such as verification and testing.

Constructive procedures aim at assurance that the development is carried out according to a systematic process usually described in standards or guidelines. These procedures try to minimize the possibility of errors. A major standard to be mentioned here is the ISO/IEC/IEEE 15288:2015 (ISO/IEC/IEEE, 2015), which provides a common process framework for describing the life cycle of systems. It focuses on defining stakeholder needs and required functionality early in the development cycle, specifying requirements, then proceeding with design synthesis and system verification and validation, while addressing a problem in its entirety. The life cycle of a system spans the period from the system conception through to its retirement. The current version of the standard is the product of a coordinated effort by the IEEE and ISO/IEC, and replaces the ISO/IEC 15288:2008 (second edition), which was technically revised in conjunction with a corresponding revision of the ISO/IEC/IEEE 12207 (for software life-cycle processes).

Model engineering is regarded as an integral part of systems engineering. INCOSE defines model-based systems engineering (MBSE) as the formalized application of modeling, to support system requirements, design, analysis, verification, and validation throughout the life cycle of systems (INCOSE, 2007). Therefore, the constructive procedures should also be carried out in an integrative fashion. This includes adapting the systems engineering life-cycle process to involve models as artifacts and supporting it with further standards, guidelines, methods, and techniques for model engineering, which also considers further modeling guidelines, model checking, and qualification of tools that are developed for model engineering and further code generation.

Analytical procedures include verification activities. As indicated in Fig. 2, model verification in the model-based development process involves methods and techniques that ensure the modeling flaws have been detected and avoided. Code verification then includes testing the autogenerated code and assuring it conforms to the model.

3 CONSTRUCTIVE PROCEDURES IN MODEL QUALITY ASSURANCE

3.1 Process Improvement

Process improvement is crucial in systems development. Capability Maturity Model Integration (CMMI) is a state of the art, well-employed framework which provides essential practices for process improvement (SEI, 2010). CMMI is one of the quality management tools which target continuous process improvement. There are three types of CMMI: CMMI for development, CMMI for services, and CMMI for acquisition. CMMI for development consists of four categories: project management, engineering, support, and process management. Each category deals with certain process areas. For example, the engineering process areas are requirements development (RD), technical solution (TS), product integration (PI), verification (VER), and validation (VAL). Each process area has specific goals and specific practices to reach those goals. CMMI assesses capability and maturity levels based on how well organizations perform in achieving specific goals and adhering to practices in particular process areas. Integrated model quality assessment requires identifying goals and specifying practices for model engineering within CMMI. Mahmoodi et al. (2017) conducted such a study for simulation engineering. They assessed CMMI Engineering process areas for simulation life-cycle processes in order to optimize its full potential in the simulation domain. This tailored CMMI integrates IEEE Recommended Practice for Distributed Simulation Engineering and Execution Process (DSEEP) (IEEE, 2010) requirements into CMMI engineering process areas to reach a higher level of process coverage and quality in simulation systems engineering, comprehensively covering all its process areas. Similarly, a study should explore the requirements of model engineering and attempt to integrate them into frameworks like CMMI.

3.2 Modeling Guidelines

Modeling guidelines are important for maintainability. They increase the readability of the model, and facilitate expendability, testing, and reuse (Stürmer et al., 2008). Through the use of modeling guidelines, the experience relating to good and bad modeling practices can be collated. They are useful for modelers as a reference for quality assurance tasks (Hu et al., 2012). Guidelines are usually inspected and verified by modeling experts and made available to a wider group of modelers. They prevent regularly occurring problems in model design and can reduce the amount of reworking significantly (Eisemann, 2006).

Among others (Ferrari et al., 2009; Eisemann, 2006; Erkkinen, 2005; Ohata and Komori, 2009), a particularly good example of well-employed guidelines are those specified by the MathWorks Automotive Advisory Board (MAAB) (MAAB, 2015). The guidelines appear in various categories: model layout, arithmetical problems, exception handling and tool or project-specific constraints.

Table 1—taken from Stürmer et al. (2008)—elaborates on these categories.

Model layout guidelines ensure readability, maintainability, and portability. An example layout guideline is depicted in Fig. 3. It mandates all sum blocks must be rectangular. It also states that the size of a sum block should be selected to avoid overlapping input signals.

Further examples can be listed as follows:

- Signals should not cross each other or other blocks.
- In-ports should be located on the left-hand side (LHS) whereas outports should be on the right-hand side (RHS).
- Every element of a model should be connected.

While guidelines about arithmetic operations address typical problems such as avoiding division by zero, exception handling guidelines ensure the robustness of the model. Two

TABLE 1   Categories of Modeling Guidelines

Category	Aim/Goal
Model layout	Increase readability, maintainability, portability
Arithmetical problems	Prevent typical arithmetical problems (e.g., division by zero, fixed-point limitations)
Exception handling	Increase robustness of the model
Tool-specific considerations	Address tool-specific considerations, e.g., ensuring that the model can be tested with model testing tools such as MTest (Model Engineering Solutions, 2007) and TPT (Piketec GmbH, 2017)
Project-specific guidelines	Naming conventions

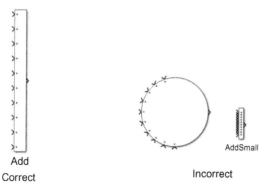

FIG. 3   An example layout guideline from MAAB (2015).

noteworthy example tools that support model engineering are MTest (Model Engineering Solutions, 2007) by Model Engineering Solutions GmbH and TPT (Piketec GmbH, 2017) by Piketec GmbH. There are usually modeling guidelines associated with the use of such tools. MTest and TPT both require that the interface of model modules should be well defined. Finally, there may also be project-specific guidelines, such as naming conventions.

3.3 Enforcement of Modeling Guidelines and Model Repair

Modeling guideline checking involves a static model analysis. It verifies that the model design adheres to selected modeling guidelines. The manual effort required to check guideline compliance is too high. Such checks can be turned into constructive procedures, only if tool support is available.

The two most popular tools in the industry used for enforcing modeling guidelines for Simulink or TargetLink models are MES Model Examiner® and Simulink Model Advisor®. MES Model Examiner®(Stuermer et al., 2014) is the most common tool used for checking Simulink and TargetLink modeling guidelines in the area of safety-critical software development and is particularly tailored to the requirements of the automotive sector. Simulink Model Advisor®(Popinchalk, 2008) is another example of a modeling guideline checking tool. It automatically checks a MATLAB/Simulink model or its subset for some common mistakes. It supports MAAB Guidelines, as well as other model checks associated with safety standards, such as DO-178B/DO-331 or ISO 26262.

Stürmer et al. (2008) identified the automation of model repairing as the crucial challenge. He proposed MATLAB Simulink and Stateflow Analysis and Transformation Environment (MATE) (Stürmer et al., 2007) which complements the analysis capabilities of commercial model checking tools with constructive model quality improvement. The capabilities provided by MATE in 2007 included automatic repair functions for straightforward repair functions; interactive repair functions for ones that require user feedback; design pattern instantiation such as if-then-else or switch-case constructs and model "beautifying" operations for a better model layout.

3.4 Model Refactoring

Model refactoring or reconstruction is listed as one of the key model engineering technologies. Zeigler and Zhang (2015) described it as adjusting the internal structure without changing the external functions of the models with a view to optimization of model performance, understandability, maintainability, and adaptability.

Refactoring has been used in classical software development as an evolutionary modernization technique, in order to incrementally alter the structure of an artifact to achieve a better quality, while keeping its behavior unchanged. Fowler described refactoring as a process of cleaning up the code to improve its design after it has been written (Fowler and Beck, 1999). With refactoring, code is tidied up in order to keep its shape. Chikofsky and Cross (1990) classified refactoring as one of the appearance of restructuring, which is essentially a transformation from one form to another at the same abstraction level while maintaining functionality and semantics.

While common practice is employed over commercial modeling and simulation tools in model engineering activities within the context of model-based design, such as MATLAB/ Simulink, refactoring is a labor intensive and repetitive task. Durak (2016) encouraged model developers to develop modification scripts following a well-established methodology. He proposed a model refactoring approach that is accessible, maintainable, and adoptable by modelers based on pragmatic model transformations for Scilab/Xcos, which is an open source model-based design and simulation environment (Campbell et al., 2010).

The proposed approach by Durak (2016) provides an application programming interface (API) to conduct in place model-to-model transformations for refactoring. The transformation operation involves matching the LHS pattern in the model being transformed and replacing it with the RHS pattern in place (Czarnecki and Helsen, 2006). Pattern specification metamodels for the LHS can be obtained by subjecting the original language metamodel to relaxation, augmentation, and modification (Kühne et al., 2009). Accordingly, Durak proposes an Xcos refactoring metamodel (Fig. 4) for the LHS pattern specification derived from the Xcos metamodel employing relaxation and augmentation. Regular expressions are proposed in order to define the constraints as the values of attributes in LHS pattern structure as an augmentation. Not all the fields of the Xcos metamodel are suitable for constraint definition; the Xcos metamodel is simplified for refactoring purposes as a relaxation. Furthermore, all the data types of the parameter values are specified as strings in order to enable the application of regular expressions in the simplification.

It is suggested that the RHS pattern—that is, the replace pattern—be specified using the same structure as the model conforming to the Xcos metamodel (Fig. 5).

The specification problem in model transformation addresses the definition of the precondition and the postcondition, namely the LHS and the RHS. Czarnecki and Helsen (2003) proposed variables, patterns, and logic that are used to specify LHS and RHS. Variables are defined as the elements from the source and target. Patterns are defined as model fragments with zero or more variables. Lastly, the logic refers to the constraints on the model elements. Durak (2016) suggested the variables of the transformation as the objects of the Xcos diagram with their attributes. Afterward, patterns can be introduced as the composition of these elements. Logic specification that employs regular expressions is recommended.

Following the aforementioned approach for the specification of the patterns, Durak (2016) proposed an API for both atomic operations and the overall transformation process. Atomic functions include finding, adding, deleting, and replacing a block and, similarly, finding, adding, deleting, and replacing a link. Furthermore, getting the list of connected blocks and the list of connecting links between the blocks is also proposed as an atomic function. While these atomic model transformation functions provide the modeler with the building blocks for developing their own algorithms to manipulate or transform their models, the composite model transformation functions find subdiagram, add subdiagram, delete subdiagram, replace subdiagram, and an overall find and replace are proposed for complex refactoring tasks (Table 2).

It would be helpful to demonstrate the proposed approach by using a refactoring scenario. Sometimes, as exemplified in Fig. 6, rather than using a gain block, which multiplies its input signal with the constant value defined as the parameter of the block, modelers use explicitly product blocks to multiply a signal with a constant value. Although it is mathematically correct, gain blocks enhance readability by reducing the number of blocks.

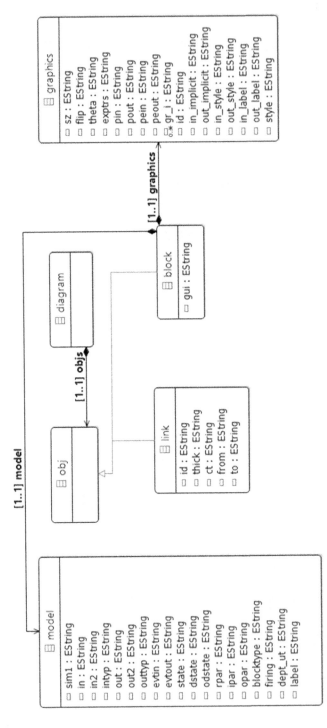

FIG. 4   Xcos refactoring metamodel. *Redrawn from the metamodel presented in Durak, U., 2016. Pragmatic model transformations for refactoring in Scilab/Xcos. Int. J. Model. Simul. Sci. Comput. 7(01), 1541004.*

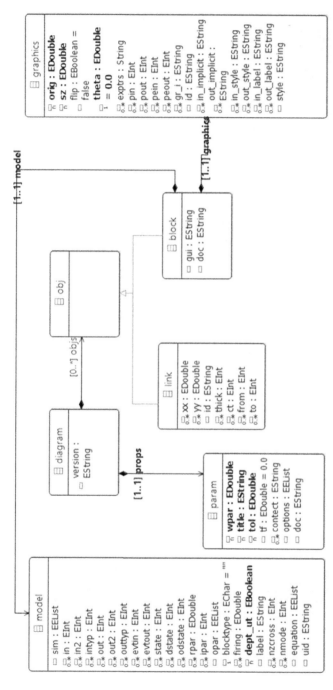

FIG. 5 Xcos metamodel. Redrawn from the metamodel presented in Durak, U., 2016. Pragmatic model transformations for refactoring in Scilab/Xcos. Int. J. Model. Simul. Sci. Comput. 7(01), 1541004.

TABLE 2 Composite Transformation Functions

Function Name	Input Arguments	Output Arguments
Find subdiagram	Diagram	List of indexes for matching blocks
	Pattern to be searched	
	List of indexes for the constraining objects	List of indexes for matching links
	List of constraining attributes List of constraints	
Add subdiagram	Diagram	Updated diagram
	Pattern to be added	
Delete subdiagram	Diagram	Updated diagram
	List of indexes of the blocks to be deleted	
Replace subdiagram	Diagram	Updated diagram
	List of indexes for the blocks to be replaced	
	New structure	
Find and replace	Diagram	Updated diagram
	Pattern to be searched (LHS)	
	List of indexes for the constraining objects (Logic)	
	List of constraining attributes (Logic)	
	List of constraints (Logic)	
	pattern to be inserted (RHS)	

Adapted from Durak, U., 2016. Pragmatic model transformations for refactoring in Scilab/Xcos. Int. J. Model. Simul. Sci. Comput. 7(01), 1541004.

To fix the issue, manual review and refactoring take time and cannot guarantee full coverage. For this refactoring task, the modeler develops a script using the proposed API. The sample listing for the script is given in Listing 1. It calls find_subdiagram to identify the blocks that match the specified pattern. The blocks and the associated links are first removed from the diagram. Then a gain block is added and required links are constructed. The script searches for a matching subdiagram and conducts the replacement operation until no match is found. The resulting model is depicted in Fig. 7.

3.5 Tool Qualification

Referring to the aforementioned categorization for sources of error in model engineering, tool qualification targets possible tool errors and potential incorrect usage of a tool. The constructive procedures for model quality assurance mandate standards for the qualification of tools in modeling engineering as they are identified as a possible source of errors that may influence code quality undetected.

ED-215/DO-330 Software Tools Qualification Considerations (RTCA/EUROCAE, 2012c) is the extension for ED-12C/DO-178C which introduces a software tool for life-cycle processes (Fig. 8).

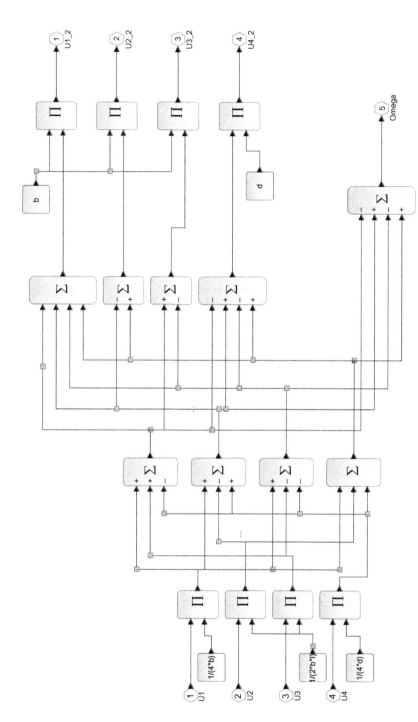

FIG. 6 A model excerpt with the sample issue. *Redrawn from the Scilab/Xcos model presented in Durak, U., 2016. Pragmatic model transformations for refactoring in Scilab/Xcos. Int. J. Model. Simul. Sci. Comput. 7(01), 1541004.*

```
1.  // import the searched sub-diagram
2.  importXcosDiagram('productwithconstantblock.xcos')
3.  searched_subdiagram = scs_m
4.  // import the diagrams to be refactored
5.  importXcosDiagram ('quadrotor.xcos')
6.  dl = get diagrams (scs_m)
7.  // no particular selection constraints
8.  attributelist =[]
9.  constraintlist=[]
10. // for every diagram
11. for i = 1: size ( "dl" )
12.   //get the list of matched blocks and links
13.   [matched_block_list matched_link_list]=find_subdiagram(dl(i),...
14.     searched_subdiagram, attribute_list, constraint_list)
15.   // get the index of const and product
16.   while size (matched_block_list)>0
17.     for j =1: size (matched_block_list)
18.       if dl(i).objs(matched_block_list(j)).gui=="CONST"
19.         const_index = j;
20.       elseif diagram_list(i).objs(matched_block_list(j)).gui=="PRODUCT"
21.         product_index = j
22.       end
23.     end
24.     // get the value for the gain
25.     gain_value = dl(i).objs(matched_block_list(const_index)).model.rpar
26.     // delete the const block
27.     delete_block(dl(i), matched_block_list(const_index))
28.     gain_block = GAINBLK("define") // create a gain block
29.     gain_block.model.rpar = gain_value // set the gain value
30.     // replace it with the product block
31.     replace(d(i), matched_block_list(product_index), gain_block)
32.     // check for any other match
33.     [matched_block_list matched_link_list]=find_subdiagram(dl(i), ...
34.       searched_subdiagram, attribute_list, constraint_list)
35.   end
36. end
```

LISTING 1   Listing for sample refactoring script. *Adapted from Durak, U., 2016. Pragmatic model transformations for refactoring in Scilab/Xcos. Int. J. Model. Simul. Sci. Comput. 7(01), 1541004.*

ED-215/DO-330 enables qualification based on the conformance to its requirements which are listed in the form of objectives and activities. The required qualification level of a tool, namely the tool qualification level (TQL), is based on its use and its potential impact on the software life-cycle process. There are five qualification levels—from TQL1 to TQL5. TQL 1 is the most rigorous that demands coverage of all objectives and requirements while TQL 5 is the least rigorous with minimum objective and requirement coverage. Determination of TQL is described in RTCA/EUROCAE (2012b) in detail.

ISO 26262 also requires the evaluation of each software tool and asks for a corresponding tool qualification (ISO, 2011). Evaluation is conducted to determine a tool confidence level (TCL) based on tool impact (TI), and tool error detection (TD) level. While TI is the possibility

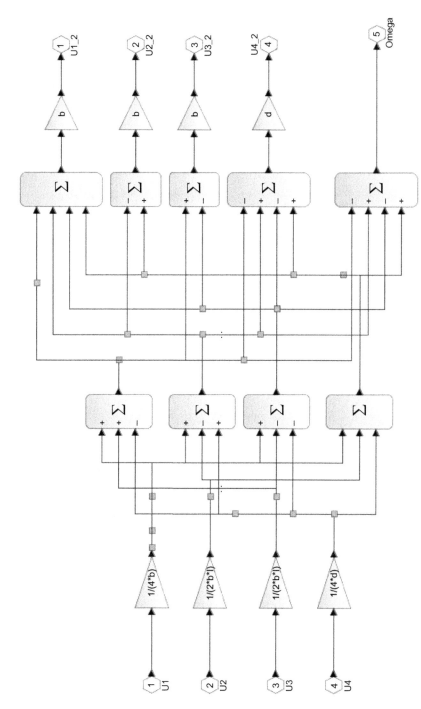

FIG. 7 Excerpt from the model after refactoring. Redrawn from the Scilab/Xcos model presented in Durak, U., 2016. Pragmatic model transformations for refactoring in Scilab/Xcos. Int. J. Model. Simul. Sci. Comput. 7(01), 1541004.

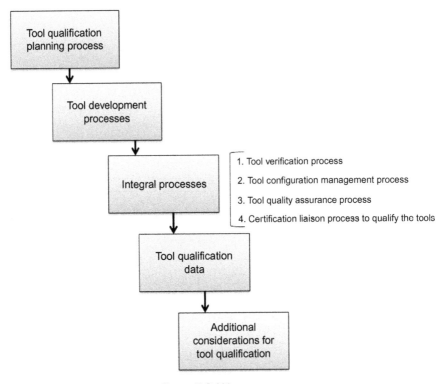

FIG. 8   The tool life-cycle processes according to DO-330.

of tool errors leading to a safety requirement violation, TD is the likelihood of detecting or preventing the error (Slotosch et al., 2012).

Tool qualification is applicable to software development tools, such as compilers or code generation tools, and software verification tools such as static analyzers or simulation models. Within the context of model engineering, tool qualification is a concern for simulation models, model checking and repair tools, however particularly for code generators and model-based testing tools. Various efforts regarding tool qualification have been made. In 2004, Stürmer and Conrad (2004) presented a test suite-oriented approach for certifying code generators. Later Conrad et al. (2010) presented tool qualification efforts in the MathWorks ecosystem that address both ISO 26262 and ED-12B/DO-178B. Recently, Wagner et al. (2017) introduced a qualification of a modeling guideline checkers for airborne systems.

4 ANALYTICAL PROCEDURES IN MODEL QUALITY ASSURANCE

4.1 Verification in Model-Based Development

Testing is the major verification method conducted as an analytical quality assurance activity. In traditional software engineering, the focus is on testing the functionality of the code and ensuring its correct behavior; whereas in model-based development, the model is tested against its requirements and the code can be verified against the executable model by means of dynamic testing. This difference has significant advantages; because code behavior can be tested before code implementation has even started, resulting in a very early verification of functional requirements. However, in model-based testing, both model and code are stimulated with the same input then the two outputs are compared with respect to certain acceptance criteria.

As depicted in Fig. 9, testing in model-based development is carried out at different stages of the development process:

- *Model-in-the-Loop (MiL)* checks the validity of the model with respect to the functional requirements within the development environment. This simulation is executed on the host PC. The simulation results are used as a reference (expected values) for the following software verification steps.
- *Software-in-the-Loop (SiL)* analyzes the generated code against possible arithmetic problems (e.g., over-/underflow), and to measure code coverage. The designed model used during MiL is compiled and executed on the host PC with the same stimuli used for MiL.

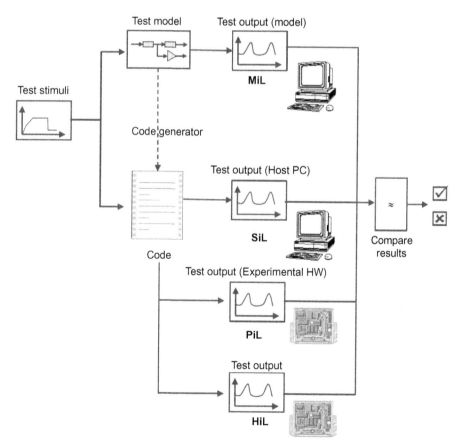

FIG. 9 Testing in model-based development (Stürmer et al., 2005).

- *Processor-in-the-Loop (PiL)* verifies the code behavior on the target processor and measures code efficiency (profiling, memory usage, etc.). The generated code is (cross-) compiled using the project's target compiler. Afterward, the code is executed on experimental hardware, which contains the same processor as the target system (such as an evaluation board) but contains additional resources for storing and exchanging test data and test results.
- *Hardware-in-the-Loop (HiL)* checks the software on the target hardware with its electrical interfaces. The software embedded into the target hardware is connected to a real-time simulation system simulating the plant and is executed.

4.2 Model-Based Testing

Model-based testing is described as a proposal for automating test case generation from a test specification, also called the "test model", instead of implementing test cases manually (Zander, 2009). It further enhances the flexibility and adaptability of the testing infrastructure by automating the test case design (Utting and Legeard, 2010).

Model-based development proposes that a formal system model is derived based on the system requirements. In the next step, executable model components can be generated from the formal system model. As presented in Fig. 10 which is adapted from Roßner et al. (2012), in model-based testing, the same system requirements are used to derive a test model that is able to generate a single test case or a test suite for a system under test (SUT). Test cases describe the intended behavior of the SUT that needs to be tested. The idea is that test cases are abstracted in a test model, and then a model-based testing tool is employed to generate a set of concrete test cases from that model.

A test case is composed of an input stimulus to be fed into a SUT, called test inputs, and the expected behavior of SUT. The expected behavior is determined using a test oracle which also contains a judgment unit to decide the verdict. Schmidt et al. (2016) claimed that a specification of a set of test models on an abstract level is desirable for model-based testing of complex, modular models with diverse testing objectives. They aim to reduce the complexity of test models on the implementation level. The test case is formalized using the experimental frame concept, first introduced by Zeigler (1976), as a specification of a limited set of circumstances under which a system or model has to be observed. In Fig. 11, the structure of a test case is proposed as an EF structure, consisting of a generator, an acceptor and a transducer, coupled with the system under investigation; in this case, the model under test (MUT) (Fig. 11).

FIG. 10 Model-based testing. *Based on Roßner, T., Brandes, C., Goetz, H., Winter, M., 2012. Basiswissen Modellbasierter Test. Dpunkt. verlag.*

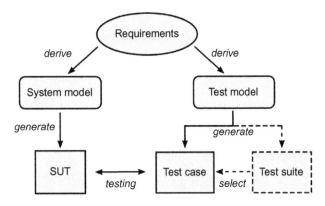

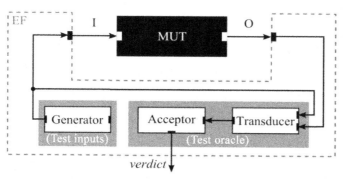

FIG. 11   Test case formalized as an experimental frame (Schmidt et al., 2016).

The generator produces the test inputs. The test oracle is made up of an acceptor and a transducer. The transducer calculates measures in the form of performance indices, comparative values, statistical values, etc. that can be assessed by the acceptor. The acceptor corresponds to a decision unit that decides the success of a test case.

The approach proposed by Schmidt et al. (2016) is based on the idea of having a collection of all possible model components to compose a test case, namely a model base (MB) and utilizing a transformation framework to automatically construct an executable test model. They employed the System Entity Structure and Model Base (SES/MB) framework (Zeigler et al., 2000) for an interactive or automatic generation of an executable simulation model (Fig. 12). The SES ontology, which was specifically developed to represent a family of modular, hierarchical systems, was used to model a family of test cases. Pruning which resolves variabilities and derives a distinct system structure with corresponding parameter settings, is used to derive a specific test scenario. Translation is then employed to collect test model elements from the MB and generate an executable test case. Based on the SES/MB toolbox (Pawletta et al., 2014), the proposed infrastructure was prototyped in MATLAB/Simulink and utilized in testing of flight simulation models (Durak et al., 2015). A screenshot from the infrastructure implementation is given in Fig. 13.

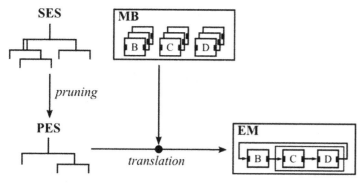

FIG. 12   System entity structure and model base framework (Schmidt et al., 2016).

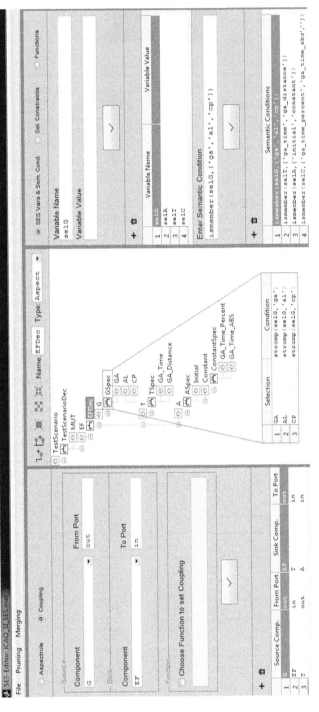

FIG. 13   SES/MB Toolbox for model-based testing (Schmidt et al., 2016).

5 CONCLUSION

As the modeling and simulation-based approach is becoming common practice in systems development, integrated approaches to assure the quality of models are being used more and more. Model quality assessment and improvement methodologies and practices are becoming more crucial in model engineering. Furthermore, safety-critical domains such as the aeronautics and automotive industries are recognizing model-based development as common practice and are endorsing standards to assure model quality.

This chapter provides an overview of model quality assessment and improvement methods and techniques. After introducing the indicators of model quality and sources of error in model engineering, an integrated quality assurance approach is promoted for model engineering activities in model-based development. In this context, the chapter provides a background for various constructive and analytical model quality assurance procedures such as modeling guidelines, model checking and refactoring, or model-based testing. It extends the discussions and exemplifies the tools and techniques about the procedures with the references from the recent literature and examples from authors' own research where appropriate.

While the chapter focusses on model engineering in context of model-based development, the quality assurance procedures introduced in this chapter are well-transferable to other model engineering applications.

References

Amundson, I., Shipton, L., Liu, A., Nowak, M., 2015. Toward efficient model-based development of aerospace applications. In: Proc. 15th AIAA Aviation Technology, Integration, and Operations Conference, Dallas, TX.

Broy, M., Kirstan, S., Krcmar, H., Schätz, B., Zimmermann, J., 2013. What is the benefit of a model-based design of embedded software systems in the car industry? In: Information Resources Management Association (Ed.), Software Design and Development: Concepts, Methodologies, Tools, and Applications. IGI Global, Hershey, PA.

BTC Embedded Systems, 2017. Products Overview. Available from: https://www.btc-es.de/en/products/overview.html. [(Accessed 20 July 2017)].

Campbell, S., Chancelier, J., Nikoukhah, R., 2010. Modeling and Simulation in Scilab/Scicos With Scicoslab 4.4. Springer, New York.

Chikofsky, E.J., Cross, J.H., 1990. Reverse engineering and design recovery: a taxonomy. IEEE Softw. 7 (1), 13–17.

Conrad, M., Munier, P., Rauch, F., 2010. Qualifying software tools according to ISO 26262. In: Proc. MBEES, pp. 117–128.

Czarnecki, K., Helsen, S., 2003. Classification of model transformation approaches. In: Proc. 2nd OOPSLA Workshop on Generative Techniques in the Context of the Model Driven Architecture, Anaheim, CA.

Czarnecki, K., Helsen, S., 2006. Feature based survey of model transformation approaches. IBM Syst. J. 45 (3), 621–645.

D'Ambrogio, A., Durak, U., 2016. Setting systems and simulation life cycle processes side by side. In: Proc. IEEE International Symposium on Systems Engineering, Edinburgh, Scotland.

Durak, U., 2016. Pragmatic model transformations for refactoring in Scilab/Xcos. Int. J. Model. Simul. Sci. Comput. 7 (01)1541004.

Durak, U., Schmidt, A., Pawletta, T., 2015. Model-based testing objective fidelity evaluation of engineering and research flight simulators. In: Proc. AIAA Modeling and Simulation Technologies Conference. Dallas, TX.

dSPACE (2017). TargetLink®. Available from: https://www.dspace.com/go/targetlink (Accessed 20 July 2017).

Estrada Jr., R.G., Dillaber, E., Sasaki, G., 2013. Best practices for developing DO-178 compliant software using model-based design. In: Proc. AIAA Infotech@Aerospace Conference, Boston, MA.

Eisemann, I.U., 2006. Modeling guidelines for function development and production code generation. In: Proc. Embedded World Conference, Nuremberg, Germany.

Eisemann, U., 2016. Applying model-based techniques for aerospace projects in accordance with DO-178C, DO-331, and DO-333. In: Proc. 8th European Congress on Embedded Real Time Software and Systems (ERTS 2016), Toulouse, France.

Erkkinen, T., 2005. Model Style guidelines for production code generation. In: Proc. SAE 2005 World Congress & Exhibition, Detroit, MI.

Ferrari, A., Bacherini, S., Fantechi, A., Zingoni, N., 2009. Modeling guidelines for code generation in the railway signaling context. In: Proc. 1st NASA Formal Methods Symposium, Hampton, VA.

Fowler, M., Beck, K., 1999. Refactoring: Improving the Design of Existing Code. Addison-Wesley Professional, Reading, MA.

Hu, W., Loeffler, T., Wegener, J., 2012. Quality model based on ISO/IEC 9126 for internal quality of MATLAB/Simulink/Stateflow models. In: Proc. 2012 IEEE International Conference on Industrial Technology (ICIT). IEEE, Athens, Greece, pp. 325–330.

IEEE, 2010. 1730–2010—IEEE Recommended Practice for Distributed Simulation Engineering and Execution Process (DSEEP). IEEE.

INCOSE, 2007. Systems Engineering Vision 2020, Version 2.03. International Council on Systems Engineering, Seattle, WA (INCOSE-TP-2004-004-02).

International Organization for Standardization, 2011. ISO 26262-1: 2011 Road Vehicles Functional Safety. ISO.

ISO/IEC/IEEE, 2015. ISO/IEC/IEEE 15288:2015 International Standard—Systems and Software Engineering—System Life Cycle Processes. IEEE.

Kühne, T., Mezei, G., Syriani, E., Vangheluwe, H., Wimmer, M., 2009. Explicit transformation modeling. In: Proc. 12th International Conference on Model Driven Engineering Languages and Systems (MODELS). Springer, Denver, CO, pp. 240–255.

Mahmoodi, S., Durak, U., Gerlach, T., Hartmann, S., D'Ambrogio, A., 2017. Tailoring CMMI engineering process areas for simulation systems engineering. In: Proc. Clausthal-Göttingen International Workshop on Simulation Science, Göttingen, Germany.

MAAB, 2015. Control Algorithm Modeling Guidelines Using MATLAB®, Simulink®, and Stateflow® V3.1. The Mathworks Inc.

Model Engineering Solutions, 2007. MES Test Manager® (MTest). Available from: https://www.model-engineers.com/en/mtest.html. [(Accessed 20 July 2017)].

Ohata, A., Komori, S., 2009. JMAAB plant modeling guidelines and vehicle architecture. In: Proc. 2009 ICROS-SICE International Joint Conference. IEEE, Fukuoka, Japan, pp. 484–487.

Pawletta, T., Pascheka, D., Schmidt, A., Pawletta, S., 2014. Ontology-assisted system modeling and simulation within MATLAB/Simulink. Simul. Notes Europe 24 (2), 59–68.

Piketec GmbH, 2017. Time Partition Testing (TPT). Available from: http://www.piketec.com/. [(Accessed 20 July 2017)].

Popinchalk, S., 2008. Introduction to Model Advisor. [Blog] Guy on Simulink. Available from: http://blogs.mathworks.com/simulink/2008/11/04/introduction-to-model-advisor/. [(Accessed 20 July 2017)].

Roßner, T., Brandes, C., Goetz, H., Winter, M., 2012. Basiswissen Modellbasierter Test. Dpunkt.verlag, Heidelberg, Germany.

RTCA/EUROCAE, 2012a. E-218/DO-331 Model-Based Development and Verification Supplement to ED-12C and ED-109A. EUROCAE.

RTCA/EUROCAE, 2012b. ED-12C/DO-178C Software Considerations in Airborne Systems and Equipment Certification. EUROCAE.

RTCA/EUROCAE, 2012c. ED-215/DO-330 Software Tool Qualification Considerations. EUROCAE.

Schmidt, A., Durak, U., Pawletta, T., 2016. Model-based testing methodology using system entity structures for MATLAB/Simulink models. Simulation 92 (8), 729–746.

SEI, 2010. CMMI® for Development, Version 1.3, Improving Processes for Developing Better Products and Services. Software Engineering Institute.

Slotosch, O., Wildmoser, M., Philipps, J., Jeschull, R., Zalman, R., 2012. ISO 26262-Tool chain analysis reduces tool qualification costs. In: Proc. Automotive-Safety & Security, Karlsruhe, Germany.

Stürmer, I., Conrad, M., 2004. Code generator certification: a test suite-oriented approach. In: Proceedings of Automotive-Safety & Security, Stuttgart, Germany.

Stürmer, I., Kreuz, I., Schäfer, W., Schürr, A., 2007. The MATE approach: Enhanced Simulink and Stateflow model transformation. In: Proc. Math Works Automotive Conference, Dearborn, MI.

Stürmer, I., Dziobek, C., Pohlheim, H., 2008. Modeling guidelines and model analysis tools in embedded automotive software development. In: Proc. Dagstuhl-Workshop MBEES:Modellbasierte Entwicklung Eingebetteter Systeme IV, Braunschweig, Germany.

Stürmer, I., Pohlheim, H., 2012. Model quality assessment in practice: how to measure and assess the quality of software models during the embedded software development process. In: Proc. Embedded Real Time Software and Systems, Toulouse, France.

Stürmer, I., Pohlheim, H., Rogier, T., 2010. Berechnung und visualisierung der modellkomplexität bei der modellbasierten entwicklung sicherheits-relevanter softwar. In: Proc. Automotive-Safety & Security, Stuttgart, Germany.

Stürmer, I., Weinberg, D., Conrad, M., 2005. Overview of existing safeguarding techniques for automatically generated code. In: Proc. ACM SIGSOFT Software Engineering Notes, vol. 30, pp. 1–6.

Stuermer, I., Eisemann, U., Salecker, E., 2014. Distributed development of large-scale model-based designs in compliance with ISO 26262. SAE Technical Paper 2014-01-0313. https://dx.doi.org/10.4271/2014-01-0313.

Tolk, A., Hughes, T.K., 2014. Systems engineering, architecture, and simulation. In: Gianni, D., Tolk, A., D'Ambragio, A. (Eds.), Modeling and Simulation-Based Systems Engineering Handbook. CRC Press, Boca Raton, FL, pp. 11–42.

Utting, M., Legeard, B., 2010. Practical Model-Based Testing: A Tools Approach. Morgan Kaufmann.

Wagner, L., Mebsout, A., Tinelli, C., Cofer, D., Slind, K., 2017. Qualification of a model checker for Avionics Software verification. In: NASA Formal Methods Symposium. Springer, Cham, Switzerland, pp. 404–419.

Wang, R., Dagli, C.H., 2008. An executable system architecture approach to discrete events system modeling using SysML in conjunction with colored petri net. In: Proc. 2nd Annual IEEE Systems Conference, Montreal, Canada.

Zander, J. (2009). Model-Based Testing of Real-Time Embedded Systems in the Automotive Domain. TU Berlin. Available from: https://doi.org/10.14279/depositonce-2126 (Accessed 20 July 2017).

Zeigler, B.P., 1976. Theory of Modeling and Simulation. Wiley Interscience, New York.

Zeigler, B.P., Praehoffer, H., Kim, T.G., 2000. Theory of Modelling and Simulation, second ed. Academic Press, Elsevier San Diego, CA.

Zeigler, B.P., Zhang, L., 2015. Service-oriented model engineering and simulation for system of systems engineering. In: Yilmaz, L. (Ed.), Concepts and Methodologies for Modeling and Simulation. Springer, pp. 19–44.

Zhang, L., Shen, Y., Zhang, X., Song, X., Tao, F., Liu, Y., 2014. The model engineering for complex system simulation. In: Proc. 26th European Modeling & Simulation Symposium, Bordeaux, France, pp. 10–12.

Validation of DEVS Models Using AGILE-Based Methods

L. Capocchi, J.F. Santucci

SPE Laboratory (UMR CNRS 6134), University of Corsica, Corte, France

1 INTRODUCTION

This chapter deals with validation via simulations of discrete event system specification (DEVS) models at the early phases of the design process. DEVS is a widely used formalism in the framework of discrete-event simulation of complex systems. The validation of models is traditionally a step, which is relegated at the end of the design process: once the models have been defined and coded, experiments are conducted in order to validate them using simulation. However, this traditional way to perform validation of models is often an expensive and time-consuming activity, and the resulting quality of the models is still poor. Consequently, new approaches for coping with these challenges are necessary. The same remarks can be formulated when dealing with software testing. Considering software testing, one emerging trend is stronger integration of testing as early as possible in the design process of a program. For that reason software engineering has proposed new design and test as Agile methods, which include test-driven development (TDD) (Fraser et al., 2003) and behavioral-driven development (BDD) (Solis and Xiaofeng, 2011) methods.

In order to go on with the analogy between modeling and simulation (M&S) and software engineering, one can imagine applying BDD and TDD Agile methods to the design and test of DEVS models. Our main objective is to develop an approach that is able to use different software testing techniques stemming from software engineering (Agile test methods to be more specific) that are applied to design of DEVS models in order to improve quality assurance of the resulting DEVS models while proposing inexpensive and no time-consuming activity.

Our approach consists in applying the BDD method for the design of DEVS models. In order to achieve this goal, we have carefully performed a correspondence between the BDD method when applied to software and the BDD method which has to be applied to design of DEVS models.

Model Engineering for Simulation
https://doi.org/10.1016/B978-0-12-813543-3.00011-1

Therefore, the problem is to perform a BDD method when defining DEVS models. Defining such a method in the DEVS M&S context requires the resolution of the following basic problems:

1. To define a semiformal format for the behavioral specification of the test for any atomic models involved in a DEVS model.
2. To define how to generate parameters for a test from a specification document of the tests of DEVS models.
3. To define how to perform the previously defined tests using simulations.

To solve these problems we propose:

1. to define a semiformal format from the natural specifications of FD-DEVS as proposed by Zeigler (see Zeigler and Sarjoughian (2012) for a description);
2. to use the specificity of the Python language to generate a test (by means of the decorator programming concept); and
3. to perform the tests by combining software engineering programming concepts such as decorators, patch, and mocking objects in order to perform the tests using DEVS simulations in the framework of the DEVSimPy (Capocchi et al., 2011) environment.

DEVSimPy is a collaborative general user interface implemented in Python language allowing us to experiment with new approaches inside the DEVS formalism. To ensure this, we use DEVSimPy plug-ins in order to be more generic.

The next part of the chapter discusses related work on similar problems in validation of DEVS models inspired by the software engineering techniques. Section 3 briefly introduced the DEVS formalism and the DEVSimPy framework before the presentation of the main notions involved in Agile test methods. In Section 4 an overview of the proposed approach is briefly presented. Section 5 presents in details the different aspects involves in DEVS model validation using Agile methods. Section 5.1 is devoted to the definition of the semiformal format chosen for the user to write the behavioral specification of the tests as required in a BDD method. In Section 5.2 we describe how we have been able to generate the test parameters, which will be used to validate the models. Section 5.3 presents how the tests are performed using simulations within the DEVSimPy framework by integrating decorators, patches, and mocking objects into DEVS models. In Section 5.4 the integration of the tests performed at the internal level (behavior of atomic models) into coupled models is described. This integration is called external testing. Section 5.5 is dedicated to present the architecture of the DEVSimPy plug-in allowing to implement the proposed approach in previous sections. The last part will permit to conclude and to give a brief overview of future work we envision.

2 RELATED WORK

Today, it is essential to take into account the verification and validation (V&V) processes in the field of the M&S of systems (Pace, 2004). Concerning the M&S of discrete-event models, many research works have been develop to validate a model of a system using simulations. Basically, simulations are preformed and compared against the requirements. The model is validated if all of the considered simulation results comply with the modeler's expectations.

Hollmann et al. (2012, 2014) propose to apply the model-based testing (MBT techniques like the test template framework in the model simulation process. The idea of their work is to enable modelers to define a set of testing criteria (rules) to conduct simulations of DEVS models in order to validate them. The specification of these rules can be realized using formal language and then to contribute to the automatization of the validation by simulations process. Note that the proposed technique does not run the simulations and requires to be integrated in the simulation process.

Labiche and Wainer (2005) highlight the open research area of the V&V of DEVS models and give open research paths in the field of DEVS modeling V&V. They propose to explore new techniques to incorporate automated testing facilities mostly used in the Software Engineering field to the testing framework based on DEVS Experimental Framework and the CD++ toolkit. Saadawi and Wainer (2013) introduce an extension of the DEVS formalism called Rational Time-Advance DEVS (RTA-DEVS) allowing the formal checking of real-time systems using standard model-checking algorithms and tools. The correctness of models is guaranteed by verifying DEVS models using the timed automata theory and tools.

Zengin et al. (2010) and Zengin and Ozturk (2012) give a good introduction of V&V of DEVS simulation models and apply their approach in the DEVS-Suite environment. Based on technique developed by Forrester and Senge (1980), they used a case example Open Shortest Path First DEVS (OSPF-DEVS) simulator to perform V&V tests.

Olamide and Kaba (2013) propose a framework allowing a model-based verification using the simplified model checking tools. It is based on comparison of trajectories and events of DEVS-Driven Modeling Language simulation models with real system. The authors claim that "This framework provides a model refinement iterative procedure that helps to enhance the DEVS Simulation Model, correct errors, or adapt to changing contextual requirements" and validate their approach on the case study of a GSM telecommunication system by checking and refining the model.

Jianpeng et al. (2014) propose a model-driven methodologies to specify a unified model-driven design and validation approach to service-oriented architecture. The methodology consists first to extend of the DEVS modeling language to support nondeterministic state transition and use this extension as a model transformation intermediary to bring together Model-Driven Service Engineering with Service-oriented architecture Modeling Language and M&S methodology based on DEVS.

Most of the previous approaches briefly presented earlier are based on software engineering techniques (Labiche and Wainer, 2005; Hollmann et al., 2012, 2014). There are many manners to test a DEVS model (atomic or coupled model). Li et al. (2011) focus on the validation of DEVS formalism implementation. The approach described in Byun et al. (2009) concerns the correctness of a given simulation. It proposes a framework allowing to check all the possible paths involved in a given simulation model. In Hu et al. (2007), test agents are presented. This chapter proposes a definition of a test agent, which is connected to the I/O of a given model and is used to point out the behavioral information concerning the model. The test is performed using this information. However, the proposed approach is not a generic one since test agent should be specific to a given model. In this chapter, the proposed solution is derived from the software engineering domain but with exploitation of Agile test methods.

3 BACKGROUND

3.1 DEVS Formalism

The DEVS formalism is fully described in the companion volume to this book (Bernard et al., 2018). Here we only review aspects needed for this presentation. DEVS provides a means of specifying a mathematical object called a system. Basically, a system has a time base, inputs, states, outputs, and functions for determining next states and outputs given current states and inputs. The DEVS formalism is a simple way in order to characterize how discrete-event simulation languages may specify discrete-event system parameters. It is more than just a means of constructing simulation models. It provides a formal representation discrete-event systems capable of mathematical manipulation just as differential equations serve this role. Furthermore by allowing an explicit separation between the modeling phase and simulation phase, the DEVS formalism is one of the best ways to perform an simulation of systems using a computer.

In the DEVS formalism, one must specify: (i) basic models from which larger ones are built, and (ii) how these models are connected together in hierarchical fashion. An atomic model allows specifying the behavior of a basic element of a given system. Connections between different atomic models can be performed by a coupled model. A coupled model tells how to couple (connect) several component models together to form a new model. This latter model can itself be employed as a component in a larger coupled model, thus giving rise to hierarchical construction.

An atomic DEVS model (AM in Fig. 1) can be considered as an automaton with a set of states and transition functions allowing the state change when an event occur or not. Regarding the initial version of the DEVS formalism called "classical with port," when no events occur, the state of the atomic model can be changed by an internal transition function called δ_{int}. When an external event occurs, the atomic model can intercept it and change its state by applying an external transition function called δ_{ext}. The life time of a state is determined by a time advance function called t_a. Each state change can produce output message via an output function called λ. A simulator is associated with the DEVS formalism in order to exercise instructions of coupled model to actually generate its behavior. The architecture of a DEVS simulation system is derived from the abstract simulator concepts (Bernard, 1976) associated with the hierarchical and modular DEVS formalism.

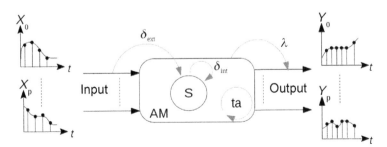

FIG. 1   DEVS atomic model in action.

Finite and Deterministic Discrete Event System Specification (FD-DEVS) (Pace, 2004) has been inherited from classic DEVS in order to model and analyze discrete-event systems in both simulation and verification ways. With regard to natural language specification, FD-DEVS offers a support for XML translation or graphic representation.

3.2 DEVSimPy Software

There are many tools, which provide a user interface dedicated to help the user to define DEVS models and to perform simulations. A nonexhaustive list can be done: PowerDEVS (Bergero and Kofman, 2011), DEVSim++ (Kim et al., 2010), DEVS-Suite (Kim et al., 2009), VLE (Quesnel et al., 2007), DEVSimPy (Capocchi et al., 2011), CD++Builder (Bonaventura et al., 2013), MS4Me (Zeigler and Sarjoughian, 2012), etc. Special attention will be given to DEVSimPy (stand for DEVS simulator in Python), which is a collaborative M&S software.

DEVSimPy (Python Simulator for DEVS models) (Capocchi et al., 2011) is a user-friendly interface for collaborative M&S of DEVS systems implemented in Python language. Python is a programming language known for its simple syntax and its capacity to allow modelers to implement quickly their ideas (Langtangen, 2005). The DEVSimPy project uses the Python programming language and provides a GUI based on PyDEVS (Bolduc and Vangheluwe, 2001) API in order to facilitate both the coupling and the re-usability of PyDEVS models. This API is used in the excellent multimodeling GUI software named ATOM3 (De Lara and Vangheluwe, 2002), which allows to use several formalism without focusing on DEVS. DEVSimPy is an open source project under GPL V3 license and its development is supported by the University of Corsica Computer Science research team. It uses the wxPython graphic library, which is a wrapper of the most popular WxWidgets C library (Julian et al., 2005).

The main goal of the DEVSimPy environment is to facilitate the modeling of DEVS systems using the GUI dynamic libraries and the drag and drop functionality. With DEVSimPy, models can be stored in a dynamic library in order to be reused and shared (left panel in Fig. 2). The creation of dynamic libraries composed with DEVS components is easy since the user is coached by dialogues and wizard during the building process. With DEVSimPy, complex systems can be modeled by a coupling of DEVS models and the simulation is performed in a automatic way. Moreover, DEVSimPy allows the extension (or the overwrite) of their functionalities in using special plug-ins managed in a modular way. The user can enabled/disabled a plug-in using a simple dialog window. We propose in this chapter to use the extension capability offered by DEVSimPy plug-ins in order to implement a model-based validation of DEVS models.

3.3 Agile Test Methods

Agile methods (Cockburn, 2002) are the increasingly common practice throughout the lifecycle to develop a software iteratively. These methods may include: the TDD (Fraser et al., 2003) method and its extension/revision, that is the BDD (Solis and Xiaofeng, 2011) method. TDD is a software development methodology, which essentially states that for each unit of software, a software developer must: (i) first define a test set for the unit, (ii) then implement the unit, (iii) finally verify that the implementation of the unit makes the tests

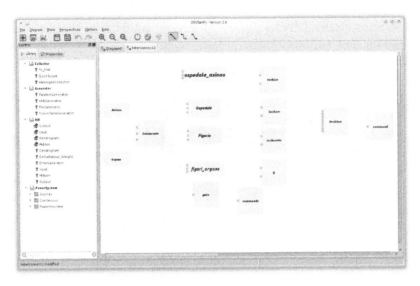

FIG. 2   DEVSimPy general interface.

succeed. BDD is a specialized version of TDD, which focuses on behavioral specification of software units. It is based on: (i) the use of examples to describe the behavior of a given application or code units; (ii) the use of feedback and regression testing from previous examples; (iii) the use of "Mocks" (Mock, 2012) replacing the code modules, which have not yet been written.

The main steps of the BDD method can be summarized by the following points:

1. BDD test of any units of software should be specified in a document written in a semiformal format composed by a set of scenario.
2. The specification document has to be read and each scenario of the document is breaking up into meaningful clauses. Each individual clause is transformed into some parameters defining a given test.
3. The framework then executes the tests of each scenario.

In this chapter, we present a set of main BDD characteristics, which are used to implement the test of DEVS model.

4 PROBLEM DESCRIPTION

The context of the proposed work relates to the cycle of software development. Traditionally software development corresponds to a logical and intuitive approach described in Fig. 3.

The basic approach of such traditional cycle is that we code first and then we perform the tests.

As explained in Section 1, we have developed an analogy between software engineering design process and DEVS model design process. From a software engineering point of view,

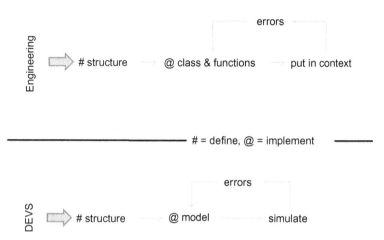

FIG. 3   Traditional cycle of model development.

the structure of the software is first defined, then the implementation of the related classes and functions is performed, and finally the implementation is executed in order to find errors (as it can be seen in the engineering part in Fig. 3). The same kind of development cycle is used for DEVS model development: the structure of models is first defined, models are then implemented and the obtained implementation is simulated in order to find potentials errors (as it can be seen in the DEVS part in Fig. 3). If errors are raised during simulation, they have to be fixed one by one and verified each time by simulation. We also can deduce numerous problems from a validity and productivity points of view from Fig. 3.

- The models cannot take into account all the situations and the predictions have a low reliability because of the uncertainty observability of the produced results.
- The behavior of the developed models has high probability to be erroneous and it will certainly be necessary to adjust certain parameters or certain functions.
- The maintenance, the refactoring, or the evolution of a model cannot guarantee the preservation of the behavior.
- The same model implemented in various environments is completely different. It can raise problems from the perspective of a standardization.

In Section 5 we focus on the developed solution in order to propose an embedded mechanism allowing to automatically take into account the test part at the beginning phase of the design process of DEVS models.

5 PROPOSED SOLUTION

As described earlier, the traditional cycle for DEVS model design raises numerous problems. The proposed approach is summarized in Fig. 4.

We have developed a solution by analogy with the BDD approach defined in the software engineering domain. As shown in Fig. 4 (engineering part) this approach consists in firstly

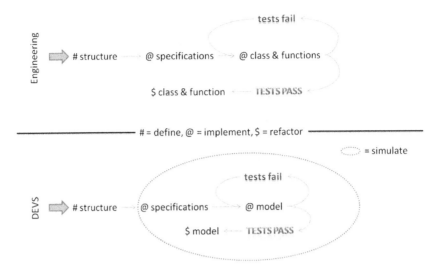

FIG. 4   Proposed behavior-driven development cycle.

defining the structure of the software to be designed and immediately after to write the test specifications. Then the designer can implement and at the same time test the class and functions of the software to be designed. The same kind of development cycle is proposed for the design of DEVS models as it is shown on the DEVS part of Fig. 4. The only difference is that the specifications writing and the model implementation are performed using an M&S framework. The simulation engine is used in order to perform the tests. Concerning the possible relation with the DEVS experimental frame, DEVS defines this notion as an entity, which provides inputs to a simulation model and decides its outputs. Experimental frames can be specified with the same formalism used to specify the simulation model itself. Our approach uses the DEVS experimental frame in order to perform the test of the models by using simulation.

Implemented behavior in a DEVS model is directly tested. This brings numerous advantages:

1. The produced code is reliable.
2. The basic elements of the DEVS model can be tested one after the other.
3. Even if there is an evolution of the implementation of the DEVS model, the behavior remains the same.

The proposed approach has been implemented in the framework of the DEVSimPy environment. The design and test process has been introduced in this Python programming language-oriented simulation environment. In order to propose a generic implementation we choose to define a DEVSimPy plug-in dedicated to:

- the automatic generation of test scenario and
- the execution of the test scenario using the simulation kernel.

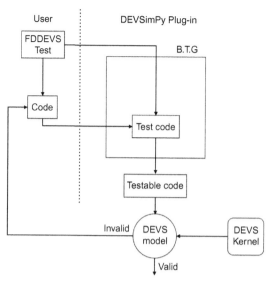

FIG. 5   DEVS design and test proposed methodology.

The plug-in involves the definition of a Behavioral Test Generator engine (called BTG) allowing to transform the test specifications into a test code, which will be executed in order to validate the DEVS model under test. Fig. 5 describes explicitly how a user is going to develop the design and test proposed methodology.

The user has first to write the FD-DEVS test specifications. Then he/she has to implement part of (or completely) a model corresponding to the previously written specifications. The DEVSimPy plug-in will allow to:

- generate testable code from the FD-DEVS specification (using the BTG engine);
- integrate the testable code into the DEVS model already defined by the user; and
- simulate the integration of the testable code and the already defined DEVS model part.

If the result of the simulation points out an invalid DEVS model definition, the user has to rewrite the DEVS model and again execute the plug-in.

5.1 Test Scenario Specification

Specification is an important point when dealing with a BDD approach for DEVS formalism. It gives a manner to describe the test patterns associated with the behavior of a DEVS model, which has to be defined and implemented. Instead of defining a new language we choose to select an already defined language, which allows to describe DEVS modeling scheme under a semiformal natural language. The pseudo-natural language described in Zeigler and Sarjoughian (2012), which is inspired from the grammar detailed in Hong and Kim (2006), has been adapted in order to offer a language for specifying test patterns of a future DEVS model. In order to use this specification language, we have defined a parser for the

FD-DEVS grammar with simple parse (SimpleParse, 2006) tool helping. The proposed grammar follows the hierarchy depicted in Fig. 6.

From this diagram the grammar shown in Listing 1 can be generated at an EBNF format.

L I S T I N G 1   DEVS SPECIFICATION OF THE GENERATOR MODEL

```
to start hold in generate for time 10!
after generate output Job!
from generate go to generate!
when in generate and receive Stop then go to passive!
```

L I S T I N G 2   EBNF SPECIFICATION LANGUAGE GRAMMAR

```
string           := [a-zA-Z],[a'zA-Z0-9]*
number           . = [1-9], [0-9]*

states           := states_fnc / state_name
states_fnc   := initial_states / passive_states / hold_states
initial_states      := c''to start '', (passive_states / hold_states)
passive_states   := c''passive in '', state_name, '' ''?, ''!''
hold_states      := c''hold in '', state_name, c''for time '', number, '' ''?, ''!''
state_name       := string / CURRENT_STATE / NEXT_STATE
CURRENT_STATE:= string
NEXT_STATE       := string

message          := OUTPUT_MSG / INPUT_MSG
OUTPUT_MSG       := (number / string)+
INPTU_MSG        := (number / string)+

functions        := (int_transition / output_fnc / ext_transition)
int_transition   := c''from '', CURRENT_STATE, c'' go to '', NEXT_STATE, '' ''?, ''!''
output_fnc       := c''after '', state_name, c'' output '', OUTPUT_MSG, '' ''?, ''!''
ext_transition   := c'' when in '', state_name, c'' and receive '', INPUT_MSG, c'' go to '',
NEXT_STATE, '' ''?, ''!''
```

An example of a generator DEVS model specification using the ENBF format is given in Fig. 7. Fig. 7 shows that the information are involved by the specification of Listing 2.

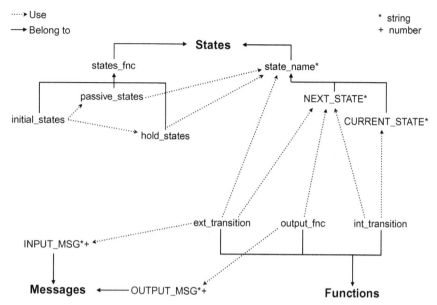

FIG. 6   Diagram of the specification language grammar.

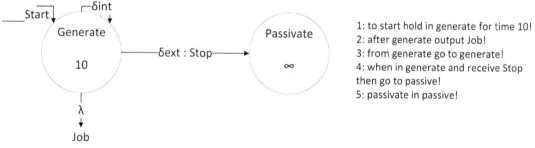

FIG. 7   State automaton and FD-DEVS specification of the generator model.

5.2 Test Scenario Generation

The purpose of the test scenario generation is to transform the previous specification (that can be considered as an abstract model in the MBT activity) into test scenario (derived test cases in the field of MBT) and integrate them into the DEVS simulation models. In order to realize this transformation, we have defined a BTG engine which is a parser. Its goal is to transform the specification (which are expressed using the language presented in Section 5.1 and named "Spec" in Fig. 8) into an adapted test code. Fig. 8 describes how the parser is used when a user has to develop the code corresponding to a DEVS model (an atomic model in this case).

The transformation is based on important information deduced from the specification and called "critical data." These "critical data" are determined by the BTG and injected as test code

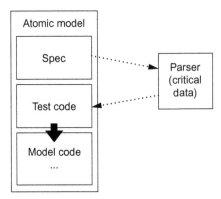

FIG. 8   The parser role is to transform specification into test code.

into the DEVS model to be implemented. Two cases have to be considered: (1) if the DEVS model has been already coded by the user, the injection will be performed using the software engineering decorator notion (Python, 2003); (2) if the DEVS model has not already been implemented, the injection will be performed using the mocking objects (Mock, 2012) notion. These two cases are illustrated by the following two pedagogical examples. An example of a resulting test code corresponding to case 1 is given in Fig. 9.

In this example the user has already written the code of an atomic DEVS model. It corresponds to the behavior, which has been specified in Fig. 7 (see Section 5.1). The example highlights how the internal transition is modified using decorators in order to implement the test scenarios. In the presented example the test scenario, which has been injected, corresponds to the behavior involved by the line 3 of Fig. 7 in the generator DEVS model specification: "from generate go to generate." This means that the injected test scenario allows to check that the state remains the same after the execution of the internal transition when the model is in the "generate" state.

Fig. 10 gives an example of the use of mocking objects, which corresponds to case 2.

```
# intTransition(generate) should do: from generate go to generate
def dec_intTransition(intTransition):
    def new_intTransition():
        realStatus = intTransition.__self__.state["status"]
        if realStatus == "generate":
            status = "generate"
            intTransition()
            if realStatus == status:
                print "intTransition[generate] --> OK"
            else:
                print "Error in intTransition function:\
                status should be %s and we have %s"
                %(status, realStatus)
        return intTransition
    return new_intTransition
```

FIG. 9   FDDEVS to decorator example.

```
    # original function: Empty
    model.extTransition()

    # [received message]: prepared message]
    criticals_data = {(1,2): 1, (2,3): 2}
    def criticalsData(*args):
        return criticals_data[args]

    # Patch of the extTransition method with side-effect
    model.extTransition = MagickMock(name="extTransition", side_effect=criticalsData)
    model.extTransition(1,2)
    => 1
    model.extTransition(2,3)
    => 2
```

FIG. 10   Simple method patching example.

In Fig. 10 we suppose that the external transition of the DEVS model to be tested has not been already implemented. From the specification we have been able to deduce the critical data, which are expressed as follows: for inputs 1 and 2, the output should be 1 while for inputs 2 and 3, the outputs should be 3. The use of patches and mocking objects is highlighted by the introduction of a MagicMock object (from line 10 to 14 of Fig. 10). The external transition ("extTransition" function is patched with the previously mentioned MagicMock object). This example points out how the behavior of DEVS functions can be defined using mocking objects when a DEVS model has not been totally implemented.

5.3 Test Execution

Fig. 11 describes how the DEVSimPy plug-in:

- is integrated into the simulation kernel; and
- is able to select between the injection of decorators and patches (mocking objects).

In order to select which kind of injection has to be performed during the simulation, we use an important property of the Python programming language called dynamic introspection: it is possible to know if a python object is completely implemented or not at any time of the execution of a given python script. Fig. 11 details how this property is used in order to dynamically (during the simulation phase) detect if a given DEVS function has been (or has

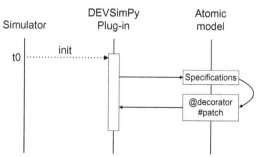

FIG. 11   Sequence diagram of the DEVSimPy plug-in.

not been) already implemented. At time t0 (initialization of the simulation phase), the DEVSimPy plug-in is executed:

1. It refers to the specifications associated with the DEVS model being implemented and tested.
2. It executes the BTG engine, which allows to transform the previous specifications into decorators or patches according to model introspection.
3. It ends by returning into the DEVSimPy simulation kernel in order to go on with the simulation.

The definition of this DEVSimPy plug-in as described earlier permits a user to perform design and test of DEVS models as presented in Fig. 5.

Execution of the scenario will lean on decorators and patches. Here, the simulation is the heart of the test. The point is that with this strategy, we really test a behavior and not just structure or variable.

To illustrate this, we have implemented a plug-in on DEVSimPy environment, which can take care of this strategy. With aspect-oriented programming (AOP) helping, the plug-in gets, on the fly, models on the start of simulation and decorates transitional functions with implemented decorators in specifications. When developers create their models, they have to do two things:

1. write specifications (Spec); and
2. implement the model, just to pass the created test.

These two steps are looped and the second step leads to a major problem. How to execute tests on the DEVS model while implementing them by simulating the execution of the test scenario?

We have to generate, according to specifications, decorators or patches. For example, you have to write your first step specification of external transition and now you want to implement it in your atomic model but you know that without having implemented the rest, it is impossible for you to simulate it! So, we can fix this problem! At the moment t0 of the simulation, the plug-in is initialized and retrieves specifications from atomic models. When it is done, the parser transforms specifications (Spec) into test code (decorator or patch according to model introspection) and then the plug-in links test code to the model.

We have seen the global work of the example plug-in that realizes the new vision of behavioral testing with DEVS but we have not seen the progress of the simulation in a user point of view.

5.4 External Testing

Previous sections dealt only with internal testing that focus on the behavior of the model like states and transition functions. Another part of the test is to see a coupled model under test as a black box in which events are sent on input ports and the output ports are observed and compared with the requirements.

In order to be able to point out an error raised in a coupled model M1, we propose the following process (see Fig. 12: from a given specification (called spec1), we are able to test the

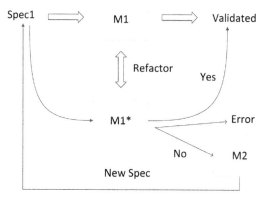

FIG. 12 Methodology for external testing.

model M1. When the model has been validated, we can refactor it. Two possible solutions appear:

- a new model M1* is validated by spec1; and
- the new model M1* raises errors.

If the re-factoring model raises errors, two possibilities have to be considered: (1) a simple re-factoring error or (2) the spec1 no longer corresponds to M1. In this last described case, new specifications have to be defined (according to the M2 model). In the former case we just have to test the M1* model. To test the model like a black box, we only have to execute it and look what happen. We have an input dataset already described in the specification.

5.5 DEVSimPy Plug-In Architecture

The plug-in architecture is open for extensions and at the same time extremely linked to EBNF grammar presented in Section 5.1. In order to explain each part of the plug-in, we begin with the main classes which have been defined.

The generator interface is the base of all delegated generator objects. It implements global functions to all components and other abstract functions that need to be specific to a generator. They are the following:

- *update_fnc(self)*: to update the function name string;
- *conditional_structure(self, data)*: to define conditional structure of the returned code specific to the component; and
- *specif_checker(self)*: return a string which contains different equality in order to check specification.

These functions serve to keep the plug-in open to possible extensions, an API to extend the grammar. If grammar needs to be extended, all we need to do is to add its grammar words (see Fig. 6) and make a new related class that extends the *GeneratorInterface*.

Fig. 13 shows all available classes.

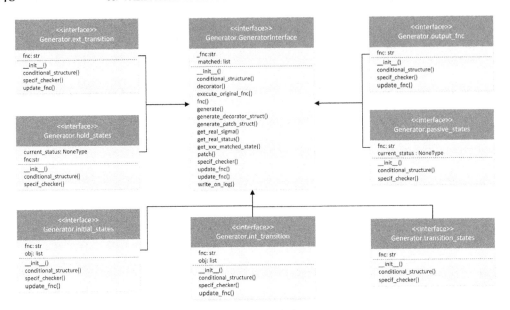

FIG. 13   Class diagram of the generator interface and related objects.

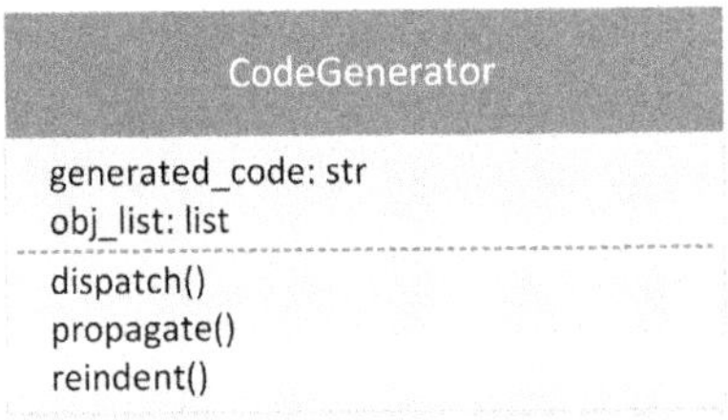

FIG. 14   *CodeGenerator* class diagram.

The responsible class of test code management is *CodeGenerator* (Fig. 14). Two important methods are called:

- *dispatch*: populate obj_list with dynamic required objects and call propagate.
- *propagate*: call generate method for each object in obj_list.

The dispatch method implements a dependency injection—by populating dynamically obj_list with instances of required generators objects—in order to limit dependency in the code but not in the execution. Thanks to the dependency injection, the plug-in operating protocol is easily updated. Methods call each other thanks to the *GeneratorInterface* but behave differently depending on the calling object. For example, *generator* method call *decorator* and *patch* method that generates different codes depending on whether the object is supposed to generate the test for ext_transition or int_transition.

6 CONCLUSION AND PERSPECTIVES

This chapter introduced an approach allowing to perform the test of DEVS models at the very early phases of the design. The main idea is to apply the concepts which have been defined in the software engineering domain in the framework of the Agile community. We described how the BDD design and test methodology stemming from the software engineering domain can be applied to the design of DEVS models. We presented how we have adapted the three main steps of the BDD methodology:

1. Definition of a semiformal language allowing test specification using the DEVSSpecL language.
2. Generation of the test scenarios using the definition of a BTG using the notion of decorators and patches.
3. Execution of the test scenario using the concepts of dynamic introspection and mocking objects.

The resulting Design and Test DEVS approach has been validated in the framework of the DEVSimPy environment. The three previous steps have been integrated into a DEVSimPy plug-in. We have also described how to integrate the test of atomic model into a process allowing to deal with coupled model (external testing). The future work will concentrate in the validation using the design of more complex DEVS models.

References

Bergero, F., Kofman, E., 2011. PowerDEVS: a tool for hybrid system modeling and real-time simulation. Simulation 87 (1–2), 113–132. https://dx.doi.org/10.1177/0037549710368029.

Bernard, P.Z., 1976. Theory of Modeling and Simulation. Academic Press, London.

Bernard, P.Z., Lisandru, M., Ernesto, K., 2018. Theory of Modeling and Simulation, third ed. Academic Press, London.

Bolduc, J.S., Vangheluwe, H., 2001. PythonDEVS: A Modeling and Simulation Package for Classical Hierarchal DEVS. In: Rapport technique, MSDL, Universitè de McGill.

Bonaventura, M., Wainer, G.A., Castro, R., 2013. Graphical modeling and simulation of discrete-event systems with CD++Builder. Simulation 89 (1), 4–27. https://dx.doi.org/10.1177/0037549711436267.

Byun, J.H., Choi, C.B., Kim, T.G., 2009. Verification of the DEVS model implementation using aspect embedded DEVS. In: Proceedings of the 2009 Spring Simulation Multiconference, vol. 151, San Diego, pp. 1–7.

Capocchi, L., Santucci, J.-F., Poggi, B., Nicolai, C., 2011. DEVSimPy: a collaborative python software for modeling and simulation of DEVS systems. In: Proceedings of 2011 20th IEEE International Workshops on Enabling Technologies: Infrastructure for Collaborative Enterprises (WETICE), pp. 170–175.

Cockburn, A., 2002. Agile Software Development. Addison-Wesley Longman Publishing Co., Inc., Boston, MA.

De Lara, J., Vangheluwe, H., 2002. ATOM3: a tool for multi-formalism and meta-modelling. In: Ralf-Detlef, K., Herbert, W. (Eds.), In: Proceedings of the 5th International Conference on Fundamental Approaches to Software Engineering (FASE '02). Springer-Verlag, London, pp. 174–188.

Forrester, J.W., Senge, P.M., 1980. Test for building confidence in system dynamics models. TIMS Stud. Manag. Sci. 14, 209–228.

Fraser, S., Beck, K., Caputo, B., Mackinnon, T., Newkirk, J., Poole, C., 2003. Test driven development (TDD). In: Proceedings of the 4th International Conference on Extreme Programming and Agile Processes in Software Engineering, Genova, pp. 459–462.

Hollmann, D.A., Cristiá, M., Frydman, C., 2012. Adapting model-based testing techniques to DEVS models validation. In: Proceedings of the 2012 Symposium on Theory of Modeling and Simulation—DEVS Integrative M&S Symposium. Society for Computer Simulation International, San Diego, CA, p. 8.

Hollmann, D.A., Cristiá, M., Frydman, C., 2014. A family of simulation criteria to guide DEVS models validation rigorously, systematically and semi-automatically. Simul. Model. Practice Theory 49, 1–26. https://dx.doi.org/10.1016/j.simpat.2014.07.003.

Hong, K.J., Kim, T.G., 2006. DEVSpecL: DEVS specification language for modeling, simulation and analysis of discrete event systems. Inf. Softw. Technol. 48, 221–234.

Hu, X., Zeigler, B.P., Hwang, M., Mak, E., 2007. DEVS systems-theory framework for reusable testing of I/O behaviors in service oriented architectures. In: Proc. 2007 IEEE International Conference on Information Reuse and Integration, Las Vegas, IL 394–399. https://dx.doi.org/10.1109/IRI.2007.4296652.

Jianpeng, H., Linpeng, H., Renke, W., Bei, C., Xuling, C., 2014. Model-driven design and validation of service oriented architecture based on DEVS simulation framework. Int. J. Serv. Comput. 2 (2), 17–29.

Julian, S., Kevin, H., Stefan, C., 2005. Cross-Platform GUI Programming with Wxwidgets (Bruce Perens Open Source). Prentice Hall PTR, Upper Saddle River, NJ.

Kim, S., Sarjoughian, H.S., Elamvazhuthi, V., 2009. DEVS-suite: a simulator supporting visual experimentation design and behavior monitoring. In: Proceedings of the 2009 Spring Simulation Multiconference (SpringSim '09). Society for Computer Simulation International, San Diego, CA, article 161, p. 7.

Kim, T.G., Sung, C.H., Hong, S.Y., Hong, J.H., Choi, C.B., Kim, J.H., Seo, K.M., Bae, J.W., 2010. DEVSim++ toolset for defense modeling and simulation and interoperation. J. Def. Model. Simul. https://dx.doi.org/10.1177/1548512910389203.

Labiche, Y., Wainer, G., 2005. Towards the verification and validation of DEVS models. In: Proceedings of Open International Conference on Modeling & Simulation, pp. 295–305.

Langtangen, H.P., 2005. Python Scripting for Computational Science (Texts in Computational Science and Engineering). Springer-Verlag, New York, NY.

Li, X., Vangheluwe, H., Lei, Y., Song, H., Wang, W., 2011. A testing framework for DEVS formalism implementations. In: Proceedings of the 2011 Symposium on Theory of Modeling & Simulation: DEVS Integrative M&S Symposium, Boston, MA, pp. 183–188.

Mock—Mocking and Testing Library, 2012. Retrieved on August 14, 2018 from http://www.voidspace.org.uk/python/mock/index.html.

Olamide, S.E., Kaba, T.M., 2013. Formal verification and validation of DEVS simulation models. In: AFRICON, 2013, pp. 1–6 9–12.

Pace, D.K., 2004. Modeling and simulation verification challenges. Johns Hopkins APL Tech. Dig. 25 (2), 163–172.

Python, 2003. Decorators for Functions and Methods. Retrieved on August 14, 2018 from http://legacy.python.org/dev/peps/pep-0318/.

Quesnel, G., Duboz, R., Ramat, E., Traoré, M.K., 2007. VLE: a multimodeling and simulation environment. In: Proceedings of the 2007 Summer Computer Simulation Conference (SCSC '07). Society for Computer Simulation International, San Diego, CA, pp. 367–374.

Saadawi, H., Wainer, G., 2013. Principles of discrete event system specification model verification. Simulation 89 (1), 41–67. https://dx.doi.org/10.1177/0037549711424424.

SimpleParse, 2006. SimpleParse A Parser Generator for mxTextTools v2.1.0. Retrieved on August 14, 2018 from http://simpleparse.sourceforge.net/.

Solis, C., Xiaofeng, W., 2011. A study of the characteristics of behaviour driven development. In: Proceedings of 2011 37th EUROMICRO Software Engineering and Advanced Applications (SEAA)pp. 383–387.

Zeigler, B.P., Sarjoughian, H.S., 2012. Guide to Modeling and Simulation of Systems of Systems. Springer, London.

Zengin, A., Ozturk, M.M., 2012. Formal verification and validation with DEVS-suite: OSPF case study. Simul. Model. Pract. Theory 29, 193–206. https://dx.doi.org/10.1016/j.simpat.2012.05.013.

Zengin, A., Köklükaya, E., Ekiz, H., 2010. Verification and validation of the DEVS models. In: Proceedings of 2nd International Symposium on Sustainable Development, Sarajevo.

DEVS Activity Tracking Based on Model Engineering and Simulation

L. Capocchi*, A. Muzy†, J.F. Santucci*

*SPE Laboratory (UMR CNRS 6134), University of Corsica, Corte, France †I3S Laboratory (UMR CNRS 6070), Team Bio-Info, Batiment Algorithme, Sophia Antipolis, France

1 INTRODUCTION

Over the past decade, numerous efforts have been made to define *activity* in modeling and simulation (Kung and Sölvberg, 1986; Liu and Meersman, 1992; Muzy et al., 2011, 2013; Hu and Zeigler, 2013; Muzy and Hill, 2011). Recently, the concept of *activity* has been introduced for discrete-event system specification (DEVS) (Zeigler et al., 2000) models as the number of transition function executions and can be used to profile DEVS models. Furthermore, in order to analyze models, DEVS designers can correlate the number of transition functions (which has some language complexity) with the CPU time consumed by a DEVS transition function. These two kinds of metrics, the number of transition executions and CPU, are available only during the simulation (simulation-based metrics). However, having a means to compute a priori *analytic activity* of components may be useful in order to anticipate the computation of the DEVS model complexity. Among the list of potential metrics, those stemming from software engineering are particularly of interest for Model Complexity analysis because: (i) they are most popular metrics in software engineering; (ii) they have strong implications for software testing; and (iii) they can be used as an estimation of the time required for the execution of the DEVS transition functions. A model complexity is measured in terms of the time and space required to simulate it. Zeigler et al. (2000) claim that the complexity of a model depends on the amount of detail in it, which in turn, depends on the size (number of components), resolution (number of states per component), and interaction (number of couplings per component). In order to validate the interest of analytic and simulation activity metrics in the framework of Model Complexity analysis, we have defined and implemented a DEVS activity pattern profiling based on both simulation and analytic metrics.

In Zeigler et al. (2000), the concept of *activity* is specifically introduced for DEVS models as the number of transition functions executions over some time interval. Furthermore, from this concept of activity several activity metrics can be defined and used to profile DEVS models. Profiling consists of collecting statistics about the models activity during a simulation. Usually, memory usage, duration, frequency of calls, CPU usage, and occupation are collected during the simulation. These metrics are available only during the simulation (called in this chapter as *simulation activity*). However, having a means to compute a priori activity of components (called in this chapter as *analytic activity* or *static activity*) may be worth when simulating a model (or parts of it) for the first time when we are looking to accelerate the simulation process. Then, during the simulation, analytic activity can be corrected using simulation measurements.

In this chapter, we introduce two kinds of analytic metrics: the first one derived from McCabe Cyclomatic Complexity metric (MCC) (McCabe, 1976) to compute analytic activity and the second one derived from program performance measurements (Park and Shaw, 1990; Shaw, 1989). Both analytic and simulation activity metrics have been implemented through a plug-in of the DEVSimPy (DEVS Simulator in Python language) environment and applied to DEVS models. DEVSimPy (Capocchi et al., 2011) is being developed at the Science Pour L'Environment (SPE) laboratory of the University of Corsica and is an open source project under GPL v3 license. The DEVSimPy environment is based on the PyDEVS API (Bolduc and Vangheluwe, 2001; Tendeloo, 2018) and aims at facilitating researches in the SPE group in order to introduce and to validate new concepts around DEVS formalism.

This chapter is organized as follows: Section 2, dedicated to background and related work, allows to present the issues associated with DEVS Model Complexity analysis (pointing out the need of both analytic and simulation metrics). This second section also introduces the DEVS formalism and the DEVSimPy environment. Section 3 deals with the simulation metrics based on activity notions and introduces the two kinds of analytic metrics that we have developed. Section 4 deals with the implementation of these metrics in a DEVSimPy plug-in called activity tracking. Section 5 is dedicated to the validation based on two complex case studies: asynchronous electrical machine and IEEE 802.3 CSMA/CD protocol. The obtained results, which highlight the relationship between the different metrics, are presented in detail. Finally, some conclusions and perspectives are given.

2 BACKGROUND

2.1 The Activity-Tracking Paradigm in the DEVS Formalism

The activity notion for a DEVS system is commonly referred as the number of transition functions executions (Muzy and Hill, 2011). Usually the activity notion is referred as quantitative activity (QA). The activity concept on the other hand has been used to connect information processing and energy consumption as proposed in Muzy et al. (2011). In this case it is called weighted activity (WA) and allows for example when counting the number of transitions to compute the weight of a transition proportionally to the time spent in state before the transition. In both cases (for the QA and WA notions) the activity is a notion defined at the simulation level and computed during the simulation (see Fig. 1). However, activity of a DEVS system can be tracked using two approaches: considering the modeling level and/

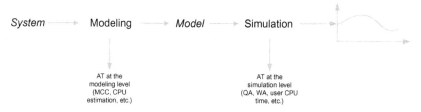

FIG. 1 AT paradigm at the modeling and simulation levels.

or the simulation level (Muzy et al., 2011). Fig. 1 shows the position of the activity tracking (AT) process in these two levels.

We can distinguish two kinds of AT metrics: (i) at the modeling level, the metrics are computed without any DEVS simulations (they are called analytic metrics) and (ii) at the simulation level, the metrics are computed by performing the DEVS simulation process (they are called simulation-based metrics). From both modeling and simulation levels, tracking the activity of a DEVS model can be considered as:

1. counting the number of state-to-state transitions in a model over some time interval—commonly referred to as QA (Muzy and Hill, 2011);
2. counting the number of weighted state transitions over some time interval—commonly referred to as WA (Hu and Zeigler, 2013);
3. computing the MCC (McCabe, 1976) of transition functions—which can be used to measure the complexity of a model;
4. estimating the CPU using a metric based on software performance measurements; and
5. measuring the user CPU time in order to measure the time spent on the processor running the code of the transition functions.

These five metrics will be presented in Section 3.

2.2 Activity Concept in System Engineering

Activity is a concept and refers to the state transition distribution in the components of a system (Hu and Zeigler, 2013). Activity metrics have been used to speed up simulation in the form of AT, which focuses computational resources on component based on their activity level. The concept of activity can be exploited in the field of the model engineering and more specially in system engineering with System of Systems (SoS). SoS is a composition of systems, which component systems have legacy properties. Activity can be used as engineering methodology in order to estimate the *complexity* of an SoS model for its construction, management, and maintenance. Depending on the definition of complexity, the activity analysis can improve the structural and behavioral properties of SoS during the design process.

2.3 DEVS Formalism and DEVSimPy Framework

The DEVS formalism was introduced by Zeigler in the 1970s (Zeigler, 1976) for modeling discrete-event systems in a hierarchical and modular way. DEVS is fully described in the companion volume to this book (Zeigler et al., 2018). Here we only review aspects needed for this presentation. DEVS formalizes what a model is, what it must contain, and what it does not contain (experimentation and simulation control parameters are not contained in the model). Moreover, DEVS is universal and unique for discrete-event system models. Any system that accepts events as inputs over time and generates events as outputs over time is equivalent to a DEVS model. DEVS allows automatic simulation on multiple different execution platforms, including those on desktops (for development) and those on high-performance platforms (such as Clusters or High Performance Computer). With DEVS, a model of a large system can be decomposed into smaller component models with couplings between them. DEVS formalism defines two kinds of models: (i) atomic models that represent the basic models providing specifications for the dynamics of a subsystem using function transitions and (ii) coupled models that describe how to couple several component models (which can be atomic or coupled models) together to form a new model. This hierarchy inherent to the DEVS formalism can be called a description hierarchy by allowing the definition of a model using a hierarchical decomposition.

An atomic DEVS model can be considered as an automaton with a set of states and transition functions allowing the state change when an event occur or not. When no events occur, the state of the atomic model can be changed by an internal transition function noted δ_{int}. When an external event occurs, the atomic model can intercept it and change its state by applying an external transition function noted δ_{ext}. The lifetime of a state is determined by a time advance function called t_a. Each state change can produce output message via an output function called λ. A simulator is associated with the DEVS formalism in order to exercise instructions of coupled model to actually generate its behavior. The architecture of a DEVS simulation system is derived from the abstract simulator concepts associated with the hierarchical and modular DEVS formalism.

DEVSimPy[1] (Python Simulator for DEVS models) (Capocchi et al., 2011) is a user-friendly interface for collaborative M&S of DEVS systems implemented in the Python[2] language. DEVSimPy is an open source project under GPL V3 license and its development is supported by the University of Corsica "Pasquale Paoli" Computer Science research group. The DEVSimPy project uses the Python programming language for providing a GUI (based on wxPython[3] graphic library) for the PyDEVS and PyPDEVS[4] (Van Tendeloo and Vangheluwe, 2014) (the parallel DEVS implementation of PyDEVS) APIs. DEVSimPy has been set up to facilitate both the coupling and the re-usability of the PyDEVS classic DEVS models and the PyPDEVS Python Parallel DEVS models. Moreover, the DEVSimPy architecture is based on an MVC (Model-View-Controller) pattern coupled with the oriented aspect programming concept, which renders the user interface and the simulation kernel (PyDEVS or PyPDEVS) independent. User can select the desired simulation kernel on DEVSimPy models, which are compatible with (a wrapper is delegate for this) the Py(P)DEVS simulators. In this way, when the simulator code sources are updated, the DEVSimPy GUI part is not affected. It should be

---

[1] See https://github.com/capocchi/DEVSimPy.

[2] See http://python.org.

[3] See http://www.wxpython.org.

[4] See http://msdl.cs.mcgill.ca/projects/DEVS/PythonPDEVS.

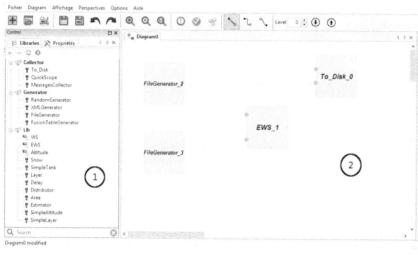

FIG. 2   DEVSimPy general interface.

noted that PyDEVS and PyPDEVS are also used in the multimodeling GUI software named ATOM3 and its promising successor ATOMPM (Van Mierlo et al., 2015).

With DEVSimPy models can be stored in a library in order to be reused and shared (1 in Fig. 2). More specifically, a DEVSimPy model is a compressed file composed by a python file (behavioral specifications according to Py(P)DEVS specifications) and a text file (graphical view according to wxWidgets API). When a model is instantiated, the corresponding compressed file is extracted and (i) the graphical representation is built from the text file and (ii) the behavior is instantiated from the Python file. Thus the view and the behavior of a model are split and an user can change a behavior of a model (by changing the python file) without changing its graphical view. The same strategic storing approach was taken in Fard and Sarjoughian (2015) where the data for every atomic model are stored in two flat files (Domain and Diagram files). A set of DEVSimPy models constitutes a shared library due to the fact that all models can be loaded or updated from an external location such as from a file server (Dropbox, GoogleDrive, GitHub, etc.), which could be also considered as a kind of "online model store." However, it should be stated that this concept of model store can significantly extended with the concept of model repositories as discussed in Chapter 16 of Zeigler and Sarjoughian (2013) or in Sarjoughian and Elamvazhuthi (2009).

Nevertheless, a python file with Py(P)DEVS specifications is embedded by DEVSimPy that transforms it into an object with a default graphical view when it is dropped in the interface (2 in Fig. 2). The creation of dynamic libraries composed with DEVS components is easy since the user is coached by dialogues and wizards during the building process. With DEVSimPy, complex systems can be modeled by a coupling of DEVS models (2 in Fig. 2) and the simulation is performed in an automatic way. Moreover, DEVSimPy allows the extension (or the override) of their features in using special plug-ins managed in a modular way (Santucci and Capocchi, 2013). The philosophy of DEVSimPy is to be an open, extensible, and participative environment for the developers. A plug-in manager is proposed in order to expand the functionalities of DEVSimPy allowing their enabling/disabling through a dialog window. For example, the plug-in "Blink" is proposed to visualize the activity of models during the simulation. It is based on a step-by-step approach and illuminates each active model with a color, which depends on the executed transition function. Section 4 presents the AT DEVSimPy plug-in with a specific focus on its setting and use.

3 DEVS MODEL ACTIVITY TRACKING METRICS

3.1 Simulation-Based Activity Metrics

Simulation-based metrics are considered during the execution of a program (simulation process). Concerning the activity of DEVS models, the CPU user time computed and updated during the simulation can be weighted by the quantitative activity.

Muzy and Hill (2011) define the QA of a system as "the number of discrete-events received by the system, over a simulation time period." According to Hu and Zeigler (2013), a measure of activity can be considered as a measure of information processing by counting over some time interval the number of state-to-state transitions in a model. In other words, it is the number of transitions in a given time interval. The QA is a notion defined at the modeling level but quantified (tracked) during the simulation. Muzy and Zeigler (2012) give the following definition of the total activity for an atomic DEVS model AM_i in a simulation time interval:

$$QA_{AM_i} = QA_{AM_i}^{\delta_{int}} + QA_{AM_i}^{\delta_{ext}} \tag{1}$$

where $QA_{AM_i}^{\delta_{ext}}$ (respectively, $QA_{AM_i}^{\delta_{int}}$) is the external (respectively, internal) activity. The external (respectively, internal) activity is defined as a natural number equal to the sum of DEVS external (respectively, internal) transitions δ_{ext} (respectively, δ_{int}) execution. In Muzy and Hill (2011), the activity of a coupled DEVS model CM is defined as the sum of the total activity of its N atomic models QA_{AM_i} with $i \in \{1, ..., N\}$:

$$QA_{CM} = \sum_{i \in 1, ..., N} QA_{AM_i} \tag{2}$$

Let consider the definition of activity given in Eq. (1) (respectively, Eq. 2) as the definition of the QA for an atomic (respectively, coupled) model.

CPU user time is the time spent on the processor running your program's code (or code in libraries) while system CPU time is the time spent running code in the operating system kernel on behalf of your program. We are interested here in the CPU user time for an atomic DEVS model AM_i:

$$CPU_{AM_i} = CPU_{AM_i}^{\delta_{int}} + CPU_{AM_i}^{\delta_{ext}} \tag{3}$$

According to Hu and Zeigler (2013), WA for a component $i \in D$ is the sum of weighted state transitions over some time interval T:

$$WA = WA_{int} + WA_{ext} \tag{4}$$

where WA_{ext} (respectively, WA_{int}) is the external (respectively, internal) transition-weighted activity computed as follows:

$$WA'_{ext} = WA_{ext} + w_{ext}(x,s) \tag{5}$$

$$WA'_{int} = WA_{int} + w_{int}(s) \tag{6}$$

where $w_{ext}(x, s)$ (respectively, $w_{int}(s)$) is the external (respectively, internal) transition weighting function, which must be implemented by the modelers.

An example of the implementation of these function has been given in Hu and Zeigler (2013) by allocating the weight of a transition proportionally to the time spent in state before the transition. As it has been defined for the QA, the WA for a coupled model CM is equal to the sum of the WA of all atomic models included in D (the set of references to lower-level components):

$$WA_{CM}(T) = \sum_{i \in 1,...,N} WA_{AM_i} \tag{7}$$

3.2 Analytic-Based Activity Metrics

Getting a good activity metrics of models, in order to evaluate their performance for example, during the simulation is very difficult. These activity metrics can be defined from metrics, which have been defined in software engineering. A software metric is usually used to determine the degree of maintainability of software products. However, software metrics may also be used to predict the execution time consumption of functions or methods of an object. In DEVS modeling and simulation context, these kinds of metrics can be employed to evaluate the analytic and simulation activity of models.

3.2.1 McCabe Complexity

Among the list of recommended metrics proposed in most of software engineering subfields (Halstead complexity (Halstead, 1977), McCabe complexity (McCabe, 1976), Coupling (Stevens et al., 1974; Beck and Diehl, 2011), etc.), MCC (McCabe, 1976) has been chosen because: (i) it is one of the most popular metrics in software engineering; (ii) it has strong implications for software testing; and (iii) it can be used as an estimation of the time required for the execution of the transition functions.

MCC depends only on the decision structure of a program. The cyclomatic number of a directed graph G, where each node corresponds to a block program, is a graph-theoretic complexity:

$$MCC(G) = e - n + p$$

where e is the number of edges of the graph, n the number of nodes of the directed graph, and p the number of connected components (exit nodes). This number depends on the number of linearly independent paths, that is, to the decision (if statement, conditional loops, etc.) structure of a program.

MCC usually correlates with the amount of work required to test a program; therefore, it is used to have a measure of the test complexity of a program. However, in DEVS modeling and simulation context, MCC can be used at atomic function level considering the whole functions as a "block program." Indeed, the higher the number of independent decision paths, the more the system is expected to be an event hub of high CPU activity: if the MCC of an atomic function is high, there is a significant probability that the time spent in the function execution will be high too. Moreover, the MCC can be considered as a metric, which provides useful feedback to the DEVS designers during the modeling phase.

For DEVS model, we can defined the McCabe Activity (MCA) of an atomic model AM_i as:

$$MCA_{AM_i} = MCC_{\delta_{ext}} + MCC_{\delta_{int}} \qquad (8)$$

3.2.2 *Program Performance Measurements Metrics*

In software engineering, profiling ("program profiling" and "software profiling") is a form of program analysis that measures, for example, the time complexity of a program, the usage of particular instructions, or the frequency and duration of function calls (Park and Shaw, 1990; Shaw, 1989). Profiling is achieved by instrumenting the program source code. Such kind of techniques performs the investigation of a program's behavior using information gathered as the program executes. A code profiler is a performance analysis tool that, most commonly, measures only the frequency and duration of function calls. Generally program execution time is measured from program initiation at presentation of some inputs to termination at the delivery of the last outputs. Several different measures of software performance are of interest: (i) worst-case execution time—the longest execution time for any possible combination of inputs; (ii) best-case execution time—the shortest execution time for any possible combination of inputs; and (iii) average-case execution time for typical inputs. We have to point out that defining a metric based on code execution is language dependant. We present in Section 4 how using the Python language we have been able to use the *timeit* module (Beazley and Jones, 2013). Indeed, during the coding phase, one can want to know how long it takes for a particular function to run. This topic is known as profiling or performance tuning. Python has a couple of profilers built into its Standard Library like the *timeit* module. This module uses a platform-specific method to get the most accurate run time of a specific function by running the code *n* number of times and returning a list of times it took to run (best case, worst case, and average case).

4 IMPLEMENTATION OF ANALYTIC AND SIMULATION ACTIVITY

Built upon definitions of activity metrics given in the previous section, DEVSimPy implements a new plug-in called AT. This plug-in increases the handling of the recent definition of the activity metrics thus opening new perspectives for the use of AT in DEVS formalism. DEVSimPy plug-in AT is generic and can be applied to any DEVS models. It does not require any modification of the DEVS simulation algorithm and does not require any additional methods in DEVS models to operate.

The *DEVSimPy AT* plug-in works the following way:

- The user enables the plug-in (see Fig. 3) and chooses the set of DEVSimPy atomic models for AT (see Fig. 4). Before the simulation, the DEVS models are scanned in a recursive

FIG. 3 Activation of AT plug-in in DEVSimPy preferences.

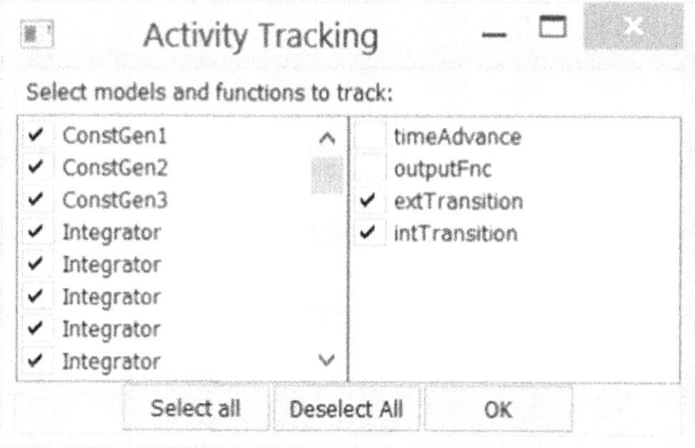

FIG. 4 AT plug-in configuration.

way to collect all atomic models selected by the user in the plug-in interface. The external and internal transition functions of all selected models are decorated with a new method aimed at introducing at AT computation of these functions. A decorator function adds a new attribute to the DEVS object in a dynamic way (offered by the Python language combined with the use of oriented aspect programming) for each transition function. This new attribute is a dictionary key with the simulation time associated at the CPU of the tracked transition function. Moreover, knowing the code of the transition functions for each selected atomic DEVS model, the associated MCA metric can be performed before the simulation. We also compute how long it takes to execute the transition functions using the python *timeit* module. In the same way, the coupling metrics can be computed from coupling relationships between models inside all coupled models.

- The user can now perform the simulation of the model during which the QA metric is measured by counting the number of executions of DEVS transition functions. From the previous dictionary, QA is measured by counting the number of its keys after the simulation. In the same way the simulation time that a model waits for the coordinator to give it a *message is also computed (difference between simulation times stored in dictionary).
- Finally, when the simulation is over, the plug-in offers tables resuming all of the computed metrics.

While the simulation is running, the plug-in offers dynamically, among others, the QActivity, WActivity, CPU, MCA, Worst_case, and Best_case metrics for each tracked model.

5 VALIDATION

5.1 Asynchronous Electrical Machine Use Case

The model used for the experiments contains 49 atomic models, 15 coupled models, 203 coupling, and 3 levels of encapsulation between coupled models. It models an asynchronous electrical machine (Yazidi et al., 2010) employed for the diagnosis of eolian motors. We simulate this model during 1 second and we compute for each tracked model (via the AT plug-in) the three metrics: QA, WA, and MCA. The two first ones are computed over the simulation time while the MCC analytic metric is obtained before the simulation starting.

Fig. 5 depicts the 49 ordered models (identified by ID) according to the MCA metric. We can note that there is a significant gap between the 16 models with the highest MCA value (75) and the 9 models with the lowest MCC value (2). In Fig. 6, for each component $i \in D$, normalized activity metrics

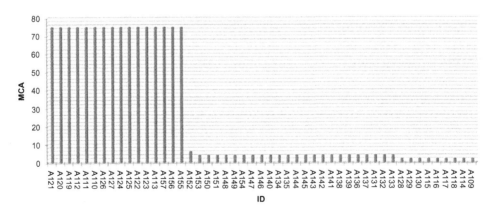

FIG. 5   MCA analytic metric computation.

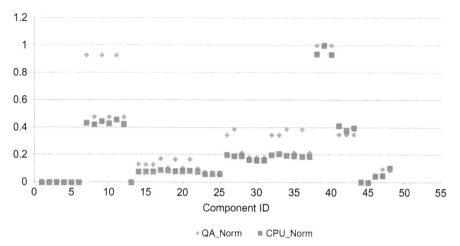

FIG. 6   Normalized QA and normalized CPU metrics.

$$NQA_i = \frac{QA_i}{QA_{\max}} \quad \text{with} \quad QA_{\max} = \max\{QA_i \mid i \in D\}$$

is compared with normalized CPU metrics

$$NCPU_i = \frac{CPU_i}{CPU_{\max}} \quad \text{with} \quad CPU_{\max} = \max\{CPU_i \mid i \in D\}$$

It can be seen that normalized activity corresponds mostly to normalized CPU.

Fig. 7 depicts the distribution of normalized $\frac{NMCA}{NCPU}$ ratio. It can be seen that normalized NMCA constitutes a good prediction for most of components. Correct prediction concerns

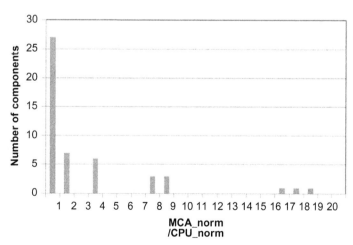

FIG. 7   Distribution of NCPU and NMCA comparison.

27 components over the 49 ones. NMCA overestimates the other components. This means that for some components, NMCA predicts they will have more activity. However, during the simulation, these overestimation could be dynamically corrected.

One can ask: "Why would we still care about correcting the analytic activity as soon as we have found the simulation activity, instead of simply using the dynamic activity?" Analytic activity can depend on initial state but also on input event (for fire spread a cell can receive water or not impacting its simulation activity). Therefore, a component with a high analytic activity could have a low level of activity at the beginning of the simulation. However, the level of activity can increase during the simulation, which was anticipated by the high-level analytic activity. The analytic activity allows to anticipate the high-level activity and to take faster decision/prediction (e.g., for load balancing in a case of distributed simulation).

5.2 The IEEE 802.3 CSMA/CD Protocol Use Case

This section introduces a pedagogical example allowing to validate the proposed approach. This case study concerns the IEEE 802.3 CSMA/CD (Carrier Sense, Multiple Access with Collision Detection) protocol (Zeigler, 1976). The CSMA/CD protocol is designed for networks with a single channel and specifies the behavior of stations with the aim of minimizing simultaneous use of the channel (data collision). The basic structure of the protocol is as follows: when a station has data to send, it listens to the medium, after which, if the medium was free (no one transmitting), the station starts to send its data. On the other hand, if the medium was sensed busy, the station waits a random amount of time and then repeats this process.

In Fig. 8, we have presented the DEVS automata model of the station model. A station starts by sending its data. If there is no collision, then, after λ time units, the station finishes sending its data. On the other hand, if there is a collision, the station attempts to retransmit the packet where the scheduling of the retransmission is determined by a truncated binary exponential

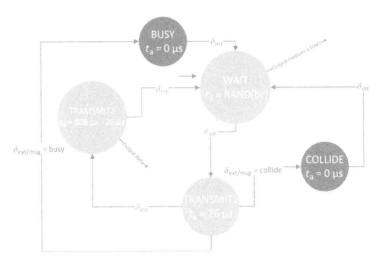

FIG. 8   Station model automata.

back-off process. The delay before retransmitting is an integer number of time slots (each of length slot time). The number of slots that the station waits after the nth transmission failure is chosen as a uniformly distributed random integer. Once this time has elapsed, if the medium appears free the station resends the data (event send), while if the medium is sensed busy (event busy) the station repeats this process.

The medium is initially ready to accept data from any station (event send). Once a station starts sending its data there is an interval of time (at most σ), representing the time it takes for a signal to propagate between the stations, in which the medium will accept data from the other station (resulting in a collision). After this interval, if the other station tries to send data it will get the busy signal (busy). When a collision occurs, there is a delay (again at most σ) before the stations realize there has been a collision, after which the medium will become free (represented by the event CD). If the stations do not collide, then when a station finishes sending its data (event end) the medium becomes idle. The medium state automata is summarized in Fig. 9.

Fig. 10 depicts the DEVSimPy model of the case study. "Station_1" and "Station_2" can be considered as the model AM1 and the "Medium_3" as the model AM2. Simulation has been performed during 1000 seconds (simulation time).

Fig. 11 shows the setup interface of AT plug-in in DEVSimPy for the DEVS model of the case study. Once the user has enabled the plug-in activation, he/she has to select both the atomic models, functions (δ_{ext}, δ_{int}, λ, or t_a) and type AT (analytic or simulation) under investigation.

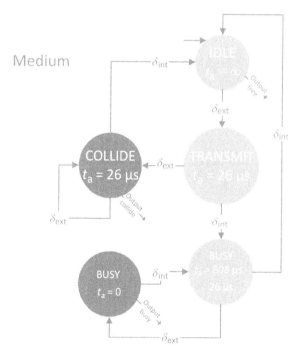

FIG. 9 Medium model automata.

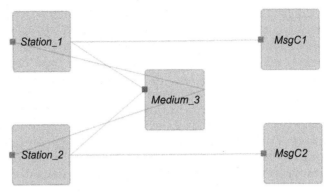

FIG. 10   The CSMA/CD protocol DEVSimPy model.

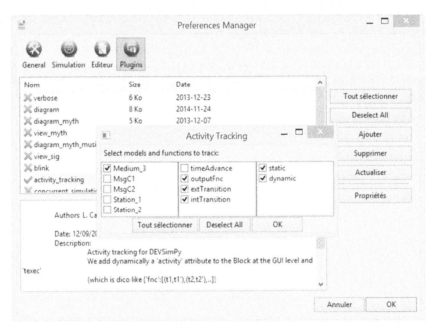

FIG. 11   The setup of the AT DEVSimPy plug-in.

Fig. 12 gives a summary of obtained results concerning simulation-based activity metrics. The results of the three metrics introduced in Section 3.1 are presented: QA (QActivity column), the WA (WActivity column), and the CPU user time (CPUs column).

Fig. 13 gives the obtained results concerning the analytic-based metrics. The two metrics presented in Section 3.2: the MCC metric (MCC column) and the program performance measurements metrics (Worst_case and Best_case columns).

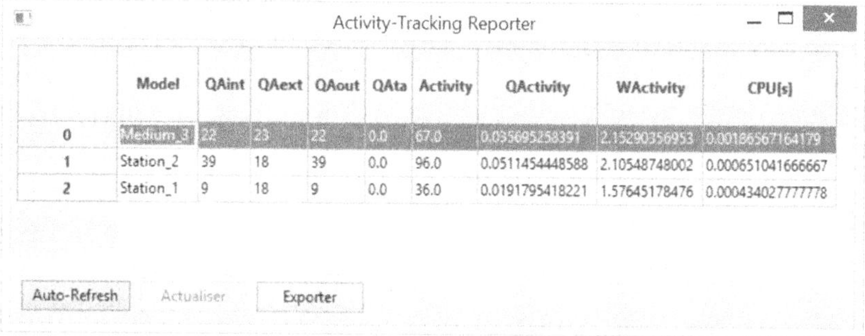

Activity-Tracking Reporter

	Model	QAint	QAext	QAout	QAta	Activity	QActivity	WActivity	CPU[s]
0	Medium_3	22	23	22	0.0	67.0	0.035695258391	2.15290356953	0.00186567164179
1	Station_2	39	18	39	0.0	96.0	0.0511454448588	2.10548748002	0.000651041666667
2	Station_1	9	18	9	0.0	36.0	0.0191795418221	1.57645178476	0.000434027777778

Auto-Refresh Actualiser Exporter

FIG. 12 Simulation activity results.

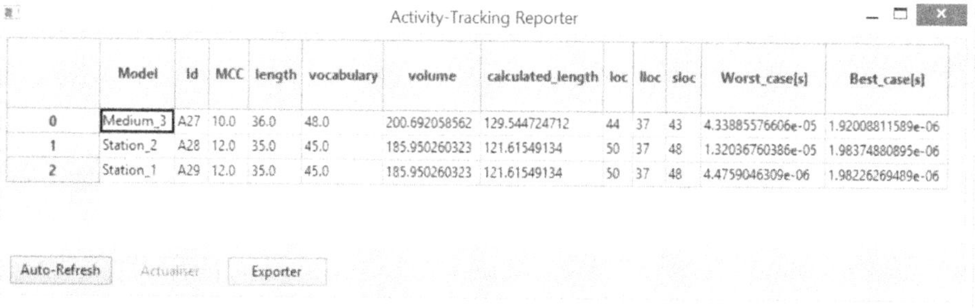

Activity-Tracking Reporter

	Model	Id	MCC	length	vocabulary	volume	calculated_length	loc	lloc	sloc	Worst_case[s]	Best_case[s]
0	Medium_3	A27	10.0	36.0	48.0	200.692058562	129.544724712	44	37	43	4.33885576606e-05	1.92008811589e-06
1	Station_2	A28	12.0	35.0	45.0	185.950260323	121.61549134	50	37	48	1.32036760386e-05	1.98374880895e-06
2	Station_1	A29	12.0	35.0	45.0	185.950260323	121.61549134	50	37	48	4.4759046309e-06	1.98226269489e-06

Auto-Refresh Actualiser Exporter

FIG. 13 Analytic activity results.

6 CONCLUSION AND FUTURE WORK

In this chapter, analytic and simulation activity metrics have been presented with a special attention about their relationships. Concerning the analytic activity metric, the new MCC activity metric (MCA) of a DEVS model as well as a software performance measurement (*timeit* python module based) metric are presented. For simulation activity metrics, the well-known QA and WA metrics have been used in the context of profiling the activity distribution over components. Both MCA and QA metrics have been compared with the user CPU time of the DEVS transition functions are introduced. The MCA, QA, WA, CPU, and timeit-based metrics have been implemented through a plug-in of the DEVSimPy (DEVS Simulator in Python language) environment and applied to a case study. It is clear that MCA and the timeit-based metrics constitute a good approximation allowing a priori estimation of component simulation overheads.

The main perspective concerns the improvement of the simulation algorithm using the activity metrics. Structure modification of abstract simulators will be guided by a combination of both activities associated with a parallel and distributed simulation algorithm. DEVSimPy framework already integrates PyPDEVS API thus exploiting parallel and distributed features. We plan also to use an activity prediction model, which requires that the user provides (domain-specific) knowledge about how a specific model will use computational resources when simulated with a specific simulator. Furthermore, in order to avoid the break of the model-simulator abstraction (due the fact that the modeler needs some knowledge about the simulator's operation), we plan to explore the use of Domain-Specific Language to automated the construction of the activity prediction model.

We also work on PDEVS protocol performance prediction using activity patterns with Finite Probabilistic DEVS. We propose to model the PDEVS protocol using a Markov Continuous Time Model, which is set up using two parameters: (i) the probability of a state giving an output according to input patterns and (ii) the rate of the coordinator's release of imminent. These parameters are computed using activity metrics described in this chapter and inserted into the PDEVS modeling scheme in order to predict the performance of a PDEVS protocol in the framework of distributed simulations (Zeigler, 2017).

References

Beazley, D., Jones, B.K., 2013. Python Cookbook. O'Reilly Media, Inc., Boston, MA. ISBN 1449340377, 9781449340377.

Beck, F., Diehl, S., 2011. On the congruence of modularity and code coupling. In: Proc. of the 19th ACM SIGSOFT Symposium and the 13th European Conference on Foundations of Software Engineering, ESEC/FSE '11. ACM, New York, NY, pp. 354–364.

Bolduc, J.S., Vangheluwe, H., 2001. The Modelling and Simulation Package PythonDEVS for Classical Hierarchical DEVS. McGill University. http://msdl.cs.mcgill.ca/projects/projects/DEVS/. MSDL-TR-2001-01.

Capocchi, L., Santucci, J.-F., Poggi, B., Nicolai, C., 2011. DEVSimPy: a collaborative python software for modeling and simulation of DEVS systems. In: 2011 20th IEEE International Workshops on Enabling Technologies: Infrastructure for Collaborative Enterprises (WETICE), pp. 170–175.

Fard, M.D., Sarjoughian, H.S., 2015. Visual and Persistence Behavior Modeling for DEVS in CoSMoS, eighth ed., vol. 47. The Society for Modeling and Simulation International, San Diego, CA, pp. 227–234.

Halstead, M.H., 1977. Elements of Software Science. Operating and Programming Systems SeriesElsevier, New York, NY. http://opac.inria.fr/record=b1084731 ISBN 0-444-00205-7.

Hu, X., Zeigler, B.P., 2013. Linking information and energy-activity-based energy-aware information processing. Simulation 0037-549789 (4), 435–450. https://dx.doi.org/10.1177/0037549711400778.

Kung, C.H., Sölvberg, A., 1986. Activity modeling and behavior modeling. In: Proc. of the IFIP WG 8.1 Working Conference on Information Systems Design Methodologies: Improving the Practice. North-Holland Publishing Co., Amsterdam, pp. 145–171. http://dl.acm.org/citation.cfm?id=20143.20149.

Liu, L., Meersman, R., 1992. Activity model: a declarative approach for capturing communication behavior in object-oriented databases. In: VLDB '92. Proc. of the 18th International Conference on Very Large Data BasesMorgan Kaufmann Publishers Inc., San Francisco, CA, pp. 481–493. http://dl.acm.org/citation.cfm?id=645918.672342.

McCabe, T.J., 1976. A complexity measure. IEEE Trans. Softw. Eng. 0098-55892 (4), 308–320. https://dx.doi.org/10.1109/TSE.1976.233837.

Muzy, A., Hill, D.R.C., 2011. What is new with the activity world view in modeling and simulation? Using activity as a unifying guide for modeling and simulation. In: Proc. of the 2011 Winter Simulation Conference (WSC), pp. 2882–2894.

Muzy, A., Zeigler, B.P., 2012. Activity-based credit assignment (ACA) in hierarchical simulation. In: Proc. of the 2012 Symposium on Theory of Modeling and Simulation—DEVS Integrative M&S Symposium, TMS/DEVS '12. Society for Computer Simulation International, San Diego, CA, pp. 5:1–5:8. http://dl.acm.org/citation.cfm?id=2346616.2346621.

Muzy, A., Jammalamadaka, R., Zeigler, B.P., Nutaro, J.J., 2011. The activity-tracking paradigm in discrete-event modeling and simulation: the case of spatially continuous distributed systems. Simulation 87 (5), 449–464. https://dx.doi.org/10.1177/0037549710365155.

Muzy, A., Varenne, F., Zeigler, B.P., Caux, J., Coquillard, P., Touraille, L., Prunetti, D., Caillou, P., Michel, O., Hill, D.R., 2013. Refounding of the activity concept? Towards a federative paradigm for modeling and simulation. Simulation 89 (2), 156–177. https://dx.doi.org/10.1177/0037549712457852.

Park, C., Shaw, A.C., 1990. Experiments with a program timing tool based on source-level timing schema. In: Proc. 11th Real-Time Systems Symposium, pp. 72–81.

Santucci, J.F., Capocchi, L., 2013. Implementation and Analysis of DEVS Activity-Tracking With DEVSimPy. In: vol. 1. ITM Web of Conferences, p. 01001. https://dx.doi.org/10.1051/itmconf/20130101001.

Sarjoughian, H.S., Elamvazhuthi, V., 2009. CoSMoS: a visual environment for component-based modeling, experimental design, and simulation. In: Proc. of the 2nd International Conference on Simulation Tools and Techniques, Simutools '09. ICST (Institute for Computer Sciences, Social-Informatics and Telecommunications Engineering), ICST, Brussels, Belgium, pp. 59:1–59:9. https://dx.doi.org/10.4108/ICST.SIMUTOOLS2009.5744.

Shaw, A.C., 1989. Reasoning about time in higher-level language software. IEEE Trans. Softw. Eng. 0098-558915 (7), 875–889. https://dx.doi.org/10.1109/32.29487.

Stevens, W.P., Myers, G.J., Constantine, L.L., 1974. Structured design. IBM Syst. J. 13 (2), 115–139.

Tendeloo, Y.V., 2018. PyPDEVS package. Retrieved on August 15, 2018 from http://msdl.cs.mcgill.ca/people/yentl.

Van Mierlo, S., Mustafiz, S., Barocca, B., Van Tendeloo, Y., Vangheluwe, H., 2015. Explicit modelling of a parallel DEVS experimentation environment. In: Proc. of the Symposium on Theory of Modeling & Simulation, DEVS '15Society for Computer Simulation International, San Diego, CA, pp. 860–867.

Van Tendeloo, Y., Vangheluwe, H., 2014. The modular architecture of the Python(P)DEVS simulation kernel (WIP). In: Proc. of the Symposium on Theory of Modeling & Simulation—DEVS Integrative, DEVS '14. Society for Computer Simulation International, San Diego, CA, pp. 14:1–14:6. http://dl.acm.org/citation.cfm?id=2665008.2665022.

Yazidi, A., Henao, H., Capolino, G.-A., Betin, F., Capocchi, L., 2010. Inter-turn short circuit fault detection of wound rotor induction machines using bispectral analysis. In: 2010 IEEE on Energy Conversion Congress and Exposition (ECCE), pp. 1760–1765.

Zeigler, B.P., 1976. Theory of Modeling and Simulation. Academic Press, London.

Zeigler, B.P., 2017. Using the parallel DEVS protocol for general robust simulation with near optimal performance. Comput. Sci. Eng. 1521-961519 (3), 68–77. https://dx.doi.org/10.1109/MCSE.2017.52.

Zeigler, B.P., Sarjoughian, H.S., 2013. Guide to Modeling and Simulation of Systems of Systems. Springer-verlag, London.

Zeigler, B.P., Kim, T.G., Praehofer, H., 2000. Theory of Modeling and Simulation, second ed. Academic Press, Inc., Orlando, FL. ISBN 0127784551.

Zeigler, B.P., Muzy, L., Kofman, E., 2018. Theory of Modeling and Simulation, third ed. Academic Press, London.

13

Generic Concept and Architecture for Efficient Model Management

Günter Herrmann*, Axel Lehmann†, Robert Siegfried‡

*Institut für Technik Intelligenter Systeme, Neubiberg, Germany †Institut für Technische Informatik, Universität der Bundeswehr München, Neubiberg, Germany ‡aditerna GmbH, Riemerling, Germany

1 INTRODUCTION—DEMANDS FOR M&S MANAGEMENT

For an increasing range of application domains modeling and simulation is becoming more and more a key enabling technology for various purposes, such as the following:

- for education and training
- for analysis and evaluation of systems, processes
- as planning tool for procurement and
- for decision support.

All these models, simulations, and data (M&S) are developed and applied with a specific purpose in mind, in a well-defined context and under specified restrictions for application. Therefore, these M&S should be subsequently used only within this context.

There is an increasing demand for reducing M&S design, development, and maintenance resources and costs as well as considering new requirements arising from demands to apply an M&S as part of distributed or network-centric environment. Often continuously changing context conditions or an increasing complexity of applications impose new requirements on the easy accessibility of permanently updated information of an M&S life cycle–its design, implementation, usage and adaptations, etc. In addition, due the complexity and multiple applications of many models, simulations and their data, measures for quality and credibility assurance are receiving more and more importance. Like in many engineering disciplines, application, respectively, adaptation of systems engineering principles should be considered for all phases of M&S lifetime (see, e.g., Kossiakoff et al., 2011).

In order to cope with all these demands, it is required to specify and document in a central repository for each M&S:

- scope of the application (training, analysis, decision support, procurement);
- traces of all activities regarding M&S usage, maintenance, verification and validation results, redesigns, or adaptions in respect of all M&S lifecycle phases;
- coupling with or integration of other models or submodels as well as with real systems forming a distributed simulation platform;
- experiences obtained by users in context of each application and reusability opportunities.

In order to match these demands, the development as well as documentation of the models, their simulation application as well as their input and output data (M&S) has to satisfy the following criteria:

- All phases of M&S—from task specification, conceptual and formal model design, implementation of the simulation platform, as well as corresponding data analyses and interpretation—have to be executed according to a well-defined, standardized M&S development process. This development process has to include detailed guidelines on the required phase products and corresponding documentation.
- Simulation models have to be comparable in order to decide whether a specific coupling or integration is feasible and makes sense.
- A standardized set of phase products and the corresponding documentation enables the evaluation of opportunities for coupling or integration of different models for a specific purpose and within a specific scenario.
- The variety of potential users requires the provisioning of different views on the documentation of M&S. Finally, the user should be able to decide upon the information presented to him whether a specific model, simulation, or even data is suitable for an intended specific purpose and how to apply them (Neches et al., 1991; Arnold et al., 2005).
- Models, simulations, data, and scenarios have to be managed and provided to users in a suitable, controlled way.

These demands may be met by applying organizational and technical measures. In accordance with basic system engineering principles, this chapter proposes a sound and solid conceptual framework of a model management system (MMS) (Herrmann et al., 2017) which is of crucial importance for the success of such an integrated approach for M&S design, development, maintenance, and operation.

2 REQUIREMENTS AND DESIGN PRINCIPLES FOR AN MMS

As mentioned above, management and permanent accessibility to updated and valid documentation regarding an M&S development, its applications, and users during all stages of its lifetime are crucial for its effectiveness as well as for efficient quality control and credibility assurance. As prerequisite, M&S design, development, application, and maintenance processes should follow well-experienced principles of systems engineering (e.g., Kossiakoff et al., 2011). In this regard, a model management supporting system (MMS) must satisfy three fundamental categories of requirements which are complementary to one another:

- The MMS must integrate the domain-specific concepts for model development, documentation and quality assurance, and VV&A (Wang and Lehmann, 2007a, 2010; Wang et al., 2009a; Sargent, 2011; Lehmann and Wang, 2017).
- The MMS must offer project organization support for carrying out the domain-specific concepts.
- The MMS must support ever-increasing complexity requirements. Its fundamental concepts must therefore intrinsically support high flexibility and upgradeability of the architecture and its use.

On the basis of the experience in other technical fields, we therefore decided to consider *Model Engineering* as an engineering discipline like any other fields, and to manage model conception, development, usage, maintenance, and quality control as a pure engineering process.

2.1 Domain-Specific Concepts

The main MMS design motivation for the domain-specific concepts for model development and VV&A consists in subjecting the model development process to a structured approach and to systematically document, according to predefined templates, each of the individual phases of this structured approach, as well as the products of each of these individual phases.

Fig. 1 shows the seven phases of the model development process as well as the seven corresponding phase products, see also Wang and Lehmann (2007b) and Wang and Lehmann (2010). The process starts with the *Preliminary phase* and ends with the *Interpretation phase*, possibly after iterations caused by error detections, requiring adjustments or new versions. Each of these phases, as well as each of the products of each phase must be documented according

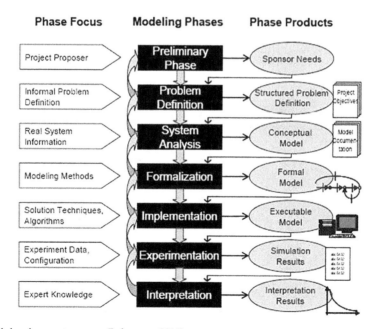

FIG. 1   Model development process (Lehmann, 2014).

to predefined, standardized but tailorable templates (Wang et al., 2009b). As an example, in Fig. 1, the phase product *Structured Problem Definition* is also represented as a *project objectives* report (or documentation) which is the concrete product resulting from the Problem Definition phase. All phase products of phase i form the basic specification or prerequisite for the work to be performed in phase $i+1$. In Wang and Lehmann (2010), we have also specified precisely the roles of team members within each phase of an M&S project. This proposed concept of M&S roles guarantees that responsibilities as well as contributors of the project are known and visible for all project members. Fig. 2 shows an overview of the model documentation as well as the closely related concepts and products which are produced during the model development process.

The specialized aspects and functions of the MMS are however not only restricted to model development, they are really intended to cover the entire life cycle of the model. In particular, the aspects of model use, reuse, and replacement must be supported. Additional documentation and quality assurance concepts have been developed for these further model life cycle phases as well. As starting points for these documentation and quality assurance measures, the family of products developed during the model development phases was enlarged to include products relevant to further phases of the model life cycle. Examples for these additional products include experiment descriptions and scenario definitions.

The use cases in these further phases of the model life cycle include, for instance, support for integrating existing data bases, for taking over documentation from existing repositories,

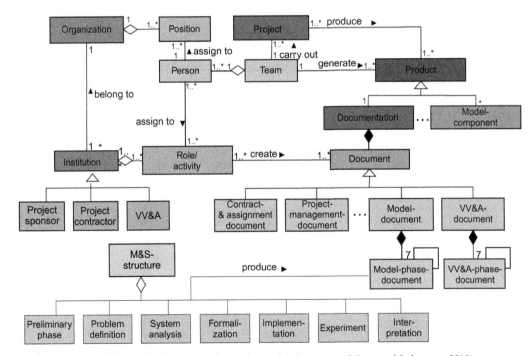

FIG. 2 Overview of the model documentation and associated concepts (Wang and Lehmann, 2010).

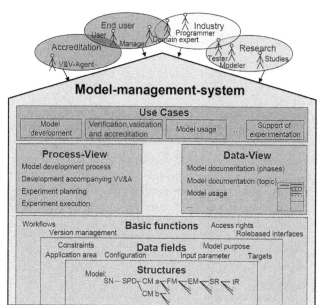

FIG. 3   High-level overview of the architecture of the proposed model management system (MMS) (Lehmann, 2014; Herrmann et al., 2017).

as well as for designing and conducting experiments. For instance, they offer component search and coupling support, the follow-up documentation of models, as well as extended search and output possibilities (queries, reports, documentation).

Fig. 3 illustrates an overview of the architecture of the proposed MMS, including the use cases which are to be supported. Especially, Fig. 3 highlights that the various user groups (like end user, developer, etc.) have very different demands and therefore will use the MMS within very different use cases. Each use case in turn combines processes and data in a coherent way, thereby providing the exact right amount of information needed by the user within the actual use case. Technically, the use cases and processes are built on top of basic (domain-independent) functionalities like workflows and version management. These functionalities operate on so-called structures (e.g., a model is a structure) which are described and documented by a multitude of associated data fields. All components shown in Fig. 3 are explained in detail in Section 3.

2.2 Project Organization Support

Clearly, the functional demands on the MMS are very closely related with the project organization demands. In fact, the project support functions (e.g., search functions, coupling assessment) are fundamental for the usability of the MMS and are certainly applicable within various use cases. The basis for the project support functions comprises low level and technical functionalities, which are either context independent or which only become useful through being linked to a specialized context. The MMS supports interactive and collaborative functions (e.g., know-how exchange, Wiki, Blog) as well as general administrative

functions such as workflow and role management, version management, history manage-
ment, problem management, change management, document management, and configuration management.

2.3 Flexibility

Fig. 4 provides an overview of the use cases resulting from the requirements analysis and
the supporting functions on which these use cases are built.

Strategic user expectations are always coupled with an MMS meeting these requirements.
Such expectations include, for instance, the use of the MMS as central model catalog within an
organization, the integration of the MMS in the existing experimental environments and simulation environments as well as enabling the constitution of competence networks. The

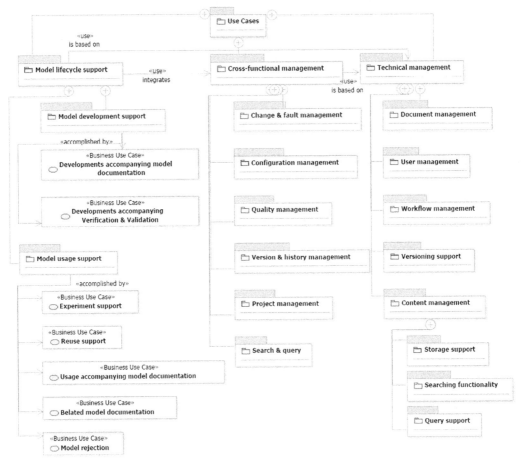

FIG. 4 Overview of the MMS use cases and of the supporting functions on which these use cases depend
(Herrmann et al., 2017).

heterogeneity of the user community as well as the future requirements and developments which will result from it has to be continually integrated into the MMS concept. Therefore, the MMS concept was developed so that the supported use cases can be flexibly and adaptively designed and also they tolerate extensions (within certain limits), that is, so that additional use cases can be seamlessly integrated into the existing MMS concept.

3 MMS CONCEPT

3.1 The MMS Meta-Concept as Design Pattern

A straight modeling of predetermined domain-specific requirements is not the goal of our MMS concept, because the prescribed ability to support numerous unfocused and incomplete future requirements precludes this straightforward but unupgradeable approach.

The usual approach to develop a partial solution for each management requirement, independent of the other requirements, is neither desired nor implementable for a future-proof MMS concept and MMS development. Such an ad hoc approach would contradict the basic premise of the coordinating and centralizing function of an MMS in which the fundamental core is a unitary and consistent domain model.

How can a common concept support this multiplicity of unfocused, still unknown future requirements? A possible solution is to develop a structurally abstract MMS meta-concept. The result of this abstraction is a design pattern for requirements and problems of a specific problem class. The advantage of this abstract design pattern is that it permits in principle a unitary representation of all requirements and problems, present and future, known and yet unknown.

In summary, the intention of the MMS meta-concept is to provide a design pattern within which the presently known documentation requirements in the domain of M&S can be implemented, while being general and flexible enough to be able to seamlessly integrate a large array of still unknown additional requirements as they might arise in the future. The MMS meta-concept is modeled using Unified Modeling Language (UML), according to Object Management Group (2009).

3.2 Problem Class

The problem class is defined by determining the ideas on which most of the requirements are based. These ideas do not exist on the abstraction level of the domain-specific problems which are very specific requirements. Rather they are the commonalities underlying most of the individual requirements.

This problem class can be identified as well as conceptually modeled on a higher level of abstraction. It can be imagined as a kind of frame in which the actual problem expressions can be fitted in their appropriate locations. The specific requirements must be understood as instantiation of this general problem class.

The resulting meta-concept allows to represent the specificity of the requirements, and in a way this meta-concept is a design pattern for a specific class of requirements and problems. The problem class of the MMS is the description of real-world entities in the application domain of M&S. The design pattern for this problem class enables the modeling of real-world structures and the description of the identified entities.

In this manner, the meta-concept of the MMS is in principle usable for all use cases which perform any kind of documentation tasks or which retrieve description data. However, semantic consistency requirements restrict the application domain of the MMS meta-concept to a limited semantic field which is definable without contradiction. This is due to the fact that for all core elements (i.e., the structures, data fields, and roles as described in the following) of the specific domain model first clarity and unicity of meaning must be guaranteed; and second disjoint meanings between these elements must be ensured. Since the meaning of a term is dependent on the context of its use, this context must be restricted according to specialized fields in order to exclude the possibility of multiple meanings. In the case of the MMS, this context is the application domain of M&S.

Requirements which go beyond the mere description of modeled entities and beyond the mere providing information about these entities (for instance, actual coupling of components) must therefore be excluded from the concept. These functionalities have to be considered separately and added using a different structural concept.

3.3 Benefits of the MMS Meta-Concept

Introducing a design pattern to describe the use cases seem to be lots of effort for unclear reasons. We believe that the strict use of the proposed design patterns helps to meet the various (and continuously evolving) requirements, thus leading to considerable benefits:

- Most notably, following the design pattern ensures that all use cases are modeled (and subsequently implemented) in the same way. This eases the comparability of use cases and allows to define globally valid quality requirements. Specific requirements may be defined once in the meta-concept, and are in turn applied to all use cases. As an example, the meta-concept states that all activities are executed by specific roles. Assigning an activity to a single person or a whole organization is therefore prohibited.
- A direct consequence of this is that the MMS is based on a consistent domain model. Especially, this implies that all parts (roles, data fields, structures, views) may be reused. In fact, they are modeled just once and have to be reused. When creating a new use case, already existing parts have to be reused and only parts, which are not yet included in the domain model have to be added. The meta-concept forces the modeler/developer to reuse the existing parts and secures the consistency of the domain model. As an example, a data field for the description of the "Problem statement" may be referenced in several use cases.
- The MMS domain model is limited to the domain of modeling and simulation. This limitation is necessary to ensure the unambiguousness of the concepts defined within the domain model.

3.4 The MMS Meta-Concept in Detail

The basis of the MMS meta-concept as design pattern is the strict separation of static elements (domain model) and dynamic elements (model management), as depicted in Fig. 5.

The strength of the MMS meta-concept as design pattern resides in its ability to define unlimited interaction possibilities with the modeled entities. These interaction possibilities

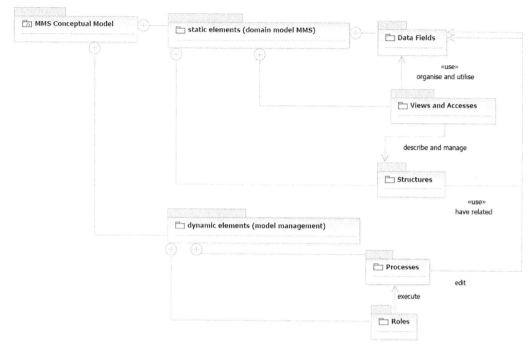

FIG. 5 The very fundamental idea of the MMS meta-concept is to separate the static and dynamic elements (Herrmann et al., 2017).

represent user-specific views in the application domain of M&S, and are modeled as USE CASES. The design pattern covers these use cases in their dynamic and static aspects. (Note: Terms in small capitals explicitly denote classes which are part of the MMS meta-concept.)

Each modeled USE CASE may and will in general extend the domain model of the MMS. This can occur because additional real-world entities must be documented or because additional description data fields must be assigned to the existing or newly modeled entities. The first priority when introducing these extensions is to use the existing STRUCTURES and DATA FIELDS. If new aspects must be introduced at all, these must be integrated as conservatively and as generically as possible in the structure model and data model, taking care of integrating them as seamlessly as possible with the existing ones.

The goal is to saturate the domain model with STRUCTURE and DATA FIELD definitions by integrating a sufficiently large number of USE CASES, so that the resulting MMS ends up being based on a consistent, comprehensive, and self-consolidated domain model for the problem class.

3.4.1 Static Elements: MMS Domain Model

Structures: The domain model of the MMS consists of a representation of the entities relevant for the problem class. These entities are modeled in their inner and outer relations as STRUCTURES.

This modeling of STRUCTURES makes it possible to explicitly work with the modeled entities like MODEL, PHASE MODEL, and SUBMODEL in the MMS. Especially, every instance of such a STRUCTURE can be described by data fields.

An example is the structure of a model as shown in Fig. 6: A MODEL is internally structured in seven PHASE MODELS, with three of them (CM, FM, EM) including at least one SUBMODEL, which itself can include sub-SUBMODELS. This example is taken from the use case "Development accompanying model documentation." A further use case, for

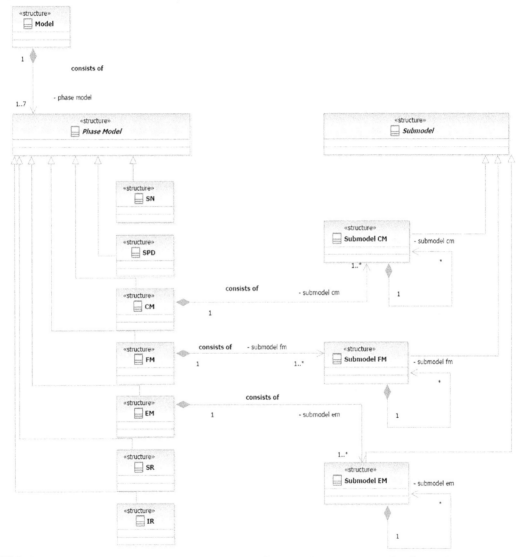

FIG. 6   All entities handled by the MMS, like a MODEL for example, are represented in the MMS meta-concept as STRUCTURES and have to be modeled explicitly with their inner and outer relations (Herrmann et al., 2017).

example, "Model usage (experimentation)" might define a new structure (e.g., "Experiment") which could reuse the already existing structure EM.

Data fields: Since the problem class of the MMS is the description of real-world entities (which are represented in the MMS meta-concept as STRUCTURES) in the application domain M&S and since a central requirement is to maintain the consistency of the data model, the STRUCTURES have been conceptually separated from their description. For this reason, the MMS structure model is supplemented by a separate data model, in which all possible description aspects in the application domain are modeled in the form of DATA FIELDS.

In principle it would be sufficient to model the DATA FIELDS as an unstructured set of classes. But to ensure a clear overview and understanding, an aspect-oriented description hierarchy was modeled internally to the MMS, the leaves of this hierarchy being the DATA FIELD classes. Fig. 7 shows an overview of the most general categories of the representation aspects as well as an example of a branch of this hierarchy.

The main benefit of using separate DATA FIELD classes is the possibility to reuse data fields for describing various structures. For example, a data field "Classification/Copyright" might be used for models as well as experiments.

Views and accesses: Only the actual instantiation of a STRUCTURE is associated with specific description data (instantiated DATA FIELDS). This association is determined by a predefined and prespecified USE CASE.

The link between the DATA FIELDS on the one hand and the STRUCTURES on the other hand are the VIEWS AND ACCESSES, which define how structures are described by the DATA FIELDS. For every USE CASE, there are one or more association possibilities of DATA FIELDS to STRUCTURES. It is possible to specify several views as, for instance, several

FIG. 7   MMS data model: Overview of the hierarchy of DATA FIELDS, including a selection of DATA FIELDS for CONSTRUCTION ASPECTS (Herrmann et al., 2017).

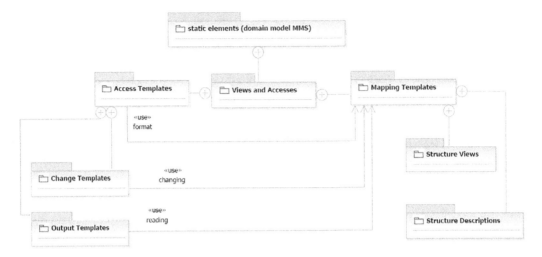

FIG. 8   MMS meta-concept: VIEWS AND ACCESSES are the linking element between the DATA FIELDS on the one hand and the STRUCTURES on the other hand (Herrmann et al., 2017).

combinations of DATA FIELDS might be desired for describing a STRUCTURE (e.g., a model). The totality of these associations constitutes the third pillar of the MMS domain model: the Views and Accesses.

As shown in Fig. 8 the description of a STRUCTURE has two aspects: Firstly, it must be clear which data fields will be used in order to describe the entity, out of the large pool of available DATA FIELDS. The DATA FIELDS are then associated to STRUCTURES as specific description templates. These content description templates are called MAPPING TEMPLATES. They are further subdivided into use case-specific mappings (STRUCTURE VIEWS) and general mappings (STRUCTURE DESCRIPTIONS) internally defined within the MMS.

A small example may clarify this. If we consider a structure (like a "model"), the structure description of this structure contains all data fields which are currently defined (e.g., "Problem statement," "Input parameters," or "Simulation results"). A single structure view refers only to a subset of these data fields. A structure view for the use case "Model documentation" might only refer to the "Problem statement" and "Input parameters." Another structure view, related to the use case of model usage, might refer to the "Simulation results."

To summarize, the structure description contains all data fields related to a specific structure. The structure views in turn refer to the specific subset of data fields which is required within the current use case.

Secondly, the DATA FIELDS must be accessible, either to read them or for data management. The simple content description is not sufficient as the user of the MMS must actually be able to read or write data. For this reason the content patterns must be cast in a specific form and the access rights and access methods of the user must be defined. The ACCESS TEMPLATES are the interfacing templates which perform the tasks of formatting the content description and controlling user access. For this purpose, each ACCESS TEMPLATE uses specific MAPPING TEMPLATES, according to the user-specific view and use case. The ACCESS TEMPLATES which are read-only are called OUTPUT TEMPLATES, those granting writing privileges are called CHANGE TEMPLATES.

While the structure views define only the subset of data fields which is required within the current use case, the ACCESS TEMPLATES define the order of these data fields and the access rights of the users. An actual system implementing this MMS meta-concept would provide a user interface which is defined by the CHANGE TEMPLATES.

MAPPING TEMPLATES therefore describe which DATA FIELDS serve the description of which STRUCTURE from which perspective. The ACCESS TEMPLATES use this content description structure and make it available in a manner suitable to the current user, so that he can actually work with these DATA FIELDS.

3.4.2 Dynamic Elements: Model Management

The selection of which views to be used and thereby actually receive data is determined individually by each USE CASE. A USE CASE is defined by the association of the process description and the participating ROLES (dynamic aspect), as well as by the used views with their related STRUCTURES and the DATA FIELDS (static aspect) describing them.

Processes: The dynamic aspects are modeled by processes which carry out the USE CASES. Such a process defines the sequencing order, the relations to generated products (in the form of MAPPING TEMPLATES and ACCESS TEMPLATES) as well as the participating ROLES. Fig. 9 shows an example of such a modeled process suitable for carrying out the model documentation which is itself part of the overall process for the USE CASE development accompanying model documentation.

It exhibits how the PHASE MODEL specific documentation versions are produced and how the model documentation (in a certain version-specific form) results from putting the PHASE DOCUMENTS together. In particular, project specific tailored Change Templates for phase specific model documentation are used here.

Roles: ROLES define task fields and responsibilities within a process (see also Fig. 2). Instantiated processes (WORKFLOWS) associate ROLES to actual people. These ROLES are similarly defined in the ACCESS TEMPLATES, which are also USE CASE specific. The ACCESS TEMPLATES statically define the access rights of ROLES to DATA FIELDS, that is, the access rights are not defined in the processes themselves. The actual execution of a process thus clearly defines who take which Roles (via the Workflow) and which Roles can access which Data Field instances, and how they access them (via the instantiated Access Templates).

Fig. 10 represents the overall architecture of the MMS meta-concept and clarifies how ROLES are integrated in it.

3.4.3 Execution of a Use Case in the MMS

The execution of a USE CASE in the MMS is carried out along the following steps:

- Triggering and execution of a WORKFLOW.
- Instantiation of the ROLES and corresponding allocation of actual persons.
- Selection and instantiation of views, respectively, ACCESS TEMPLATES.
- Instantiation (and versioning) of STRUCTURES.

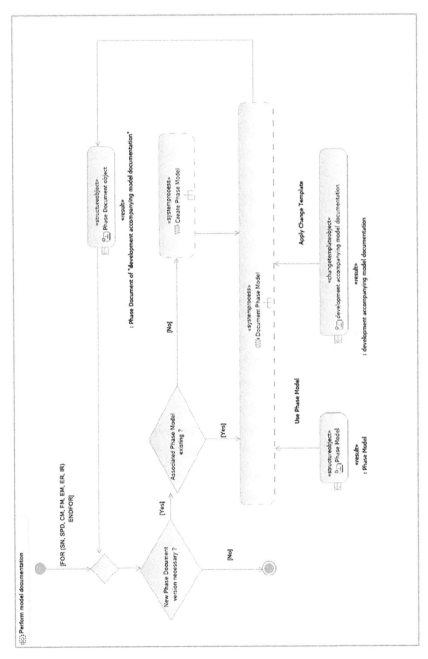

FIG. 9   Process for performing model documentation (Herrmann et al., 2017).

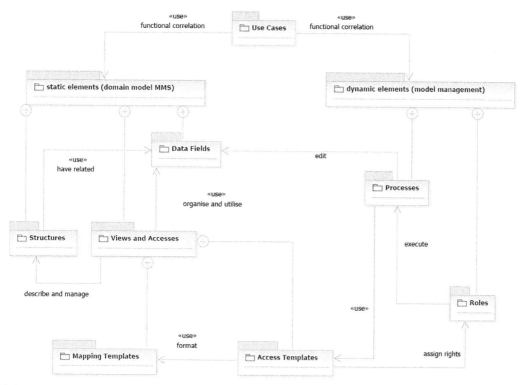

FIG. 10 Overview of the overall architecture of the MMS meta-concept (Lehmann, 2014; Herrmann et al., 2017).

- Instantiation of DATA FIELDS and allocation of these DATA FIELD instances to the STRUCTURE instances (according to the view).
- Content processing of the DATA FIELDS during the WORKFLOW execution.

The USE CASE execution results in additional STRUCTURE instances and associated description data (DATA FIELD instances) stored in the MMS dataset.

4 FORMALIZATION OF THE MMS CONCEPT

In order to enable further investigations the MMS concept had to be formalized in an unambiguous and computationally accessible way while reflecting the static elements (domain model) and dynamic elements (model management) of the MMS concept. The MMS domain model was formalized as ontology using the Web Ontology Language (OWL) (Gómez-Pérez et al., 2004). Building upon the model development process (Fig. 1) and the related products (Fig. 2), the devised MMS concepts were structured and formalized, defining the scope of the MMS ontology.

This MMS ontology is intended to represent the core data model of the MMS concept which supports the model development process (including documentation) as described in

FIG. 11   High-level overview about the OWL formalization of the static elements of the MMS meta-concept, that is, of the MMS domain model (Herrmann et al., 2017).

Section 2.1. The basic MMS ontology modeling pattern directly reflects the static elements of the MMS concept (see Fig. 11) and is therefore composed of three main components, as shown in Fig. 11:

- *Structures*: they model real-world structures like, for example, a model version, phase model, or configuration.
- *Data fields*: they represent places where the actual data of the structure description are stored.
- *Views and accesses*: they define how structures and data fields are associated with each other.

Each of the classes shown in Fig. 9 has multiple relations to other classes which are defined as OWL object properties, like, for example, *hasDocumentation* and *hasDatafields* for relating models with their documentation and these documents with the necessary data fields. All the object properties are modeled with their corresponding inverse relations, like *inverseOf_hasDocumentation* and *inverseOf_hasDatafield* in order to permit easy navigation within the data.

Just like the MMS concept itself, the MMS ontology relies strictly on the separation of structures (like models or model documentation), their description via the data fields, and the interconnection of the structure and its describing data fields. The main advantage of this modeling pattern of strict separation is that it manages and maintains a large number of concepts in a clear and simple structure away from their corresponding data in order to:

- avoid the complexities of frequently handling or moving massive amounts of data whenever the usage or structure of this data (or related concepts) changes;
- avoid multiple occurrences in the structure of the same data, which could cause data maintenance problems or data inconsistencies; and
- facilitate concept matching between different sources when seeking to integrate them into the MMS ontology.

During the formalization process and the subsequent development of a demonstrator (see next section), it turned out that the capabilities of the target system with respect to the ontology support directly influence the ontology development. In order to evaluate the limitations resulting from this constraint, two ontologies were designed. The first task consisted in developing a custom-made ontology for the demonstrator which uses only very simple language concepts. The second task was to develop an ontology which resembles the MMS concept (originally modeled in UML) as closely as possible in a system-independent way. In order to maintain computability even in this case, no language concepts higher than OWL DL were used.

It is interesting to note that the difference between these two ontologies—a system-dependent ontology using only very basic constructs and a system-independent one—are rather small. In our case, this means that a rather simple formalization standard is sufficient for formalizing the (rather complex) MMS concept.

Due to the nature of ontologies, the MMS ontology (irrespective of whether system-independent or not) covers only the static aspects of the use case "development accompanying model documentation" (i.e., structures, data fields, views, and accesses). The dynamic elements (i.e., the model management consisting of workflows, assignment of roles, etc.) have not been formalized beyond the devised UML model. In fact, the dynamic elements had to be implemented directly within the demonstrator by writing Java code for performing the necessary functionalities. It has to be remarked that this is purely an implementation issue and not a conceptual problem; upcoming releases of the underlying software (or some other software systems) might be able to make use of externally developed workflow descriptions using some standard exchange format like the XML Process Definition Language (XPDL).

5 MMS DEMONSTRATOR

5.1 Purpose of the Demonstrator

In order to show the potential and flexibility of the devised MMS concept, several well-selected parts were exemplary implemented within an existing software framework and made available for experimental usage. This MMS demonstrator mainly serves three purposes:

- Demonstration of the practical applicability of the MMS concept.
- Illustration of the benefits of formalizing the MMS concept as an ontology.
- Confirmation that the ontology has a *well-chosen* design.

By implementing a selected use case (development accompanying model documentation) including all its relevant concepts, the applicability and implementability of the underlying MMS concept should be evaluated. In addition, it should be evaluated whether the various demands on an MMS can be met. Therefore, the demonstrator was developed in close contact with the users and could be tested during an extensive 1-week workshop. As most participants of this workshop were the same as those who contributed to the requirements analysis a direct comparison between the user's expectations and their fulfillment was possible.

5.2 Scope of the Demonstrator

Due to the purpose of the demonstrator and the central importance of model documentation, the whole model development process (as described in Section 2.1) is actually represented within the demonstrator. This includes especially the following aspects:

- Representation of models, phase models, and submodels.
- Representation of model documentation along all phases of the model development process.
- Definition of workflows and roles, as well as access rights associated with specific steps and roles within these workflows.
- In addition to the phase-oriented documentation, an aspect-oriented view on the model documentation was also modeled and implemented. In contrast to the default view which is oriented along the phases of the model development process, the aspect-oriented view combines the data fields according to a few well-defined topics.

In order to provide these desired aspects all necessary structures (Model, Phase Model, Submodel as well as the respective versions of these structures) along with the required data fields were modeled. Besides the central domain-specific functionalities (i.e., supporting the model development and documentation process), the demonstrator provides additional technical functionalities required for executing this use case:

- Version management for selected concepts (Models, Documentation).
- Role- and task-specific user interface.
- User friendly templates for data input as well as several output possibilities for selected data.
- Searching and browsing within the model catalog.
- Management of further model-associated data (responsible person, model status, annotations).

As indicated, two different access templates to the model documentation have been implemented: phase-oriented and aspect-oriented.

5.3 Technical Realization

The MMS demonstrator is based on the framework WebGenesis (Fraunhofer IITB, 2018). Basically, WebGenesis is a three-tier system consisting of a database backend, an application server, and a client (typically a browser). In this context, a notable feature of WebGenesis is its capability of directly importing an OWL ontology for the use as internal data model.

At this point, the chosen approach via a formalization of the MMS concept as ontology shows its benefits: the development of the ontology can be done with powerful external tools and is (in the best case) completely independent of the actual software environment used for the demonstrator.

5.4 Evaluation of the MMS Demonstrator

The MMS demonstrator implements a selected subset of the MMS concept and shows practically how the realized functionality was built from the MMS meta-concept in a component-oriented way. Furthermore, by using the underlying technical and ontological structures, the demonstrator illustrates in a very straightforward way the extensive conceptual and technical flexibility and extensibility of the MMS with regard to future use cases. Last but not least, technical constraints of the system used for the current implementation could be identified.

The evaluation of the demonstrator took place in form of a 1-week workshop. This workshop was attended by a heterogeneous user group which carried out different example usage scenarios (individual work and teamwork with varying roles) and documented several simulation models for which real model documentation (according to the model development process) was available. All participants agreed that the modeled and implemented documentation process as well as the roles and document templates are reasonable and target into the right direction. In general, the realization of this process within the demonstrator was considered good. Due to the fact that the demonstrator did not focus especially on the usability, many improvements regarding the design of the user interface were suggested.

Based on the results from the demonstrator development of itself as well as of the evaluation workshop two main conclusions can be drawn:

1. The implementation of the MMS concept in a software system is straightforward, while at the same time remaining very closely to the MMS concept. Although the MMS concept itself is quite complex and partly a bit abstract, it is also part of the MMS concept to hide this complexity from the user. The evaluation of the demonstrator shows that this complexity can be hidden very well, especially if supported by a well-designed user interface. Furthermore, the evaluation results gained from the experiences of the users support the overall conclusion that the MMS demonstrator provides a very good foundation. With regard to the necessary domain expertise the demonstrator was easily and plausibly usable for the use case development accompanying model documentation.
2. The underlying software system WebGenesis proved to be a very good foundation as ontologies can be directly imported and used. Although WebGenesis provides lots of additional functionalities, the demonstrator showed that a fair amount of extra effort is necessary for an optimal implementation of the MMS concept (especially regarding the ontology expressivity, version management as well as user and role management).

In summary, the MMS concept can be implemented without any major changes to the basic modeling concepts. Besides the applicability of the MMS concept, the model development process also received a very positive feedback during the user evaluation. Integrated into the MMS meta-concept, the model development process is one of the many use cases which are of central importance during the whole life cycle of a model. The successful implementation of this use case demonstrates that the MMS meta-concept itself—as abstract as it may look like—is very powerful on the one hand, but on the other hand remains easy to use as it is only visible in the form of well-defined domain-specific use cases.

6 SUMMARY AND OUTLOOK

This chapter addresses urgent demands for support of an effective model, simulation and data (M&S) management and proposes a conceptual and technical approach for the implementation of a model management system (MMS). It describes both a structural concept as well as an exemplary implementation of such an M&S management tool along with supporting the complete model life cycle, thereby also integrating all further relevant products of an domain-specific M&S.

The proposed MMS is conceptually based on phase-oriented model development and accompanying documentation processes. Due to a multitude of further or currently unknown requirements regarding management functionalities as well as with regard to formal as well as technical aspects, it was necessary to develop a modular and flexible MMS concept. This was mainly achieved by developing a generic MMS meta-concept which serves as design pattern and therefore permits implementation of a wide variety of use cases within a consistent and clear framework. This formally defined MMS meta-concept with its exact specifications for instantiation of various kinds of data involved, its strict definition in regard to an implementation of extensions makes it possible to handle the enormous complexity of user demands within an application domain of M&S in a consistent manner.

The MMS meta-concept as central aspect of the overall MMS concept serves as design pattern for arbitrary documentation processes within the designated application domain of M&S. Within this design pattern specific use cases need to be modeled (like the M&S design and implementation process, see, e.g., Figs. 1 and 2) which are finally instantiated and executed by a team of developers where each of them serves a well-defined role.

It is crucial to mention that all use cases of the MMS collaborates with a single, consistent domain model which consists of the (object) structures to be handled as well as the data fields describing these (object) structures. Integration of new use cases can be done in a very consistent way without having to change the domain model, although some new use cases may require the extension of the domain model for new data fields. Each use case is accompanied by a set of user-specific views onto the relevant parts of the domain model. Actually each view associates a structure with a very specific set of data fields, thereby presenting each user exactly the information needed within the current situation (for performing a task according to his or her role. The definition of roles and access rights (either to read or write specific data fields) is modeled by using the so-called access templates which define precisely the possible actions of a role, respectively, of a user within a workflow.

The MMS concept as a whole has been modeled (implemented) in an object-oriented way using UML. Static aspects of a domain-like object structures, data fields, views, and access rights were subsequently formalized as OWL ontology. For the implementation of an MMS demonstrator, this ontology was imported to the already available software system WebGenesis. Dynamic aspects like processes and roles were directly taken from the MMS concept and implemented in Java as an extension of the underlying software system.

Finally, a user assessment of the MMS demonstrator implementation revealed that the MMS meta-concept is not just very powerful, but at the same time very comfortable to work with. This is due to the fact that users do not experience the complexity of the meta-concept but instead can work efficiently with the predefined use cases defined within the framework of the MMS meta-concept.

Acknowledgments

The authors thank the Federal Office of Bundeswehr Equipment, Information Technology and In-Service Support as well as the Federal Office of Defense Technology and Procurement for supporting this research over several years, as well as Hwa Feron and Zhongshi Wang for their valuable contributions.

References

Arnold, V., Dettmering, H., Engel, T., Karcher, A., 2005. Product Lifecycle Management Beherrschen. Springer-Verlag, Berlin, Heidelberg.

Fraunhofer IITB, 2018. Framework für Generierung und Support Web-basierter Informationssysteme. Available from:https://www.iosb.fraunhofer.de/servlet/is/18052/. (viewed March 2018).

Gómez-Pérez, A., Fernandez-Lopez, M., Corcho, O., 2004. Ontological Engineering: With Examples From the Areas of Knowledge Management, e-Commerce and the Semantic Web. Springer, London.

Herrmann, G., Lehmann, A., Siegfried, R., 2017. A generic architecture for a model-management-system (MMS). Proceedings of AsiaSim 2017, Malaysia. (to be published in: *Communications in Computer and Information Science.* Springer, 2018).

Kossiakoff, A., Sweet, W., Seymour, S., Biemer, S., 2011. Systems Engineering Principles and Practice. John Wiley & Sons, Hoboken, NJ.

Lehmann, A., 2014. Verification and validation (V&V) of models and simulations (M&S)—past, present and future. In: NATO-Lecture Series on Application of the GM-VV—the Generic Methodology for Verification & Validation of Models, Simulations and Data. NATO STO-EN-MSG-123.

Lehmann, A., Wang, Z., 2017. Efficient use of V&V techniques for quality and credibility assurance of complex modeling and simulation (M&S) applications. Int. J. Ind. Eng. Theory Appl. Pract. 24 (2), 220–228.

Neches, R., Fikes, R., Finin, T., Gruber, T., Patil, R., Senator, T., Swarout, W., 1991. Enabling technology for knowledge sharing. AI Mag. 12 (3), 36–56.

Object Management Group, 2009. UML 2.2. Available from:http://www.omg.org/spec/UML/2.2/ (viewed February 2009).

Sargent, R., 2011. In: Verification and validation of simulation models.Proceedings of the 2011 Winter Simulation Conference, Arizona, United States.

Wang, Z., Lehmann, A., 2007a. Verification and validation of simulation models and applications: a methodological approach. In: Ince, A. et al., (Ed.), Recent Advances in Modeling and Simulation Tools for Communication Networks and Services. Springer, New York.

Wang, Z., Lehmann, A., 2007b. A framework for verification and validation of simulation models and applications. In: Park, J. et al., (Ed.), AsiaSim 2007—Proceedings of Asia Simulation Conference 2007. Springer-Verlag, Berlin, Heidelberg.

Wang, Z., Lehmann, A., 2010. Quality assurance of models and simulation applications. Int. J. Model. Simul. Sci. Comput. 1 (1), 27–45.

Wang, Z., Kißner, H., Siems, M., 2009a. Applying a documentation guideline for verification and validation of simulation models and applications: an industrial case study. In: Proceedings of the 7th Industrial Simulation Conference 2009, Loughborough, United Kingdom.

Wang, Z., Lehmann, A., Karagkasidis, A., 2009b. A multistage approach for quality- and efficiency-related tailoring of modelling and simulation processes. Simul. News Eur. 19 (2), 12–20.

14

Model Management and Execution in DEVS Unified Process

José L. Risco-Martín*, Saurabh Mittal†

*Universidad Complutense de Madrid, Madrid, Spain †The MITRE Corporation, United States

1 INTRODUCTION

Cloud infrastructures provide rapid resource provision for on-demand computational requirements. Cloud simulation environments are largely client-server architectures with multiple slave nodes solving problems through Monte Carlo methods. A cloud simulation is not the same as a distributed simulation. A cloud implementation of an M&S application does not ensure that the infrastructure will scale and complexity inherent in distribution simulation is addressed. In addition, implementing distributed modeling and simulation (M&S) services in cloud infrastructure is a nontrivial problem (Cayirci, 2013). Having a cloud-based deployment does not guarantee that a distributed simulation infrastructure is a "given." Both have different architectures. However, cloud computing brings on-demand resources and technologies like virtualization and containerization. Incorporating cloud computing for distributed M&S infrastructure then is a logical thing to get the best of both technologies and capitalize on the Moore's law with minimal increase in physical hardware costs.

In this chapter, we describe a methodology to deploy a formal discrete-event dynamic system simulation infrastructure based on discrete-event systems (DEVS) formalism, known as DEVS/SOA (Mittal, 2007; Mittal et al., 2009) in a distributed cloud environment. DEVS is component-based M&S framework founded on mathematical systems theory. DEVS also supports model continuity through a simulation-based development and testing life cycle (Hu and Zeigler, 2005). This means that the mapping of high-level requirement specifications into lower-level DEVS formalizations enables such specifications to be thoroughly tested in virtual simulation environments in cloud environments before being easily and consistently transitioned to operate in a real environment for further testing and fielding.

A scalable M&S architecture has distinct M&S layers. In order to deploy in cloud environment, sufficient automation is needed at both the simulation layer and the modeling layer.

This can now be achieved by current practices in DevOps implemented using Docker technology. DevOps, a recent buzzword, provides methodologies to automate developer operations, such as compiling, building, releasing, testing through executable scripts. DevOps has recently been applied to DEVS component-based models (as "nodes") (Mittal and Risco-Martín, 2017). This automated deployment of various "DEVS nodes" under a single administrative control is defined as a DEVS Farm. A DEVS Farm can be readily deployed and put to use for distributed simulation. A cloud infrastructure is not the only means to deploy a DEVS Farm. The high-level architecture (HLA) can also serve to perform these distributed simulations as well. An example is given in another chapter presented in this book, entitled *DEVSim++ME: HLA-compliant DEVS modeling/simulation environment with DEVSim++*, which describes a model engineering environment with support for the development of HLA-compliant DEVS models for discrete-event systems. This chapter is more focused on a cloud-based M&S methodology utilizing the concept of Simulation as a Service (SaS).

DEVS Unified Process (DUNIP) leverages the above advancements along with the foundational DEVS. It was conceptualized by Mittal (2007) during his doctoral work in collaboration with José L. Risco-Martín and Bernard Zeigler (Mittal et al., 2009). From the original SOA-based implementation, it is evolved in the last 10 years to include constructs like domain-specific languages (DSLs), metamodeling, model-driven engineering, automated code-generation, model-based testing, test-suite generation and the more recent, microservices, and containerization support. DUNIP has been applied to M&S of netcentric System of Systems engineering (SoSE) for its extensibility, modularity, flexibility, and explicit semantics. This chapter provides an overview of the state of the art on DUNIP. It describes model management, model engineering, and model execution mechanisms in DUNIP.

The chapter is organized as follows. Section 2 presents the discrete-event world-view on why it is important to understand discrete-event system in a formal way, especially when applying M&S to SoSE. Section 3 provides an overview of DUNIP. Section 4 describes the modeling infrastructure within DUNIP through its two major elements: the DEVS modeling language (DEVSML) and the improved DEVSML Stack Version 3.1. Section 5 provides the simulation infrastructure details in a Cloud environment with Docker containerization support. Section 6 discusses model engineering in DUNIP using metamodeling concepts, followed by model integration and interoperability considerations in Section 7. Finally, this chapter is concluded in Section 8.

2 DISCRETE-EVENT WORLDVIEW

2.1 Overview

A simulation is an imitation of some real thing, state of affairs, or process in action. The act of simulating something generally entails representing certain key characteristics or dynamic behaviors of a selected physical or abstract system.

Simulation is used in many contexts, including the modeling of natural systems or human systems in order to gain insight into their functioning. Other contexts include simulation of technology for performance optimization, safety engineering, testing, training, and education. Simulation can also be used as a prediction tool to show the eventual real effects of alternative conditions and courses of action.

Key issues in developing a simulation technology include acquisition of valid source information about the referent, selection of key characteristics and behaviors, the use of simplifying approximations and assumptions within the simulation model, and fidelity and validity of the simulation outcomes.

A computer simulation attempts to simulate an abstract model (that is computationally represented) of a particular system. A system is part of the real world under study and that can be identified from the rest of its environment for a specific purpose. Such a system is called a real system because it is physically part of the real world. The state of a system is defined as that collection of variables necessary to describe a system at a particular time, relative to the objectives of a study.

Systems may be categorized in two types: discrete and continuous. A discrete system is one for which the state variables may change only at discrete values of time. It is also known as discrete-time system. A continuous system is one whose state is capable of changing at any instant of time. It is also known as continuous-time signal system. Few systems in practice are wholly discrete or wholly continuous, but it will be usually possible to classify a system as being either discrete or continuous (Law and Kelton, 2000). In this chapter we are focused on discrete systems.

Continuing with the notion of system, a few things are common at the systems level: (1) there is a large number of components, (2) hierarchy is used to manage the large number of components and complexity, (3) interactions among components play a vital role in system's behavior, (3) a system has a boundary, (4) a system manifests outward behavior and internal states, and (5) a system interacts with the environment.

A netcentric system is a system that utilizes standards to integrate and operate in a network centric environment. The network, which can be managed through a cluster, a data center, or the cloud, is the underlying communication mechanism. Usage of widely adopted standards facilitates integration and interoperability.

At the complex system level, things become a bit more complicated in the sense that the complex systems are inherently dynamic due to a lot of moving parts. A system may transform into a complex system. A complex system exhibits: nonlinear behavior, dramatic changes, low predictability, etc.

To model and simulate such a variety of systems (complex, centralized, or distributed), recently identified as an open system concept in Mittal and Risco-Martín (2013b), the DEVS specification has been selected as the heart of DUNIP. DEVS is a theoretical framework to define, implement, and simulated such heterogeneity in a consistent way. In the following, we briefly describe this formalism.

2.2 The Discrete-Event System Specification

DEVS is a general formalism for discrete-event system modeling based on set theory (Zeigler et al., 2000). The DEVS formalism provides the framework for information modeling, which gives several advantages to analyze and design complex systems: completeness, verifiability, extensibility, and maintainability. Once a system is described in terms of the DEVS theory, it can be easily implemented using an existing computational library. The parallel DEVS (PDEVS) approach was introduced, after 15 years of the inception of Classic DEVS (Zeigler et al., 2018). Currently, PDEVS is the prevalent DEVS, implemented in many libraries. PDEVS accounts for the confluent condition, that is, a time instant at which both internal and external events becomes imminent. In the following, unless it is explicitly noted, the use of DEVS implies PDEVS.

DEVS enables the representation of a system by three sets and five functions: input set (X), output set (Y), state set (S), external transition function (δ_{ext}), internal transition function (δ_{int}), confluent function (δ_{con}), output function (λ), and time advance function (ta).

DEVS models are of two types: atomic and coupled. Atomic DEVS processes input events based on their model's current state and condition, generates output events and transition to the next state. The coupled model is the aggregation/composition of two or more atomic and coupled models connected by explicit couplings.

Particularly, an atomic model is defined by the following equation:

$$A = \langle X, Y, S, \delta_{ext}, \delta_{int}, \delta_{con}, \lambda, ta \rangle \tag{1}$$

where

- X is the set of inputs described in terms of pairs port-value:

$$\{p \in IPorts, v \in X_p\}$$

- Y is the set of outputs, also described in terms of pairs port-value:

$$\{p \in OPorts, v \in Y_p\}$$

- S is the set of sequential states.
- $\delta_{ext} : Q \times X^b \to S$ is the external transition function. It is automatically executed when an external event arrives to one of the input ports, changing the current state if needed.
  - $Q = (s, e)s \in S, 0 \le e \le ta(s)$ is the total state set, where e is the time elapsed since the last transition.
  - X^b is the set of bags over elements in X.
- $\delta_{int} : S \to S$ is the internal transition function. It is executed right after the output (λ) function and is used to change the state S.
- $\delta_{con} : Q \times X^b \to S$ is the confluent function, subject to $\delta_{con}(s, ta(s), \varnothing) = \delta_{int}(s)$. This transition decides the next state in cases of collision between external and internal events (i.e., an external event is received and elapsed time equals time-advance). Typically, $\delta_{con}(s, ta(s), x) = \delta_{ext}(\delta_{int}(s), 0, x)$.
- $\lambda : S \to Y^b$ is the output function. Y^b is the set of bags over elements in Y. When the time elapsed since the last output function is equal to $ta(s)$, then λ is automatically executed.
- $ta(s) : S \to \mathfrak{R}_0^+ \cup \infty$ is the time advance function.

The formal definition of a coupled model is described as:

$$M = \langle X, Y, C, EIC, EOC, IC \rangle \tag{2}$$

where

- X is the set of inputs, also described in terms of pairs port-value:

$$\{p \in IPorts, v \in X_p\}$$

- Y is the set of outputs, also described in terms of pairs port-value:

$$\{p \in OPorts, v \in Y_p\}$$

- C is a set of DEVS component models (atomic or coupled). Note that C makes this definition recursive.
- EIC is the external input coupling relation, from external inputs of M to component inputs of C.
- EOC is the external output coupling relation, from component outputs of C to external outputs of M.
- IC is the internal coupling relation, from component outputs of $c_i \in C$ to component outputs of $c_j \in C$, provided that $i \neq j$.

Given the recursive definition of M, justified by closure under coupling (Zeigler et al., 2018), a coupled model can itself be a part of a component in a larger coupled model system giving rise to a hierarchical DEVS model construction.

3 DEVS UNIFIED PROCESS

3.1 Overview

DUNIP is based on an open system concept. An open system is a dynamical system that can exchange energy, material, and information with the outside world through its reconfigurable interfaces over a period of time. An open system also possesses the capability to form complex hierarchical structures enabling them to compete and cooperate at the same time. In fact, the mechanism to reorganize in a hierarchical structure is one of the basic requirements to manage complexity. The open systems are also characterized by emerging behavior and evolving structure.

In order to have an executable adaptive System of System (SoS) model, DUNIP must provide capabilities to model an open system. In addition, a process also needs to be defined that allows the development of an executable open system. Much of the open system development hinges on the variable structure capability within a component-based system. Desired characteristics of an open system modeling framework are the ability to add or remove hierarchical components, change connections among components, and lastly, modify the behavior of a component as it evolves per its surroundings. While the first two capabilities are structural in nature and have been documented in DEVS literature, the third one is behavioral modification at runtime. This capability is the most difficult to achieve. The DEVS open systems approach underlying DUNIP gives it strong formal foundation to develop M&S complex system software capable of designing emergent behaviors (Mittal, 2013).

In an SoS, systems and/or subsystems often interact with each other because of interoperability and integration requirements. These interactions are achieved by efficient communication among the systems using either peer-to-peer communication or through central coordinator in a given SoS. Since the systems within SoS are operationally independent, interactions among systems are generally asynchronous in nature. A simple yet robust solution to handle such asynchronous interactions (specifically, receiving messages) is to throw an event at the receiving end to capture the messages from single or multiple systems. Such system interactions can be represented effectively as discrete-event models. In discrete-event modeling, events are generated at random time intervals as opposed to some predetermined time interval seen commonly in discrete-time systems. More specifically, the state change of a discrete-event system happens only upon arrival (or generation) of an event, not necessarily at equally spaced time intervals. To this end, a discrete-event model is a feasible approach in simulating the SoS framework and its interaction. There are many discrete-event simulation engines that can be used in simulating interaction in a heterogeneous mixture of independent systems. The main advantage of DEVS is its effective mathematical representation and its support to distributed simulation.

DEVS formalism has been in existence for over 40 years. It has been applied to multiple domains and many of the continuous, discrete or hybrid formalisms can be reduced to the DEVS formalism (Zeigler et al., 2018). DEVS is based on Systems theory with its hierarchy of system specifications and closure under coupling properties. DUNIP is focused on DEVS and is the consummation of how DEVS can be applied to SoSs design and analysis in full systems engineering life cycle setup (Mittal, 2007). DUNIP is not a single concept but an integration of various concepts that have been developed over the years in DEVS research. These concepts have now evolved into an integrated process that facilitates complex systems M&S. Combining the Systems theory, M&S framework, and model-continuity principles, it leads naturally to a life cycle development process, originally referred as Bifurcated Model-Continuity-Based Life Cycle Methodology (Zeigler et al., 2005): a precursor to DUNIP.

DUNIP is a universal process and is applicable in multiple domains. However, the understated objective of DUNIP is to incorporate DEVS formalism as the binding factor at all phases of this development process. Fig. 1 illustrates the DUNIP, adapted from Mittal and Risco-Martín (2013b). The important concepts and the process within DUNIP are listed below:

1. *Requirements specification using DSLs*: DSLs are used to specify system requirements and definitions. This item is described in Section 4.
2. *Platform-independent modeling at lower levels of systems specification using DEVS DSL*: This is performed through the DEVSML, which is also presented in Section 4.
3. *Model structures at higher level of system resolution using system entity structures (SES)*: This item is focused on the role of SES (Zeigler and Sarjoughian, 2013) at higher levels of systems specification and a model-based repository framework in which components stored in a repository can be used for systems development, which is briefly mentioned in Section 4.
4. *Platform-specific modeling (PSM), that is, DEVS implementations on different platforms*: Sections 4 and 5 show how platform-independent DEVS models can be implemented in a platform-specific language such as JAVA, C#, or C++.
5. *Netcentric execution in a distributed setup*: Section 5 presents several frameworks for DEVS execution, with details on the DEVS simulation architecture, distributed message management and cross-platform execution of DEVS PIMs. It will show how to define a

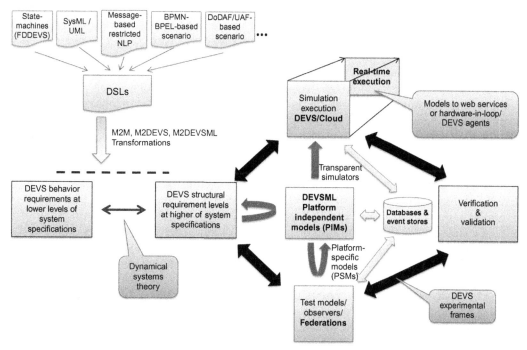

FIG. 1 The DEVS Unified Process. *(Improved from Mittal, S., Risco-Martín, J.L., 2013. Netcentric System of Systems Engineering With DEVS Unified Process. CRC Press, Boca Raton, FL, p. 712.)*

DEVS/Cloud simulation engine based on DEVS/SOA (and supported by xDEVS (Risco-Martín et al., 2017)) to achieve distributed execution.

6. *Automated test model generation using DEVS PIMs*: Automated generation of DEVS observers and test agents from DEVS platform-independent models is discussed at length in Mittal and Risco-Martín (2013b). It describes how DEVS DSL plays a critical role in achieving this capability.

7. *Interfacing of models with real-time systems*: DEVS can act as a production system and can interface with live services, hardware-in-the-loop, and live, virtual, and constructive environments (Mittal et al., 2015).

8. *Verification and validation (V&V)*: The subject of V&V is a critical aspect in developing any theory or the modeling thereof. Without valid models, the theory cannot be tested. Without verified models, model's correctness cannot be ensured. Experimental frame (EF) design for V&V is addressed in Mittal and Risco-Martín (2013b).

4 MODELING INFRASTRUCTURE

DUNIP defines the modeling infrastructure (as well as the simulation infrastructure) through the DEVSML framework (currently in its version 3.0 (Mittal and Risco-Martín, 2017)). The DEVSML 3.0 framework has two pieces: the language and the stack. In the following we describe these two pieces.

4.1 DEVSML: A Language

A DSL is a dedicated language for a specific problem domain and is not intended to solve problems outside it. For example, HTML, Verilog, VHDL, etc., are DSLs for very specific domain. A DSL can be a textual or a graphical language or a hybrid one. A DSL builds abstractions so that the respective domain experts can specify their problem well suited to their domain understanding without paying much attention to the general-purpose computational programming languages such as C, C++, Java, etc., which have their own learning curve. The notion of domain-specific modeling arises from this concept and the DSL designers are tasked with creating a domain-specific modeling language. If a DSL is also meant for simulation purposes, then one more task of mapping a specific DSL to a general-purpose computational language is also on the cards. There are many DEVS DSLs that implement a subset of rigorous DEVS formalism. One example of DEVS DSL is DEVSpecL (Hong and Kim, 2006), built on Backus-Naur Form (BNF) grammar. DSL writing tools, like Xtext, Ruby, etc., focusing directly on the Extended BNF (EBNF) grammar provide a much easier foundation to develop the Abstract Syntax Tree for Model-to-Model (M2M) transformations. The rich integration and code generation capabilities with open source tools like Eclipse give them strong acceptance in the software modeling community.

The DEVSML standard has coexisted through different DEVS DSL, like the DEVSML (Mittal and Douglas, 2012) based on Finite Deterministic DEVS (Hwang and Zeigler, 2007), an earlier developed XML-based XFD-DEVS (Mittal et al., 2012), and an expanded specification of XFD-DEVS in Mittal and Risco-Martín (2013b). Like any language, the DEVSML also has certain reserved keywords, as shown in Table 1. A DEVSML file is of the extension .fds and the specification language contains three primary element types, that is, the Atomic, the Coupled, and the Entity. While the atomic DEVS formalism has a notion of ports (input and output), the DEVSML has a notion of messages specified as Entity structures that are eventually transformed to port definitions. The DEVSML grammar is specified using Eclipse Xtext EBNF notation and is available in Mittal and Douglas (2012) and Mittal and Risco-Martín (2013b).

Once the DEVSML has been designed, the implementation of a DEVSML editor is quite straightforward, using, for example, Xtext DEVSML editors in eclipse. The current implemented framework, named DEVSML Studio (Mittal and Risco-Martín, 2016), allows the user a complete validation mechanism to check both DEVSML structure and behavior, as well as assistance with some automatic tasks, like dynamic code generation. Fig. 2 shows the DEVSML Studio atomic model rendition and Fig. 3 shows the simulation run in the Eclipse DEVSML Studio (Dunip Technologies LLC, 2015).

4.2 DEVSML Stack

The purpose of the DEVSML Stack is to integrate the transparent modeling framework with the inclusion of DSLs (like the DEVSML) through various transformations. DEVSML Stack describes how platform-independent DSLs can be transformed in this framework and eventually become operational in conjunction with the DEVS formalism.

One of the greatest advantages of DEVS is that model and simulator are completely decoupled. It allows the modeler to construct model in a platform of his choice. The ability

TABLE 1   DEVSML Keywords (Mittal and Risco-Martín, 2013b)

Package	Import	Entity
extends	coupled	models
interfaceIO	couplings	atomic
ic	eoc	eic
vars	state-time-advance	state-machine
start in	confluent	deltint
deltext	outfn	sigma
continue	reschedule	ignore-input
input-only	input-first	input-later
infinity	int	double
String	boolean	input
output	S:	S'':
this	X:[]	Y:[]

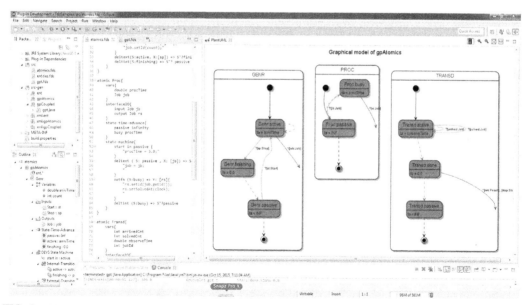

FIG. 2   DEVSML Studio showcasing the textual and autogenerated graphical DEVS state machine in Eclipse Integrated Development Environment (Dunip Technologies LLC, 2015).

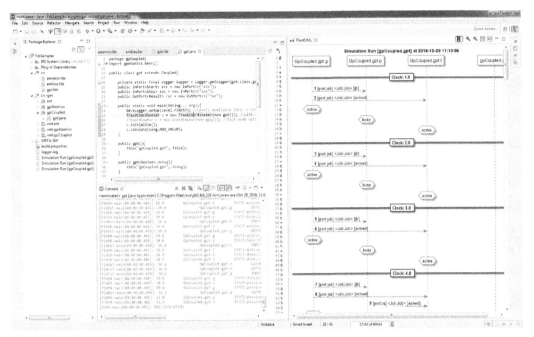

FIG. 3 DEVSML Studio showcasing the simulation log in a textual and autogenerated graphical UML Sequence Diagram in Eclipse IDE (Dunip Technologies LLC, 2015).

to execute DEVS models in multiple platforms and languages has already been achieved and demonstrated, as summarized in Moreno et al. (2009). In all these cases, the design of DEVS models was dependent on the language where the underlying assumption always has been "everything is an Object." In the new DEVSML framework, this concept has evolved to "everything is a model" (Mittal and Risco-Martín, 2013b) and there are two choices. Either the DSL designer takes the DSL directly to the execution code, which involves no transformations but only code generation to the native programming language, or she/he works with an existing framework that guarantees execution in formal systems theoretic way. If the DSL designer opts for the second option, choosing DEVS as a framework is recommended due to its rich history of model specification and simulator development. This ability provides a scalable solution and has inherent advantages for programmatic integration and M&S interoperability. Having a process to transform any DSL to DEVS components, especially to the DEVSML platform-independent specification, then has obvious advantages.

DEVSML 3.0 Stack was proposed in Mittal and Risco-Martín (2017), is improvised as Version 3.1 in Fig. 4. Starting at the bottom of Fig. 4, the execution layer of the DEVS/Cloud simulator is built upon DEVS simulators in native languages (e.g., C++, .NET, Java, etc.) that may get deployed as individual Docker containers as detailed in Section 5. The containers are built from the container images (templates) stored in the persistence layer. The next layer is the distributed communication and coordination layer that manages the containerization. This layer also incorporates the Docker scripts that build containers. Above that is the DEVS/Cloud layer that implements DEVS coordinators and simulators to perform distributed

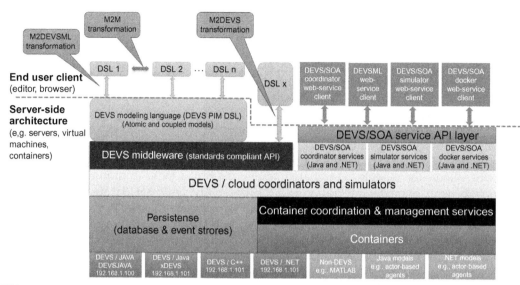

FIG. 4 DEVSML Version 3.1 stack employing model-to-DEVS transformations for model and simulator transparency in DEVS/Cloud. (*Adapted from Mittal, S., Risco-Martín, J.L., 2017. DEVSML 3.0 Stack: rapid deployment of DEVS farm in distributed cloud environments using microservices and containers. In: Proceedings of the 2017 Spring Simulation Multi-Conference (SpringSim'17).*)

simulations along with the required local databases for microservice implementation. Next is the DEVS Middleware and the DEVS/SOA Service layer that makes available the DEVS modeling and DEVS simulation services for multifarious clients. It is this Service/ Middleware layer that enables the transparent M&S framework. Finally, at the top, we have the DSLs and various service clients that utilize the DEVS M&S services. To achieve model interoperability, DEVS models can be encoded in any given language that conforms to DEVSML Application Programming Interface (API). Otherwise, DEVS wrappers can wrap the component's behavior as a DEVS model (Mittal et al., 2015). The coupled models are then specified using a platform neutral format (e.g., XML/JSON).

We need to make a clear distinction here that the DEVS modeling "language" is a DEVS modeling specification language that is anchored to DEVS simulation layer using the simulation relation in DEVS Middleware API. Consequently, a DEVSML specified model is a bona fide DEVS executable. The idea of including other DSLs at the top layer of the stack was a major addition in DEVSML Stack Version 2.0, which also added three transformations at the top layer:

1. model-to-model (M2M)
2. model-to-DEVSML (M2DEVSML)
3. model-to-DEVS (M2DEVS)

The key idea being: domain specialists (the end-user) need not delve in the DEVS world to reap the benefits of DEVS framework. The end-user as indicated in Fig. 4 will develop models in their own DSL and the DEVS expert along with the DSL designer will help develop the M2M and M2DEVSML transformation to give a DEVS backend to the DSL models. While

M2DEVSML transformation delivers an intermediate DEVS DSL (the DEVSML DSL), the M2DEVS transformation directly anchors any DSL to platform-specific DEVS. On a reverse note, a DEVS expert is ideally suited to develop DSLs in other domains as developing transformations like M2DEVS and M2DEVSML need not be negotiated with the DSL expert. A DEVS expert with DEVSML skill set can perform a dual job of both the DSL and DEVSML experts.

The addition of M2M, M2DEVSML, and M2DEVS transformations to the DEVSML Stack adds true model and simulator transparency to a net-centric M&S distributed infrastructure. The transformations yield platform-independent DEVS models (PIMs) that can be developed, compared, and shared in a collaborative process within the domain. Working at the level of DEVS DSL allows the models to be shared among the broad DEVS community that brings additional benefits of model integration and composability. DEVSML 3.0 stack allows DSLs to interact with DEVS middleware through an API. This capability enables the development of simulations that combine and execute DEVS and non-DEVS models (Moreno et al., 2009). This hybrid M&S capability facilitates interoperability. The scale is provided by the underlying distributed (or Cloud) infrastructure that is largely made of virtualization technologies and utilizes platform-as-a-service (PaaS) capabilities provided by containerization, as described in the next section. To support containers, a persistence layer is added in the proposed DEVSML Version 3.1 to account for databases and event stores. Database may store various container images and event stores help preserve runtime state in a microservices-based execution (Mittal and Risco-Martín, 2017).

5 EXECUTION (SIMULATION) INFRASTRUCTURE

The DEVSML execution infrastructure is mainly based on the Cloud capabilities provided by one of its simulation engines, following the scheme of the DEVS Virtual Machine proposed in Mittal and Risco-Martín (2013b). Then, using containers, a swarm of DEVS/Cloud simulators can be created, ready to perform distributed simulations.

Any microservices architecture is primarily an orchestration of stateless services. In a component-based M&S framework such as DEVS, a component-model has to be transformed to a stateless service and addresses two fundamental microservice architecture requirements: distributed data management and shared event stores. In the following, we discuss how the above aspects are handled at the *modeling* and *simulation* execution layers such that the model's state and inherent information can be externalized.

5.1 Modeling Layer Implementation

A DEVS model consists of ports (input and output), states, and state-variables (including model name, current state, next state, time-of-last-event, time-of-next-event, and elapsed-time) and the four characteristic functions:

$$(\delta_{\text{ext}}, \delta_{\text{int}}, \delta_{\text{con}}, \lambda(s)).$$

In a microservices-based rendition of a DEVS model, we have to partition these elements into the two buckets (of distributed data management and event stores) to achieve scalability

TABLE 2   Partitioning Atomic DEVS M&S Elements for Microservices Implementation

Atomic DEVS Model Specification	M&S Element	Data Management	Event Store	Comments
X	Model		x	Input events
Y	Model		x	Output events
name	Model	x		Model name
tl	Simulator	x		Time of last event
tn	Simulator	x		Time of next event
phase	Model	x		Current phase of the model
$v1, v2, ..., vn$	Model	x		Values and their data types

TABLE 3   Partitioning Coupled DEVS M&S Elements for Microservices Implementation

Coupled DEVS Model Specification	M&S Element	Data Management	Event Store	Comments
X	Model		x	Input events
Y	Model		x	Output events
$M1, M2, ..., Mn$	Model	x		Subcomponent names
I_Z	Model	x		Set of influencers
Z_{i-d}	Model/ simulator	x		Mapping of influencer outports to influence's inports

through stateless execution of a modular DEVS component. The DEVS state-machine needs to be separated with the operations on the state-variables inside the DEVS atomic model component. The first bucket is the model's state that is stored in state-variables, which have to be serialized in a local database for that model. This implies that all the operations on these variables through the four characteristic functions will be through the accessor functions (i.e., *get* and *set*). The second bucket is of the shared event store that is used for the *input X* and *output Y* sets. The X and Y sets are transformed into declarative events (that may be implemented as a complex data type) into a shared event store. Likewise, the DEVS atomic simulator components are partitioned as well. Table 2 shows the partitioning of atomic DEVS elements (both M&S layers). Table 3 shows the coupled DEVS elements partitioning. The above partitioning allows the model to become stateless, which then can be containerized.

5.2 Simulation Layer Implementation

The simulation layer architecture will focus on the simulator and coordinator execution and how they implement the DEVS simulation protocol in an abstract-time manner. The architecture again has to account for the above two requirements: distributed data

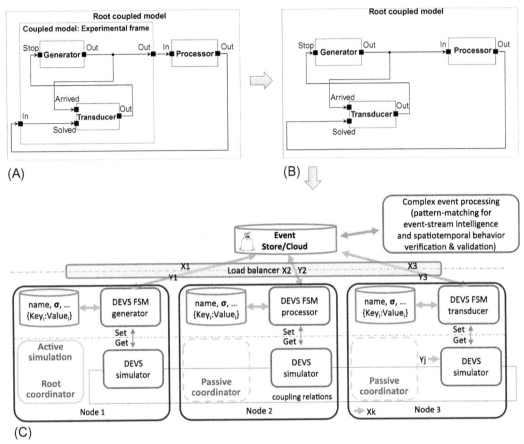

FIG. 5 DEVS/Cloud container nodes within microservices paradigm (Mittal and Risco-Martín, 2017). (A) DEVS hierarchical coupled model. (B) DEVS Flattened single-level coupled model. (C) DEVS distributed model within containerized simulators in symmetrical DEVS/Cloud architecture with Event Store/Cloud available for Complex Event Processing (CEP).

management and event stores. The DEVS simulation protocol defines the relationship between the model and its underlying simulator. The application of microservices provides resiliency at both the model and the simulator levels, to the effect that a large number of simulator instances with corresponding model instances can now be created when the data exchange between the model and the simulator is accurately partitioned (as in Tables 2 and 3).

We shall illustrate the microservices-based DEVS/Cloud simulation architecture implementation with the help of the classic EF-P example (Fig. 5). EF-P model contains two components: the EF and processor (P) models (Mittal and Risco-Martín, 2013b). The hierarchical EFP model in Fig. 5A is flattened toward a generator-processor-transducer (GPT) model depicted in Fig. 5B. Next, according to Fig. 5C, each of the three submodels in GPT, is mapped to a separate DEVS/Cloud node (container). Although, they all can also belong to a single container. One node can thus contain one or more DEVS models with their corresponding local databases to store model's data (Column 3 in Table 2). The node is selected through the simulation configuration file. For illustration purposes, Fig. 5C shows three nodes each corresponding to a single submodel in GPT. One simulator is then created for each atomic model. Due to the symmetrical architecture, each node also contains a blueprint of a coordinator. The simulation configuration file designates one of the nodes as the root coordinator. At the end, models' state and output events are always managed by their corresponding simulators, using the local databases and global event stores, respectively. Model output events are propagated through synchronous communication between the model-simulator pair, using the aforementioned `set` and `get` accessors. It should be highlighted that if the DEVS model is not flattened in the simulation configuration file, then the coupled information in Table 3 is stored as well.

The event store is implemented using various event cloud technologies such as Esper and provides a means to perform model-integrated systems engineering in which M&S itself is a part of the systems engineering (Mittal and Risco-Martín, 2013a). The event stores are usually implemented as event clouds which lend themselves to complex event processing that can do streaming analytics as well as design monitors that can detect advanced spatiotemporal patterns between the messages flowing between the components. All the containers and global event store interface through a load balancer that divides the load on each container node through various load-balancing policies.

In the following, we briefly explain the automated deployment of a DEVS/Cloud node using Docker.

5.3 Implementation of the DEVS/Cloud Support

A Docker Image is a read-only template used to instantiate Docker containers. Each image is defined with several layers that compose the final image structure. The Docker registry also called Docker Hub is a Docker Image repository. Images can be downloaded or uploaded. The Docker Hub has a considerable amount of images ready to use. Finally, a Docker Container is the runtime component of the Docker Image. Multiple containers can be instantiated from the same Docker Image in an isolated context. Docker container can be run, started, stopped, moved, and deleted. To start, Docker must be installed in the host machine. For more information, the reader should be refer to the Docker official web page (Docker, 2017).

To build our DEVS/Cloud Dockerfile, we start from a DEVS/SOA complete WAR file (Mittal et al., 2009), which includes the xDEVS simulation engine (with support for simulators and coordinators as a service) and several DEVS example models. The corresponding Docker Image must then include a minimal runtime environment that contains the DEVS/SOA dependencies: Linux OS, Oracle Java 8, and an application service such as Apache Tomcat. Next, the .war file will be deployed into the webapps directory of Apache Tomcat.

Fig. 6 depicts a Dockerfile structure. An excerpt of the source code of this file is shown in Mittal and Risco-Martín (2017). As shown in Fig. 6, the Dockerfile must start with a base image. In our case, the base image is Ubuntu. Next, MySQL and Java 8 are added. The final step is to add Apache Tomcat. Finally, more security constraints can be added to the Dockerfile. Once the Dockerfile is completed, we can show and stop container through the docker ps and docker rm commands on the host OS that has the Docker daemon running.

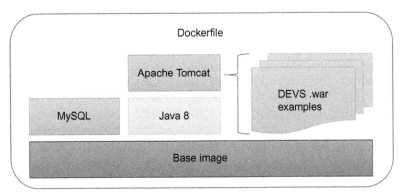

FIG. 6   Full Dockerfile structure (Mittal and Risco-Martín, 2017).

Figs. 5 and 6 show a transparent simulation infrastructure with DEVSML Stack Version 3.0 that incorporates container services for automated DEVS/Cloud deployments.

6 MODEL ENGINEERING IN DUNIP

6.1 Model-Based/Model-Driven Flavors

The "model-based (MB)" and "model-driven (MD)" terms and initials have been used in a variety of system and software-related acronyms, such as MBD, MDSD, MDD, MDA, MBSE, MDE, and many others. Although there is a consensus that these approaches suggest the systematic use of models as the primary means of a process and facilitate the use of DSLs, there is not a common understanding of the terminology (Mittal and Risco-Martín, 2013b). The definitions of the frequently referred acronyms and the objectives of those approaches are summarized following:

1. MBE: Model-based engineering (MBE) originated in the 1980s in parallel with the evolution of the Computer-Aided Design (CAD) and Model-Based Design (MBD) techniques. The main goal in MBE was to support the system development process during the design, integration, validation, verification, testing, documentation, and maintenance stages (Zeigler, 1976a,b; Wymore, 1993).
2. MBSE: In systems engineering, the application of the MBE principles is called as Model-Based Systems Engineering (MBSE) (Zeigler, 1976a; Wymore, 1984; Zeigler and Chi, 1993). MBSE provides the required insight in the analysis and design phases, enhances better communications between the different participants, and enables effective management of the system complexity.
3. MDE: Model-driven engineering (MDE) is a system development approach that uses models to support various stages of the development life cycle (Atkinson and Kuhne, 2003; Schmidt, 2006) and can be seen as a subset of MBE. MDE relies on technologies to automate model transformations thereby increasing productivity within MBE. It produces well-structured and maintainable systems because of its focus on formally defined models, metamodels, and meta-metamodels.

4. MDA: Model-driven architecture (MDA) is a software design and development approach that provides a set of guidelines for specifying and structuring models (Object Management Group, 2003), relies on the Meta Object Facility (MOF) (Object Management Group, 2006) MDA provides a natural mechanism to define models and to transform them. It prescribes the use of metamodels and meta-metamodels for specifying the modeling languages without any necessity to be domain specific.

5. MDD or MDSD: The application of the MDE principles in software engineering is called Model-Driven Development (MDD) or Model-Driven Software Development (MDSD) (Volter et al., 2006). The modern era of MDD started in the early 1990s and now offers a notable range of methods and tools. Different specifications such as MDA (Object Management Group, 2003), MIC (ISIS, 1997), Eclipse Modeling Project, and Microsoft Software Factories are some of the conceptual applications of MDD principles.

6. MIC: Model Integrated Computing (MIC) refines the MDD approaches and provides an open integration framework to support formal analysis tools, verification techniques, and model transformations in the development process (ISIS, 1997). MIC allows the synthesis of application programs from models by using customized Model Integrated Program Synthesis (MIPS) environments (e.g., Generic Modeling Environment [GME]). The meta-level of MIC provides metamodeling languages, metamodels, metamodeling environments, and metagenerators for creating domain-specific tool chains on the MIPS level.

MBE and MBSE utilize the systems V&V methodologies to the model development process relating back to the system requirements for systems test and evaluation. MDE is one area that serves both the software and systems engineering as it is domain independent. Model-Driven Systems Engineering (MDSE) gives MB/MBSE various model engineering, transformation, and tools from MDE that speed up model development through the metamodeling construct. The development of various editors based on MDE concepts involve advanced software engineering and code generation techniques. DUNIP (through its DEVSML Stack) is positioned as an MDSE that employs both the systems engineering and MDE paradigm in an agile manner (Mittal and Risco-Martín, 2013a).

6.2 Model Management in DUNIP

DEVSML and DUNIP are focused toward interoperability at the application level, specifically, at the modeling level and hiding the simulator engine as a whole. Our vision and solution development is along the lines of Model-as-a-Service (MaaS), Simulation-as-a-Service (SimaaS), and ultimately, DEVS-as-a-Service (DaaS). We would like the user or designer to code the behavior in any of the programming languages, ideally a DSL of his choice and let the DEVSML 3.1 stack develop the transformations. The DEVSML Stack is responsible for taking a DSL or a coupled DEVSML model, integrating code within their DSLs and delivering us with an executable model that can be simulated on any DEVS platform (local, virtual, distributed, or cloud).

The user can integrate his model from a model repository stored in any web location. It may contain publically available models of legacy systems or proprietary standardized models. Together they will provide more benefit to the industry as well as to the user, thereby truly realizing the model-based paradigm. In addition, the following aspects handle the model management in DUNIP:

1. Use of DSLs and guidance for model transformation: The DSLs can be platform-dependent or platform-independent. Through the model transformations (M2M, M2DEVSML, and M2DEVS), any DSL can be brought in DUNIP. Using the underlying metamodel, a DSL can be mapped to the DEVSML metamodel.
2. Alignment with Systems Theory: The alignment reflects the presence of explicit interfaces between components and strong data-type exchange mechanisms between the interfaces. This facilitates unambiguous message exchange in an explicitly connected system.
3. Support for component reusability: DEVS, and consequently DUNIP, is a component-based framework. A component in DEVS is defined by two aspects: structure and behavior. Likewise, any DSL or a DEVS-wrapped component has an explicit interface and a defined behavior. This makes it available for its inclusion in a repository with a knowledge about the structure and behavior stored in the metadata for that component.
4. Code generation, execution, and deployment mechanisms: DEVSML 3.1 Stack has a service layer that keeps the code generation (for transformations), the execution layer (simulation), and the deployment (on cloud) transparent from the end-user and model workflows. The service layer hence implements Maas, Simaas, and DaaS.

7 MODEL INTEGRATION AND INTEROPERABILITY

Interoperability is a quality that denotes the ability of diverse-independent systems to work together at a functional level. If two or more systems are capable of communicating and exchanging data between themselves to address a situation or solve a problem, the overall system manifests interoperability between these systems. The word "system" can be a general concept for an organism, component, or an agent. Interoperability facilitates model extensibility and full integration. Thus, achieving a high degree of interoperability in simulation continues to be a prime objective within the research community and this age of heterogeneity. The main reason is to assist the confluence between the large variety of legacy simulation frameworks and the latest simulation applications, not limited to augmented reality and virtual simulation.

During the last 10 years, a DEVS standard has been under development to support interoperability of DEVS models implemented in different platforms as well as with legacy simulations. Fig. 7 illustrates an architectural approach proposed to accommodate the various combinations and permutations of possible application. The basic idea was to define two sets of interfaces; the DEVS model interface and the DEVS simulator interface, as well as a DEVS simulation protocol that operates between the two.

DUNIP and its DEVSML construct (at both M&S layers) supports this conceptual architecture. At the modeling layer, DEVSML supports DEVSJAVA and xDEVS engines (Risco-Martín et al., 2017). xDEVS contains wrappers for other DEVS M&S engines like DEVSJAVA, aDEVS, or CD++. At the simulation layer, DEVSML implements xDEVS compatibility, with support for sequential, parallel, and distributed simulations.

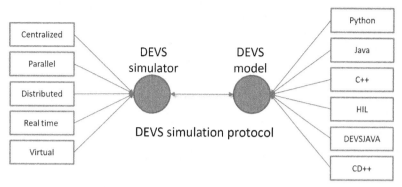

FIG. 7 Conceptual architecture of standard.

7.1 Integration at the Modeling Level

As mentioned earlier, DEVSML takes advantage of the existing DEVS model implementation interface in xDEVS (Risco-Martín et al., 2017). To support the modeling requirements implemented as in Fig. 7, the xDEVS API specifies an interface for both atomic and coupled models, which allows us to adapt the implementation to the DEVSML standard stated in Section 5. The compatibility of DEVS-to-DEVS models is tackled with the use of wrappers (Mittal et al., 2009). Fig. 8 shows how the DEVS-to-Non-DEVS interoperability is solved. To start with, the DEVSML model interface is derived from the xDEVS interface (Fig. 8). The xDEVS to DEVS mapping implements DEVS formalism specifications. These DEVS models can be implemented as full DEVS models or otherwise act as an adapter for non-DEVS integration, for example, Matlab integration (Risco-Martín et al., 2009).

Via the DEVSML simulator environment incorporated through the xDEVS framework (with all its sequential, parallel, and distributed capabilities), we are capable of modeling and simulating atomic and coupled models that share the same semantics given by the DEVS

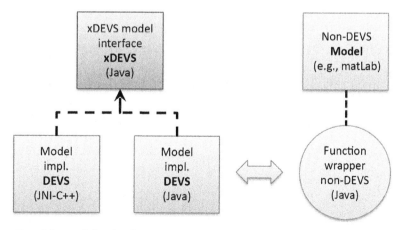

FIG. 8 Integration at the modeling level.

mathematical specifications or through DEVSML specifications. The models may differ in the computing environment implementation or deployment. As we have seen, implementations of both DEVS and non-DEVS "compliant" models (using an adequate wrapper for the case of non-DEVS models) that share a common DEVS interface can interoperate. As the DEVSML simulator remains unchanged, it is able to simulate and facilitate interoperable DEVS model implementations.

7.2 Integration at the Simulation Level

The previous methodology can be extended for the xDEVS simulation framework (left-hand side of Fig. 7). As explained earlier, xDEVS provides DEVS-based simulators as services, which are based on standard communication technologies. Each atomic or coupled component may be implemented using different simulation engines, called platforms. Fig. 9 depicts an example of multiplatform DEVS model. The DEVSML Stack acts as the framework interoperating two different simulation platforms (Fig. 5).

8 SUMMARY

The DEVS formalism, based on systems theory, provides a framework and a set of M&S tools for SoSE (Mittal and Risco-Martín, 2013b). A DEVS model is a system-theoretic concept specifying inputs, states, outputs, similar to a state machine. Critically different, however, is that it includes a time-advance function that enables it to represent discrete-event systems, as well as hybrids with continuous components, in a straightforward platform-neutral manner. DEVS provides a robust formalism for designing systems using event-driven, state-based models in which timing information is explicitly and precisely defined.

DUNIP is categorically designed to interface with service-oriented and cloud-based systems to bring an interoperable M&S environment that may include hardware-in-the-loop or software-in-the-loop (e.g., service systems). Netcentric systems can be modeled effectively

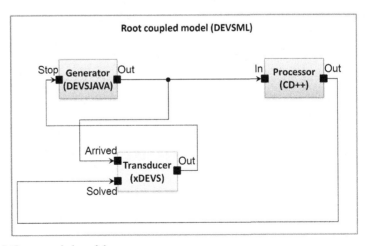

FIG. 9   Cross platform coupled model.

TABLE 4 Mapping of M&S T&E Capability Requirements and DUNIP

Desired M&S Capability for T&E	Solutions Provided by DEVS Technology in DUNIP
Support for executable architectures using M&S such as mission-based testing for Cloud-based systems	DEVS Unified Process provides methodology and Cloud infrastructure for integrated development and testing (Mittal, 2007)
Interoperability and cross-platform M&S using Cloud	Simulation architecture is layered to accomplish the technology migration or run different technological scenarios. Provide net-centric composition and integration of DEVS validated models using Cloud Computing
Automated test generation and deployment in distributed simulation	Separate a model from the act of simulation itself, which can be executed on single or multiple distributed platforms. With its bifurcated test and development process, automated test generation is integral to this methodology
Test artifact continuity and traceability through phases of system development	Provide rapid means of deployment using model-continuity principles and concepts like "simulation becomes the reality" (Hu and Zeigler, 2005)
Real-time observation and control of test environment	Provide dynamic variable-structure component modeling to enable control and reconfiguration of simulation on the fly. Provide dynamic simulation tuning, interoperability testing and benchmarking

using the DEVS formalism and consequently can leverage DUNIP. To provide a brief overview of the current DEVS capabilities within DUNIP, Table 4 outlines how DEVS can provide solutions to the challenges in netcentric design and evaluation.

The realization of netcentric DUNIP has the following pieces:

1. DEVSML Stack: the central concept
2. Distributed simulation in Cloud environment over SOA
3. Netcentric DEVS Virtual Machine (both client and server)
4. Design, development, and deployment of netcentric systems with DEVS
5. Containerization support for efficiency, scalability, and scalable deployment
6. Interfacing with event-driven architectures and live, virtual, and constructive environments that incorporate both hardware-in-the-loop and software-in-the-loop

DUNIP offers an integrated approach to bring in various DSLs in a methodical way to the DEVS ecosystem such that they could become a part of a larger SoS. This is very much a needed capability when it comes to complex multidisciplinary M&S where models from multiple domains need to be brought in together. DUNIP incorporates the latest in MDE, DevOps, and Cloud technologies to deliver a reliable M&S environment for rapid test and evaluation. Further improvements will be made as new technologies appear on the horizon.

DISCLAIMER

The author's affiliation with the MITRE Corporation is provided for identification purposes only, and is not intended to convey or imply MITRE's concurrence with, or support for, the positions, opinions, or viewpoints expressed by the author(s). Approved for Public Release. Distribution Unlimited. Case Number: PR_17-3254-10.

References

Atkinson, C., Kuhne, T., 2003. Model-driven development: a metamodeling foundation. IEEE Softw. 20, 36–51.

Cayirci, E., 2013. Modeling and simulation as a cloud service: a survey. In: Proceedings of Winter Simulation Conference.

Docker, 2017. What is Docker? https://www.docker.com/what-docker. Accessed 26 December 2017.

Dunip Technologies LLC, 2015. DEVSML Studio. http://duniptechnologies.com/jm/downloads.html. Accessed 26 December 2017.

Hong, K., Kim, T.G., 2006. DEVSpecL-DEVS specification language for modeling. Inf. Softw. Technol. 48 (4), 221–234.

Hu, X., Zeigler, B.P., 2005. Model continuity in the design of dynamic distributed real-time systems. IEEE Trans. Syst. Man Cybern. Part A 35 (6), 867–878.

Hwang, M.H., Zeigler, B.P., 2007. Reachability graph of finite deterministic DEVS. IEEE Trans. Autom. Sci. Eng. 6 (3), 468–478.

ISIS, 1997. Model Integrated Computing (MIC). http://www.isis.vanderbilt.edu/research/MIC. Accessed 26 December 2017.

Law, A.M., Kelton, W.D., 2000. Simulation Modeling and Analysis. McGraw-Hill, New York, NY.

Mittal, S., 2007. DEVS Unified Process for Integrated Development and Testing on Service Oriented Architectures (Ph.D. thesis), University of Arizona.

Mittal, S., 2013. Emergence in stigmergic and complex adaptive systems: a formal discrete event systems perspective. J. Cogn. Syst. Res. 21, 22–39.

Mittal, S., Douglas, S., 2012. DEVSML 2.0: the language and the stack. In: Proceedings of the 2012 Spring Simulation Multi-Conference.

Mittal, S., Risco-Martín, J.L., 2013. Model-driven systems engineering for netcentric system of systems engineering with DEVS unified process. In: Winter Simulation Conference.

Mittal, S., Risco-Martín, J.L., 2013. Netcentric System of Systems Engineering With DEVS Unified Process. CRC Press, Boca Raton, FL, p. 712.

Mittal, S., Risco-Martín, J.L., 2016. DEVSML Studio: a framework for integrating domain-specific languages for discrete and continuous hybrid systems into DEVS-based M&S environment. In: Proceedings of the 2016 Summer Simulation Multiconference (SummerSim'16).

Mittal, S., Risco-Martín, J.L., 2017. DEVSML 3.0 Stack: rapid deployment of DEVS farm in distributed cloud environments using microservices and containers. In: Proceedings of the 2017 Spring Simulation Multi-Conference (SpringSim'17).

Mittal, S., Martín, J.L.R., Zeigler, B.P., 2009. DEVS/SOA: a cross-platform framework for net-centric modeling and simulation in DEVS unified process. Simulation 85 (7), 419–450.

Mittal, S., Zeigler, B.P., Hwang, M.H., 2012. XFDDEVS: XML-based finite deterministic DEVS. http://www.duniptechnologies.com/research/xfddevs/.

Mittal, S., Ruth, M., Pratt, A., Krishnamurthy, D., Lunacek, M., Jones, W., 2015. A system of systems approach to integrated energy systems modeling. In: Summer Computer Simulation Conference Chicago, IL.

Moreno, A., Risco-Martín, J.L., Besada, E., Mittal, S., Aranda, J., 2009. DEVS/SOA: towards DEVS interoperability in distributed M&S. In: 13th IEEE/ACM International Symposium on Distributed Simulation and Real Time ApplicationsIEEE Computer Society, pp. 144–153.

Object Management Group, 2003. Model Driven Architecture (MDA) Guide Version 1.0.1. http://www.omg.org/mda/specs.htm.

Object Management Group, 2006. Meta Object Facility (MOF) Core Specification, Version 2.0. http://www.omg.org/spec/MOF/2.0.

Risco-Martín, J.L., Moreno, A., Aranda, J., Cruz, J.M., 2009. Interoperability between DEVS and non-DEVS models using DEVS/SOA. In: SpringSim'09: Proceedings of the 2009 Spring Simulation MulticonferenceSociety for Computer Simulation International, San Diego, CA, USA, pp. 1–9.

Risco-Martín, J.L., Mittal, S., Fabero, J.C., Zapater, M., Hermida, R., 2017. Reconsidering the performance of DEVS modeling and simulation environments using the DEVS tone benchmark. Simulation 93 (6), 459–476.

Schmidt, D.C., 2006. Model-driven engineering. IEEE Comput. 39, 25–31.

Volter, M., Stahl, T., Bettin, J., Haase, A., Helsen, S., 2006. Model-Driven Software Development: Technology, Engineering, Management. John Wiley & Sons, New York, NY.

Wymore, W.A., 1984. The Tricotyledon Theory of System Design. Springer-Verlag, New York, NY, pp. 119–132.

Wymore, W.A., 1993. Model-Based Systems Engineering. CRC Press, Boca Raton, FL.

Zeigler, B.P., 1976. Multi-Faceted Modeling and Discrete Event Simulation. Academic Press, London.

Zeigler, B.P., 1976. Theory of Modeling and Simulation. Interscience, New York.

Zeigler, B.P., Chi, S.D., 1993. Model-Based Architecture Concepts for Autonomous Systems Design and Simulation. Kluwer Academic Publisher, Boston, MA, pp. 57–78.

Zeigler, B.P., Sarjoughian, H.S., 2013. System Entity Structure Basics. Springer, London, pp. 27–37.

Zeigler, B.P., Praehofer, H., Kim, T.G., 2000. Theory of Modeling and Simulation. Integrating Discrete Event and Continuous Complex Dynamic Systems, second ed. Academic Press, London.

Zeigler, B.P., Fulton, D., Hammonds, P., Nutaro, J., 2005. Framework for M&S–based system development and testing in a net-centric environment. ITEA J. Test Eval 26 (3), 21–34.

Zeigler, B.P., Muzy, A., Kofman, E., 2018. Theory of Modeling and Simulation: Discrete Event & Iterative System Computational Foundations. Elsevier, Fairfax, VA.

GPU Parallelism-Oriented Traffic Modeling and Simulation

Xiao Song*, Yan Xu†, Gary Tan†, Fuwang Zhao*

*School of Automation Science, Beihang University (BUAA), Beijing, China †Singapore-MIT
Alliance Research & Technology (SMART), Singapore, Singapore

1 INTRODUCTION TO DEVS AND STATE-OF-THE-ART PARALLEL COMPUTING TECHNIQUES

Discrete event system specification (DEVS) is a modular and hierarchical formalism for modeling and analyzing general systems that can be discrete event systems, which may be described by state transition tables, continuous state systems, which may be described by differential equations, and hybrid continuous state and discrete event systems. In DEVS, a simulation system is composed of atomic models (AMs) and coupled models (CMs). The AMs are expressed in a basic formalism. The CMs are expressed using the CM specification—essentially providing component and coupling information.

However, DEVS falls short of addressing the following issues:

- Due to the rapid development of modeling and simulation (M&S) and software technology, some problems of DEVS have emerged. On the one hand, it is difficult to describe dynamic system behaviors of CMs using DEVS. On the other hand, advanced computer science technology including automata is widely used. As a result, the combined method is proposed by some researches. For example, visual state transition chart is combined with DEVS by extending finite state machine (FSM).
- Although DEVS introduces modeling formalisms and simulation algorithms, it does not provide implementation specifications for parallel simulation development. The DEVS can be integrated with distributed simulation specification such as high-level architecture (HLA), but it is difficult to handle the high communication cost. Therefore, it is desirable if we can design parallel DEVS executions with stand-alone computer node to avoid this cost.

Hardware accelerators, such as general-purpose graphics processing units (GPGPUs), are promising parallel platforms for high-performance computing. The GPGPU provides an inexpensive, highly parallel system to application developers. There has been growing research and industry interest in accelerating applications with GPGPUs. It has shown several advantages to run computation intensive and data-parallel applications compared to CPU platforms.

There are hundreds of applications accelerated by GPU. Various works aim to exploit the data parallelism of the applications and accelerate them with GPU. For example, the Ising model (Hawick et al., 2011) is a computation intensive model, which is used to analyze phase transitions occurring in statistical mechanics and many other systems including social networks, physical computer networks, and web page relationships on the World Wide Web (Hawick et al., 2011). The model can be set up on any graph or network where spin nodes interact with their nearest neighboring nodes according to a Hamiltonian or energy functional. It is noticeable that the calculation of the Hamiltonian actually resembles the situation where an agent tries to interact with other agents in a grid environment. Wende (2010) showed how the calculation of Hamiltonian could be resolved with Metropolis algorithm on a GPU.

When data parallelism is not obvious to utilize GPU efficiently it is necessary to transform the problem into a data-parallel one. Hong et al. (2011) introduced a GPU implementation of the level synchronous breadth first searching (BFS) algorithm to explore a graph. The CPU version of the algorithm uses queues. However, to accommodate the GPU programming, the authors changed the queue-based algorithm into an O(N) array-based algorithm. The concurrent processing on the array can make the graph exploration in parallel.

There are two network models: the fluid-based TCP model and the adaptive antenna model. These two models were converted into computation-intensive representations. With careful mapping, the data structures are positioned in the GPU memory. To be specific, the arrays of real numbers in CPU-based algorithms are represented in GPU as two-dimensional (2D) textures of floating point data values. The intermediate results are stored in GPU's texture memory to be used for subsequent passes of computation, so that the costly data transfer between the CPU and the GPU is avoided. In addition, to handle the GPU-based function calls with two or more outputs, the authors used multiple rendering targets to write multiple textures in a single pass so that good performance was achieved.

Perumalla (2008) and Perumalla et al. (2009b) introduced a method to simulate vehicle movement on a GPU by using a field-based model. This model maps the real-world road data onto a 2D lattice, with each cell in the lattice representing the possibility of turning either left/right or up/down. Cells are processed in parallel by GPU threads. Whenever there is a vehicle in a cell, its route can be calculated according to the turning probability of the cell.

Speeding up discrete event simulation is an important topic in parallel and distributed simulation. The requirement that events be processed in causal order restricts the parallelism that can be achieved. Studying GPU execution strategies for parallel discrete event simulation helps us understand the challenges and difficulties of using GPU in simulation. The major challenge is to exploit the parallelism from the application.

In another work of Perumalla (2006b), the author studied efficient implementation of discrete event simulations on the GPU. The particular application is a diffusion simulation (e.g., heat transmission and gas diffusion) which is quite computationally intensive. The author proposed a time-driven approach and an event-driven approach to simulate this scenario. The time-driven approach is inefficient, because each logical process (LP) recalculates its local state in every time advancement, but this implementation could be quite easily migrated to GPU. The event-driven approach is efficient, but it is hard to be implemented on GPU. To bridge this gap, the author proposed a hybrid approach. The clock advancement mechanism is the same as the event-driven approach in which the clock advances to the timestamp of nearest future event. However, in each time advancement multiple events are extracted and aligned to the same timestamp to be simulated concurrently, which is similar to the time-driven approach. The author fully elaborated how the extraction was done in the paper. The hybrid approach demonstrates good speedups in the scenarios with large problem size in both CPU and GPU implementations.

In the work of Park and Fishwick (2010, 2011), the authors proposed data structures for processing events in parallel. The future event list (FEL) is decomposed into sub-FELs which are then assigned to GPU threads for concurrent processing. However, in their work, the iterative data transfer between the host and the device is a bottleneck.

In summary, to take advantage of the GPU architecture, applications with higher data parallelism are more beneficial. Besides, the data structures are usually organized into arrays and the dynamic memory allocations are often avoided.

2 DEVS-BASED MESOSCOPIC TRAFFIC SIMULATION FRAMEWORK ON GPU

As shown in Fig. 1, a road network is modeled as *nodes*, *links*, *segments*, and *lanes*, which correspond to AMs in DEVS. And the traffic system can be regarded as a CM. The nodes correspond to intersections of the actual road network, while links represent unidirectional pathways between nodes. Each *link* is divided into a number of *segments*, according to geometry features. Each segment contains *lanes*. Each lane contains a number of vehicles, which are located on the lane. Each lane has capacity constraints at the upstream end and the downstream end, referred to as the input capacity and the output capacity. A queue occurs in a lane if vehicles cannot pass the lane. A spill-back occurs if a lane is blocked, which means the length of the queue on the lane is equal to the length of the lane.

According to the vehicle's status (moving or in queue), its location is updated using the following rules. If a vehicle is located in the moving part of a link, its speed is determined by a speed-density relationship on the density of the link, and its location is then updated using the speed. Example speed-density relationships can be found in the land authority publication such as the manual of Transportation Research Board (2000). In this paper, the

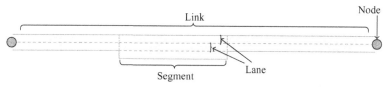

FIG. 1 Network-related terminologies used in this paper.

following equation in Yan et al. (2014) and Transportation Research Board (2000) is used to compute the speed of a link:

$$v = v_0 * \left\{ \left(1 - \left(\frac{k}{k_{jam}} \right)^{\beta} \right)^{\alpha} \right\}$$

(1)

where v_0 is the free-flow speed, k is the density, k_{jam} is the jam density, and α, β are configurable parameters to be determined for each link through calibration. In this paper, $\alpha = 1.0$ and $\beta = 0.05$.

If a vehicle is located in the queue part of a link, there are two possible conditions:

If the vehicle is at the head of the queue (at the exit of the link), it can leave the queue only if the current link has output capacity left, and the downstream link has sufficient empty space and sufficient input capacity. If both conditions are satisfied, the vehicle can pass the current link to the next time step. Otherwise, the vehicle stays at the exit of the link.

If the current vehicle is not first in the queue (i.e., there are other queuing vehicles ahead of it), it can only advance as far as vehicles in front of it do (assuming no space is left between any two consecutive queuing vehicles). The distance is then determined by how many vehicles have left the head of the queue during the same time step.

2.1 The Simulation Framework for CPU/GPU

In this chapter, we use the CPU and the GPU to enhance the performance of a traditional time-stepped mesoscopic traffic simulation enabled by the entry-time-based supply framework (ETSF), as described in Yan et al. (2014).

In most existing GPU-based simulations (Perumalla et al., 2009a; Strippgen and Nagel, 2009; Singapore-MIT Alliance for Research and Technology (SMART) & DynaMIT, n.d.), CPU code plays a role of a master thread that controls the program flow. The GPU code, on the other hand, spawns a bunch of worker threads to execute compute-intensive parts of the program in parallel, thus accelerating the overall program. Here each worker thread is mapped to an AM instance of DEVS. However, this way of implementation has two shortcomings for traffic simulations. First, the demand part, including vehicle generation, departure time choice, and path choice, is often dynamically generated in real time and thus it is not easy for the demand part to be executed on the GPU memory structure. Second, central processing unit (CPU) is not fully used, which is a waste of resources to some extent.

To tackle these problems, we design a new simulation framework making full use of CPU and GPU. In this framework, the GPU is responsible for the supply part of mesoscopic traffic simulation, which includes speed calculation, vehicle movement on a road and between roads, and queue calculation. A key feature of supply simulation is that the simulation of a road is just related to its surrounding roads, which fits GPU's data-parallel requirement. Further, the CPU is responsible for the demand part and the input/output (I/O) part of the mesoscopic traffic simulation, which includes vehicle generation, departure time choice, pre-trip route choice, en-trip route choice, and pushing simulation results to files. A key feature of demand simulation is that vehicles make decisions based on the information on the global road network. Fig. 2 shows the mesoscopic traffic simulation framework on CPU/

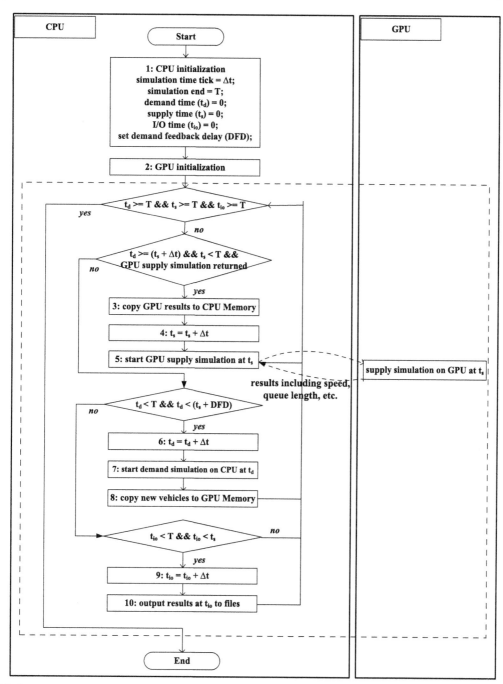

FIG. 2   Mesoscopic traffic simulation framework on CPU-GPU. The supply and demand simulation components are CMs in DEVS.

GPU, which explains the logic procedure and the simulation time management. This framework is suitable for general time-stepped mesoscopic traffic simulation (Ben-Akiva et al., 2001, 2012; Mahmassani et al., 1992).

The traditional simulation time, which controls the turnover of the system status, is divided into three components: a demand time (t_d), a supply time (t_s), and an I/O time (t_{io}). A traffic simulation is completed only if t_d, t_s, and t_{io} all reach the simulation end. In other words, the simulation is completed when all components have reached the simulation end time. In this framework, multiple time steps enable to identity the exact progress of different components in a traffic simulation. Note that, at an instantaneous time, the three time steps can be different. The time management in this framework is controlled by three rules:

- Rule 1: $t_d >= t_s$
- Rule 2: $t_s >= t_{io}$
- Rule 3: $t_d <= t_s + DFD$

First, t_d is not smaller than t_s because, only if vehicles entering the simulation at time t are generated, the supply simulation at t can start.

Second, t_s is not smaller than t_{io} because, only if the supply simulation at t is completed, the simulation results at t can be outputted to files.

The third rule involves a concept in traffic simulation: demand feedback delay (DFD), which is a multiple of the simulation time tick $\triangle t$. Vehicles generated at time t require the simulated results at $t - DFD$, for departure time choices and route choices. The minimum value of DFD is 1 s, which means vehicles have real-time instantaneous information about the global traffic status in last time step (e.g., 1 s), and DFD tends to be larger in real-world traffic systems.

In the logic procedure in Fig. 3, steps 1 and 2 initialize the required data structures on the CPU and the GPU, including the road network, traffic scenario configurations, and other parameters. After initialization, the CPU controls the simulation logic, in order to manage the simulation time and to make full use of computational resources. Without breaking the three rules in time management, the following tasks can be executed in parallel:

- Task 1: The supply simulation CM at time t_s on GPU (steps 3–5).
- Task 2: The demand simulation CM at time t_d on CPU (steps 6–8).
- Task 3: Push simulation results at time t_{io} to files (steps 9–10).

Within a loop of the logic procedure, the CPU first checks whether the GPU has finished the supply simulation at time t_s. If yes, the simulation results on the GPU (e.g., road-based speed and density) are copied to the CPU, and the supply time t_s is advanced. Then, the supply simulation at the next time step is started on the GPU. Note that the CPU will not wait for the GPU supply simulation to finish. If the supply simulation on the GPU is ongoing, the CPU checks whether the demand simulation can be started. If the simulation results required for demand simulation are available, the demand simulation will be started on the CPU. Otherwise, the CPU checks whether there are available simulation results that need to be written into files. The CPU will continue the loop until the three time steps t_s, t_d, and t_{io} all reach the simulation end. The logic of the supply simulation and the demand simulation are explained in Barcelo (2010) and Yan et al. (2014).

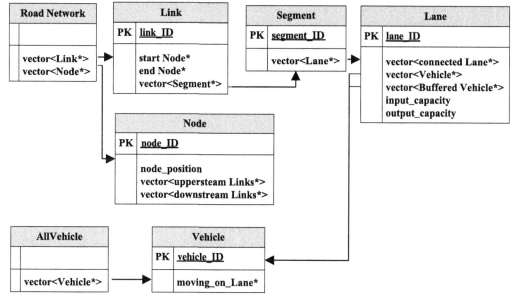

FIG. 3   Road network and vehicle modeling on the CPU.

2.2 Data Structure of Road Network and Vehicle on GPU

For parallel computing performance, the data structure in the GPU memory is completely different from CPU's, due to the thread hierarchy and the memory hierarchy in the GPU. Fig. 3 shows the road network and vehicle on the CPU memory. A road network is composed of a list of links and a list of nodes. Each link consists of a number of segments, and each node consists of a list of upstream and downstream links. Each segment consists of multiple lanes, and each lane contains a number of lane connections. Each lane also has access to vehicles that are moving on the lane.

2.2.1 Data Structure of CPU Versus GPU

From the perspective of implementation, the data structures in the CPU memory and the GPU memory have two key differences.

First, on the CPU memory, the large number of road elements and vehicles are stored in random separated memory spaces, and the objects connect with each other using pointers. While on the GPU memory, these elements are kept in arrays in a continuous memory space and different elements connect with each other using the index inside the array. The reasons for doing this on the GPU memory are to make it easy to copy the entire road network from the CPU memory to the GPU memory and, more importantly, to allow efficient coalesced memory access, which means a group of GPU threads in a warp tend to access continuous memory space.

Second, on the CPU memory, dynamic memory allocation (e.g., as done by the STL vector, which applies for additional memory space just when it is immediately required) is widely used in the data structure of a road network and vehicles, because of its flexibility and efficiency. However, on the GPU memory, dynamic memory allocation has to be replaced by static memory allocation, which utilizes a sufficiently large amount of memory space in the beginning. It is a limitation of GPU programming because it is not efficient to do random memory access. In our GPU framework, it uses "start & end indexes" to store the containing relationships.

Third, it is good to use shared memory if possible, as does the matrix multiplication example shown in NVIDIA (2014). However, it is not easy to use shared memory in our traffic simulation framework. There exist two problems. (1) In the matrix multiplication example, the submatrix is stored in shared memory because it will be used multiple times during multiplication. However, in our traffic scenario, each lane computes its speed, queue length without using the other lane's data. This means there is little need to use shared memory. (2) Considering the case in Section 4 with 100,000 vehicles, we need 99.95 MB for lanes. However, each shared memory per block is only 48 kB. Using shared memory will lead to multiple sections with various thread indexes. As such, only global memory is used in our kernel functions.

To illustrate these technical points, Fig. 4 shows the data structure of an ideal grid road network and vehicle modeling on the GPU memory.

2.2.2 Data Structure of Ideal Grid Versus Real Road Network

Compared with the ideal grid network modeling on the GPU memory, the real-world network data structure is shown in Fig. 5. This design uses indexes in data structures to save memory space. The design details are introduced as follows.

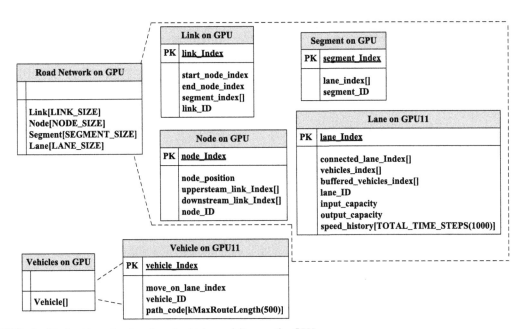

FIG. 4   Ideal grid road network and vehicle modeling on the GPU.

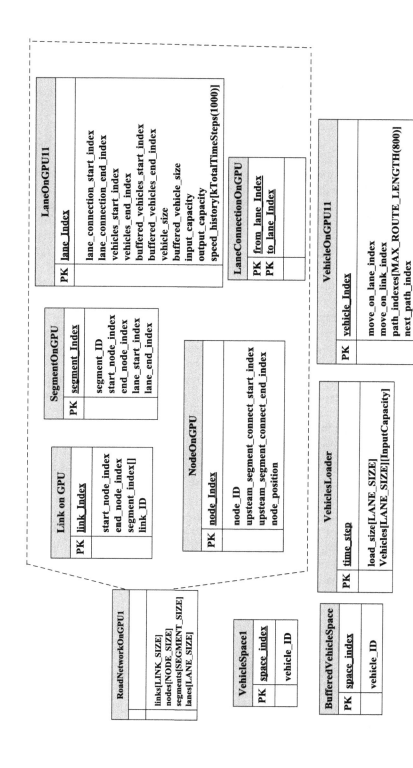

FIG. 5   Realistic road network and vehicle modeling on the GPU.

1. Realistic road network modeling is more complex. Each link of an ideal grid road network only has one segment and its lane, but there are often two or more lanes in a real network. Correspondingly, in the design of a GPU memory, a grid road network has only "lane_index" whereas there is "lane_start_index" and "lane_end_index" for each segment in a realistic road network. For example, if a segment has four lanes, then this index can be 1 and 4 (if the segment is the first segment in the network) or more likely to be (X and X+3); X is the index of the first lane in the whole network.

2. In a grid network, all lanes' lengths are equal, and the scope of vehicles' indices is thus fixed. We only need to get "vehicle_index" on the lane to know the ID of the vehicle on the lane. But in a real network, we need to get "vehicle_start_index" and "vehicle_end_index" of each lane because the length of each lane is different. For instance, about struct "LaneOnGPU," the usage of attributes "vehicle_start_index" and "vehicle_end_index" is to define global start and end index of vehicles on the network. These two attributes mean that the scope of vehicles' index on a lane, for example, if a lane is 1000 m long, and each vehicle is 5 m long, then the index is $X \sim X+199$, where X depends on other lanes in the network.

3. Considering of the complexity of real network topology, "lane_connection" is proposed in road modeling, which does not exist in the ideal grid road network because these connections are easily predetermined. The struct "LaneConnection" is devised to contain localized lane-connection information. It answers the question of "which upstream lanes are allowed to enter the downstream lane in an intersection."

4. Global path selections of vehicles are stored in CPU memory, and each vehicle has its own path. For struct "LaneOnGPU," the usage of attributes "vehicle_num" is to define the number of vehicles at a time t. "vehicle_start_index" and "vehicle_end_index" define the physical limitation of total number of vehicles; they are constant values. But "vehicle_num" changes as the simulation advances.

To further illustrate the above points, an example road network is illustrated to further explain the data structure in the GPU memory. Fig. 6A represents a real-world road network. The road network consists of a main road with one on-ramp and one off-ramp. The road network is then modeled as six nodes and five segments in Fig. 6B. The two nodes of interest are nodes 1 and 2 (as N1 and N2 in Fig. 6B). Node 1 has two upstream segments and one downstream segment; node 2 has one upstream segment and two downstream segments. The main road has two lanes; the on-ramp and off-ramp roads have one lane. The length of segment 3 (S3) is 200 m; the length of all other segments are 100 m. The length of a vehicle is 5 m.

The data structure of the road topology in the GPU memory is shown in Fig. 6C. First, continuous memory spaces (e.g., tables) are used to keep nodes, segments, lanes, etc. Second, each element has a special ID (or index), which indicates its location in the table. For example, the segment S1 is the first element in the segment table. Third, a pair of start/end indexes is used to store the relationship between nodes and segments, segments and lanes, lanes and vehicles.

Moreover, complex lane-connection rules are directly modeled in our data structure model. For example, as shown in the lane-connection table, the on-ramp segment 2 contains only one lane (lane3), and lane3 is only connected to one lane of segment 3 (lane5). It means, in this example, vehicles on segment 2 cannot pass to the other lane of segment 3 (lane4).

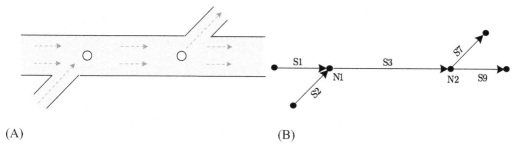

(A)   (B)

NodeOnGPU		
ID	up_start_index	up_end_index
N1	S1	S2
N2	S3	S3

SegmentOnGPU				
ID	start_node	end_node	lane_start_index	lane_end_index
S1	N*	N1	L1	L2
S2	N*	N1	L3	L3
S3	N1	N2	L4	L5

LaneConnectionOnGPU		
ID	from_Lane_ID	to_Lane_ID
LC1	L1	L4
LC2	L1	L5
LC3	L2	L4
LC4	L2	L5
LC5	L3	L5
LC6	L4	L*
LC7	L4	L*
LC8	L4	L*

LaneOnGPU				
ID	conn_start_index	conn_end_index	vehicle_start_index	vehicle_end_index
L1	LC1	LC2	V1	V20
L2	LC3	LC4	V21	V40
L3	LC5	LC5	V41	V60
L4	LC6	LC8	V61	V100
L5	LC9	LC10	V101	V140

(C)

FIG. 6 Example to show the arrays designed for node, segment, lane connection, and vehicle index for a typical road topology: (A) example road topology, (B) simulated road topology in traffic simulation, and (C) data structure in the GPU memory (* means the ID does not exist in the figure).

Meanwhile, each lane in segment 3 has a space for 40 vehicles, but the other lanes have a space for 20 vehicles. Thus, each lane in segment 3 reserves 40 vehicle ID space. An attribute "on_road_vehicle_num" is used to determine how many vehicles are on the lane at a time and which vehicle IDs are valid.

2.3 Thread and Partition of Road Network

We assume that a traffic system can be divided into numerous disjoint small partitions and assign each GPU thread to simulate several partitioned traffic models. This approach can significantly reduce the total execution time if the partitioning solution is well designed.

Normally, we have two choices to partition a traffic network. The first is to assign vehicles to threads. This approach assigns a new vehicle to the least workload thread. It is more appropriate for microscopic simulation where each vehicle has rich behaviors and more computing load.

The second is to assign road lanes or segments to threads. This approach is good for our mesoscopic simulation because of its better cache performance, as vehicles on the same road segment are processed in an order by the same thread. Besides, road segment-based properties (e.g., the average speed) can be efficiently calculated in this approach. In other words, according to current GPU architecture, the cost mainly comes from the memory access of traffic models on different threads. Thus, we think it is of high priority is to achieve coalesced memory access of vehicles on the same road segment.

As such, we choose the latter approach, where each node and all upstream road segments connected with this node correspond to each GPU thread. For instance, the road topology shown in Fig. 6B is computed with four GPU threads, as shown in Fig. 7.

In the proposed assignment method, all upstream road segments, which are connected to the same node, are processed together by the same GPU thread. Besides, all these upstream road segments are stored in a continuous structure in the GPU memory (as discussed in Section 3.2). It allows an efficient way to deal with conflicts among vehicles from different upstream road segments in order to enter into and cross the node.

2.4 Kernel Functions of Supply Simulation on the GPU

To apply the above data structure design and network partition method, the supply component of our mesoscopic simulation on GPU consists of four key functions:

(1) cpu_update ()
(2) kernel function: pre_vehicle_passing ()
(3) kernel function: vehicle_passing ()
(4) copy_simulation_results_to_cpu ()

The first function allows the CPU to change the status of the traffic simulation before starting the supply simulation at the next time step. For example, if an incident happens, the road capacity is reduced. The CPU updates the new capacity to the road network on the GPU memory before simulating the next time step. Another example is en-trip route

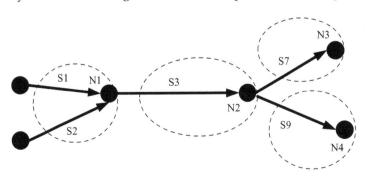

FIG. 7   Four GPU processing areas in an example road topology.

choices. En-trip route changing behavior is simulated on the CPU, and then the new routes are copied to the GPU.

The second function is a GPU kernel function updating the status of each lane (e.g., density, speed, and t_p (Yan et al., 2014)) before passing vehicles to the downstream lanes. The update unit of this kernel function is a lane, which means each lane is simulated on a GPU thread. This kernel function first loads vehicles that passed through this road at the previous time step. After that, it loads new generated vehicles into the lane. Then, the kernel function calculates the speed of the lane based on the speed-density relationship (Barcelo, 2010; Yan et al., 2014). After that, the kernel function calculates t_p. The method to calculate t_p is explained in our previous work (Yan et al., 2014). For this kernel function, the detailed procedure is shown in Algorithm 1.

The third is also a GPU kernel function scanning vehicles on the lane and passes a number of downstream vehicles to the next lane. The update unit of this kernel function is each node. As discussed in Section 3.3, each node and its upstream lanes are simulated on a GPU thread. It is because vehicles from upstream lanes might conflict with each other when crossing the node to the same next lane. One example is shown in Fig. 8. In this small road network, node 1 has two upstream lanes: lane 1 and lane 4. Thus, vehicles on lane 1 and lane 4 are processed on the same GPU thread, to remove the potential conflicts. Moreover, as shown in Fig. 7, because each lane has only one upstream node, from where vehicles might pass, there is no conflict when updating nodes in parallel. There are four rules to determine whether a vehicle can pass from a lane to the next lane, which are explained in our previous work (Yan et al., 2014). Afterward, if a vehicle crosses from the current lane to the next lane, the corresponding output

---

ALGORITHM 1

PRE_VEHICLE_PASSING

1: **Start n GPU threads (n is the number of lanes)**
2: *For each GPU thread*
3: … *for* each vehicle in the buffered space of the lane *do*
4: … … *if* there is space on the lane *do*
5: … … … transfer the vehicle from the buffered space to the lane;
6: … *if* there are vehicles in the buffered space of the lane *do*
7: … … shift vehicles in the buffered space;
8: … *for* each newly generated vehicle on the lane *do*
9: … … *if* there is space on the lane *do*
10: … … … load the newly generated vehicle to the lane;
11: … … *elseif* there is space on the buffered space of the lane *do*
12: … …… load the newly generated vehicle to the buffered space;
13: … … *else*
14: … … … cannot load the newly generated vehicle
15: … update the density of the lane;
16: … update the speed of the lane;
17: … update other attributes (e.g., t_p) of the lane.

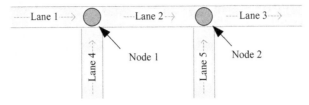

FIG. 8   A node and its upstream links are updated on the same GPU thread.

capacity of the lane, the input capacity, and empty space of the next lane are updated. Finally, the vehicle ID should be removed from the current lane and inserted into the next lane.

The details of this kernel function are further explained in Algorithm 2.

The divergence possibility of *if...else* in these two algorithms seems to be a little big. But it is almost inevitable according to their traffic processing logic. It is difficult to improve it because we must guarantee the precise results (such as lane speed and queue length) of the traffic simulation. We can find the metric of branch-taken ratio in Table 1 that the divergence ratio is indeed a little high. But we can still gain 2.37 speedup (see Fig. 14) of a Singapore network running on GPU, which is to some extent acceptable.

The last function copies the simulated results, which include the speed, density, flow, queue length, and empty space of each road, from the GPU memory to the CPU memory. As shown in Fig. 5, these data are stored in a contiguous GPU memory space in order to reduce the time cost of data transfer from the GPU memory to the CPU memory.

Further, the above two kernel functions are developed as shown in Fig. 9B. Further, the program comparison of CPU and GPU is illustrated in Fig. 9.

ALGORITHM 2

VEHICLE_PASSING

1:   **Start m GPU threads (m is the number of nodes)**

2:   *For each GPU thread*

3:   *... while the node has left capacity during the time step* **do**

4:   ... ... obtain the vehicle on upstream road segments of the node which has the maximum waiting time

5:   ... ... *if* the vehicle finishes the trip **do**

6:   ... ... remove the vehicle from the upstream road segment;

7:   ... ... *elseif* the corresponding downstream road segment has left buffered space **do**

8:   ... ... transfer the vehicle to the buffered space of the downstream road segment;

9:   ... ... update the vehicle's status;

10:   ... ... update the node's status;

11:   ... ... update the upstream road segment's status;

12:   ... ... *else*

13:   ... ... update the upstream road segment as blocked during the time step;

14:   ... end of *while*

TABLE 1 Profile of Major GPU/CUDA Kernel Functions

ID	Measurement	Grid Kernel Function 1 (G1): pre_vehicle_passing	Singapore Kernel Function 1 (S1): pre_vehicle_passing	Grid Kernel Function 2 (G2): vehicle_passing	Singapore Kernel Function 2 (S2): vehicle_passing
1	Launched GPU threads	20,352 (106 blocks)	9600 (50 blocks)	10,368 (54 blocks)	3264 (17 blocks)
2	GPU occupancy (theoretical/achieved) (%)	94/79.01	93.75/76.5	94/78.28	93.75/23.63
3	Registers (used/ available)	4224/65,536	4032/65,536	3456/65,536	2880/65,536
4	Transaction per second (load/store)	2.55/4.53	2.32/5.26	4.26/1.58	4.97/1.14
5	Branch taken ratio (%)	60.62	60.05	66.44	27.62
6	Instruction serialization (%)	0.988	1.12	3.698	1.209
7	Instruction per clock (IPC) (measurement/ maximum)	0.355/4.0	0.364/4.0	0.307/4.0	0.148/4.0
8	Warp issue efficiency (no eligible %)	90.35	91.43	93.22	96.06
9	Issue Stall Reasons (execution dependency) (%)	18.73	15.41	4.33	4.98
10	CUDA achieved GFLOPS	8.950	6.257	0.0104	0.00012

As shown in Fig. 9, in the serial program on CPU, the computation of each lane is achieved by using a loop body. The second lane starts computing when we finish the computation of the first lane. Meanwhile, to reduce the time cost and improve efficiency, the loop body and the loop control variable disappear in the parallel program on GPU. Instead, we assign each lane to a thread in GPU to compute all the lanes in parallel. Also, the __global__ prefix is added to the function that tells the compiler to generate GPU code and not CPU code when compiling this function and to make that GPU code globally visible from within the CPU.

3 EXPERIMENTS

The traffic scenario is simulated on two types of platforms: the CPU and CPU/GPU. Only the total time cost of the supply simulation during the 1000 simulation ticks is measured in this experiment. The CPU platform includes an Intel Core i5-4200H CPU @ 2.80GHZ, 8G

```
Void supply_simulation_pre_vehicle_passing(int time_step) {

        for (unsigned int lane_index = 0; lane_index<the_network->link_size; lane_index++)
{...}

}
```

(A)

```
__global__void SupplySimulationPreVehiclePassing(GPUMemory* gpu_data, int time_step,
int segment_length, GPUSharedParameter* data_setting_gpu, GPUVehicle *vpool_gpu) {

        int lane_index = blockIdx.x * blockDim.x + threadIdx.x;... ;              }
```

(B)

FIG. 9   Program comparison of the supply part: (A) computation of all the lanes on CPU and (B) computation of all the lanes on GPU.

memory, 500GB SATA 7.2K RPM. The GPU platform is a GeForce GTX 950M, which has 640 CUDA cores and 2 GB global memory. The supply simulations on the CPU and the GPU follow the same logic. The source codes are implemented using C++ on Ubuntu 12.04 and compiled using g++_4.6.3 and CUDA 6.5. The release version executable file is used to measure the time cost.

3.1 Experiment Design and Analysis

Two cases of experiments were carried out in order to evaluate the computation efficiency of CPU/GPU in artificial grid and real-world traffic scenarios.

First, an artificial large grid road network (uniform rectangular grid), which is similar to the road network topology in our companion work (Xu et al., 2014), is designed with 10,201 nodes and 20,200 unidirectional links. The length of each link is 1000 m. Each node has an index from 0 to 10,200, indicating the store location on the GPU memory. Each link also has an index from 0 to 20,199. In all, 100,000 vehicles are loaded into the road network during 1000 simulation ticks (each tick is 1 s). Vehicles are loaded into this network from nodes in the top and in the left, which are moving to the bottom and to the right. Each vehicle randomly picks a route from the pre-calculated candidate routes before starting a trip. En-trip route choice (Barcelo, 2010) is not included in this traffic scenario.

Second, a real-world Singapore expressway system traffic simulation is studied. As shown in Fig. 10, the expressway system consists of expressway segments and ramps connecting local roads with the expressway. The network has been modeled using a detailed representation of the length, geometry, and lanes of each segment. The expressway system is made up of 3179 nodes, 9419 lanes, and 3388 segments. Distribution of each segment length is shown in Fig. 11 . Most segment lengths are in the range of (300, 600). There are some short segments, which are mostly on-ramps or off-ramps, and there are also a few long segments. The demand is modeled as trips from 4106 OD pairs. Each origin is an on-ramp, where vehicles enter the

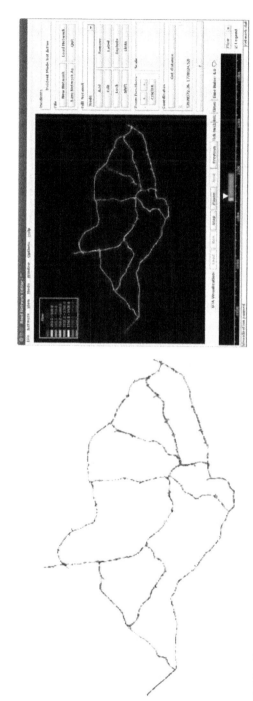

FIG. 10 Singapore expressway road network.

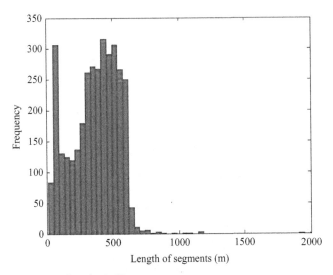

FIG. 11   Distribution of segment lengths in Singapore expressway.

expressway system from local roads, and each destination is an off-ramp, where vehicles depart the network. The configuration and calibration of the OD matrix use the same methods as in DynaMIT (Singapore-MIT Alliance for Research and Technology (SMART) & DynaMIT, n. d.; Xu et al., 2014), and the calibrated OD matrices in one period: peak hours (7:00 a.m. to 8:00 a.m.) are used in this section. In particular, there are in total 100,000 vehicles loaded into the peak traffic scenario during 1000 simulation ticks (each tick is 1 s). The routes of vehicles are pre-calculated using a path size logit model (Ben-Akiva and Bierlaire, 1999), and routes are not changed during the traffic scenarios.

3.2   Results and Analysis

The execution time results are shown in Figs. 12 and 13, and the corresponding speedup in Fig. 14. Also, the detailed profiling data of two major kernel functions are listed in Table 2.

Two groups of configurations are investigated. The first is for the artificial grid; the second is for the Singapore network. Each result is executed 20 times, and the average time cost is shown in Figs. 12 and 13. The speedup is measured by comparing the time cost of supply simulation on a GPU to the time cost of supply simulation on a CPU core.

In Fig. 14, we can observe that when the number of vehicles is 100,000, the simulation of grid on GPU obtains a speedup of 10.00 and that of the realistic network is 2.37. We can observe that, compared with the speedup on the artificial large grid road network, the speedup on the Singapore expressway is much lower. The main reasons are as follows.

First, due to the data structure difference discussed in Section 3.2, the times of memory access in the Singapore road network is more than that in the ideal grid road network. For instance, in the Singapore road network, the memory needs to access each lane's length. However, in the ideal grid road network, each lane is 1000 m, so there is no need to access memory.

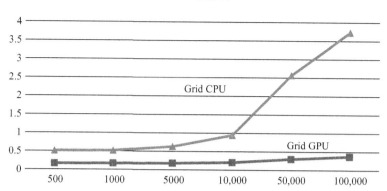

FIG. 12 Performance of grid on the CPU and the GPU. The horizontal axis is the number of vehicles; the vertical axis is the execution time in seconds.

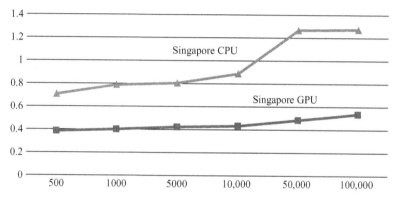

FIG. 13 Performance of Singapore Network on the CPU and the GPU. The horizontal axis is the number of vehicles; the vertical axis is the execution time in seconds.

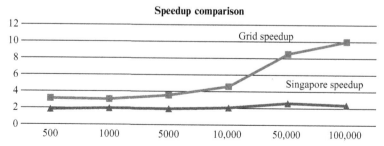

FIG. 14 Speedup comparison of grid and Singapore network. The horizontal axis is the number of vehicles; the vertical axis is the speedup.

Second, in the artificial grid road network, roads have the same length, and vehicles on roads are directly stored inside roads. However, in the Singapore expressway, roads have different lengths from 50 to 2000 m (see Fig. 11). Instead of directly storing the vehicles, the "start & end indexes" are used as introduced in Section 3.2. The "start & end indexes" make the usage of memory space more flexible. It is important to load the whole city-scale traffic simulations into the GPU memory. However, accessing a vehicle on a road requires twice the memory accesses. More memory accesses make the proposed framework less efficient.

Third, the number of GPU threads launched on the Singapore expressway is smaller than the grid road network. It makes the GPU occupancy lower.

Fourth, the artificial grid road network is more structured. For example, most nodes have two downstream links and two upstream links. However, the Singapore expressway network topology is more complicated. It makes the memory access less coalesced.

Besides the above reasons, more detailed analysis of memory copy time and execution time efficiency is studied in the following two subsections, where both cases have 100,000 cars.

Table 1 presents the profile of two kernel functions in the framework: pre_vehicle_passing and vehicle_passing. For the former, the threads are launched to compute the *lanes*. For the latter, the threads are to compute the *nodes*.

Please note that a table explaining the profiling measures illustrated in Table 1 is included in Table 2.

The analysis of these main GPU performance profiling measurements of the four kernel functions, that is, G1, G2, S1, S2, is studied as follows.

First, the simulation thread units in these two kernel functions are lanes and nodes. The block size (blockDim.x) used in these experiments is 192. In these four traffic scenarios, G1 launched 20,352 GPU threads and S2 launched 3264 threads. The former gains the highest occupancy, and the latter has the lowest occupancy.

Second, most of the occupancies of these kernel functions are high, which indicates the GPU cores are sufficiently utilized. Only the occupancy of the Singapore kernel function 2 (S2) is the lowest because the number of threads launched is the smallest. The main reason is each thread of S2 has to compute and update the status of each node, that is, has to move the vehicles to the buffered space of the downstream lane. The number of nodes is less than that of lanes. Meanwhile, the computation cost of S2 is less than that of S1, which is determined by their designed functions as discussed in Section 3.2.

Third, even after moving internal variables from the global memory to registers, registers are not a bottleneck, which means registers are enough for use in these four kernels.

Fourth, as proposed in Section 3.2, the data of lanes, nodes, and vehicles are stored into designed arrays; thus, threads in a warp can access a contiguous memory space, known as coalesced memory access. The number of memory transactions per request (both load and store) for these two kernel functions is small (<5.5), which indicates the memory access is well coalesced. In addition, G2 and S2 have less store transactions, which means they have less data to store from kernel to global memories because the vehicles of nodes are transferred to the buffered space of the downstream road lane, which means less data are transferred to the global memory to update the lanes.

Fifth, the branch-taken ratio (within threads in the same warp) varies for the kernel functions. This means the threads in these kernels do not take exactly the same branch because

TABLE 2   Brief Introduction to GPU Performance Metrics

ID	Measurement	Brief Introduction
1	Launched GPU threads	Equal to block size * the number of threads in a block. The block size is defined according to the number of parallel threads
2	GPU occupancy (theoretical/achieved)	Occupancy is defined as the ratio of active warps on an SM to the maximum number of active warps supported by the SM. Occupancy varies over time as warps begin and end, and can be different for each SM
3	Registers (used/available)	The SM has a set of registers shared by all active threads. If this factor is limiting active blocks, it means the number of registers per thread allocated by the compiler can be reduced to increase occupancy
4	Transaction per second (load/store)	This metric is the average number of L1 transactions required per executed global memory instruction, separately for load and store operations. For this metric, the lower the better. How many transactions are actually required varies with the access pattern of the memory operation and is also dependent on the compute capability of the target device
5	Branch taken ratio (%)	Total number of executed branch instructions with a uniform control flow decision
6	Instruction serialization	A flow control instruction is considered to be divergent if it forces the threads of a warp to execute different execution paths. If this happens, the different execution paths must be serialized, since all of the threads of a warp share a program counter; this increases the total number of instructions executed for this warp. For this metric, the lower the better
7	Instruction per clock (IPC) (measurement/maximum)	The average number of issued instructions and executed instructions per cycle accounting for every iteration of instruction replays. The higher numbers indicate more efficient usage of the available all resources
8	Warp issue efficiency (no eligible %)	The warp issue efficiency (no eligible) segment shows how frequently issue stalls occur. The lower of no eligible warp means more efficient the code runs on the target device
9	Issue Stall Reasons (execution dependency)	The issue stall reasons capture why an active warp is not eligible, an input required by the instruction is not yet available. Execution dependency stalls can potentially be reduced by increasing instruction-level parallelism
10	CUDA achieved GFLOPS	The number of double precision accuracy floating point instructions executed per second. The higher the better

roads have a different number of entering vehicles and a different number of passing vehicles, and the number of upstream links and number of vehicles on nodes are different. The branch-taken ratio for kernel S2 is much less (27.62%). This is expected because the divergence in Algorithm 2 (see Section 3.4) is less than Algorithm 1. Also, in S2 the number of nodes is less and the number of vehicles passing to downstream roads varies in different types of road topologies.

Sixth, the instruction serialization ratio of these two kernel functions are low (<4%). This means that the branch taken ratio does not cause performance loss in these kernels.

Seventh, the metric of instruction per clock (IPC) for these kernel functions is below 0.4, which is far below the hardware's peak value (4.0). Regarding our framework, it is mainly because almost all the data (e.g., the road network and vehicles) is stored in the global memory, which causes high memory access latency. As illustrated in Table 1, we need 546.98 MB for lanes and 75.35 MB for nodes in S1 and S2. However, each shared memory per block is only 48 kB. This requires complex data division techniques and programming skills to divide the memory into sections and handle the issue of various blocks. As such, only global memory is used in our kernel functions, and the memory access latency is not completely hidden by thousands of threads.

Eighth and ninth, the no eligible warp issue efficiency is high for all four kernels (>90%), and the execution dependency of this issue's stall reason count is <20%. This means the memory access latency is the main reason, which is the same as the low IPC, that is, almost all data are loaded from and stored into global memory. It is difficult to use highly efficient shared memory because its size is small, only 48 kB per block. If we want to use it, the 546.98 MB of lanes needs to be divided into 11,396 sections. This will lead to lots of branches taken in the kernel functions, which will inevitably results in low efficiency.

Tenth, the achieved GFLOPS for the two kernel functions are also lower than the hardware's peak, which is again related to the above-mentioned dilemma of global memory latency.

3.3 Discussions

This section discusses additional thoughts about running mesoscopic traffic simulations on the CPU/GPU platform. First, it is beneficial to run the demand simulation on the CPU, the supply simulation on the GPU, and the data communication between the CPU and the GPU in an asynchronous way. In the proposed framework, the supply simulation on the GPU is the bottleneck, and the time costs of the other two tasks are almost hidden. Second, the memory access latency is a bottleneck in the proposed mesoscopic traffic simulation framework. In mesoscopic simulation frameworks, the update of a road depends on its own road status (e.g., road density and queue status) and requires a small number of parameters from its downstream roads. There is little shared data access among nearby roads and nodes, which limits the usage of the more efficient shared memory in the GPU. We believe this is the bottleneck that prevents us from achieving a higher simulation speedup on the GPU. And the problem in general is that we should design better data structure to utilize shared memory to enable high performance.

4 CONCLUSIONS AND FUTURE WORK

The main contributions of chapter are listed as follows.

1. A comprehensive systems of systems (SoS) simulation framework, including AMs of nodes, links, and vehicles, is proposed to run demand and supply coupled components on the CPU and GPU, respectively, with designed asynchronous simulation step management. For the demand part, real-time demands are periodically generated using

calibrated data. For the supply part, road networks and vehicles are modeled and run with GPU kernel functions.

2. To enable real-world networks' critical parts, including variable lengths of segments and links, CPU linked list structures are converted to more complex GPU array structures. Moreover, to decrease global data access latency, kernel functions are elaborately designed using more registers and memory bandwidth.

3. The proposed mesoscopic traffic simulation framework is demonstrated to simulate 100,000 vehicles moving on an ideal grid network, and the supply traffic simulation on a GPU (GeForce GT 950M) gets 10.0 times speedup, compared with running the same supply simulation on a CPU core (Intel i5). However, in order to test a real-world Singapore expressway network, the speedup is only 2.37. This is mainly due to the data structure difference between the two networks.

4. The proposed traffic simulation framework on CPU/GPU offers an innovative and high-performance solution in order to reduce the computational cost of various dynamic traffic assignment (DTA) models including various components such as nodes, links, and vehicles.

To extend GPU-based traffic simulation of real networks, we think that future works should focus on the optimization of SoS component data structure alleviating memory access time. For those systems that have some already existing models, it is better to design better shared memory utilization algorithm to enable components' data sharing.

References

Barcelo, J. (Ed.), 2010. Fundamentals of Traffic Simulation. In: International Series in Operations Research & Management Science, Springer, New York.

Ben-Akiva, M., Bierlaire, M., 1999. Discrete choice methods and their application to short-term travel decisions. In: Handbook of Transportation Science, Springer, pp. 5–34.

Ben-Akiva, M., Bierlaire, M., Burton, D., Koutsopoulos, H.N., Mishalani, R., 2001. Network state estimation and prediction for real-time traffic management. Netw. Spat. Econ. 1 (3/4), 293–318.

Ben-Akiva, M., Gao, S., Wei, Z., Yang, W., 2012. A dynamic traffic assignment model for highly congested urban networks. Transp. Res. C 24, 62–82.

Hawick, K.A., Leist, A., Playne, D.P., 2011. Regular lattice and small-world spin model simulations using CUDA and GPUs. Int. J. Parallel Prog. 39 (2), 183–201.

Hong, S., Oguntebi, T., Olukotun, K., 2011. In: Efficient parallel graph exploration onmulti-core CPU and GPU. Proceedings of the International Conference on Parallel Architectures and Compilation Techniques. IEEE, pp. 78–88.

Mahmassani, H.S., Hu, T., Jaykrishnan, R., 1992. In: Dynamic traffic assignment and simulation for advanced network informatics (DYNASMART).Proceedings of the 2nd International Capri Seminar on Urban Traffic Networks, Capri, Italy.

NVIDIA, 2014. CUDA C Programming Guide. Available from: http://www.nvidia.com/CUDA.

Park, H., Fishwick, P.A., 2010. A GPU-based application framework supporting fast discrete-event simulation. Simulation 86 (10), 613–628.

Park, H., Fishwick, P.A., 2011. An analysis of queuing network simulation using GPU-based hardware acceleration. ACM Trans. Model. Comput. Simul. 21 (3), 1–22.

Perumalla, K.S., 2006b. In: Discrete-event execution alternatives on general purpose graphicalprocessing units (GPGPUs). Proceedings of the ACM/IEEE/SCS Workshop on Principles of Advanced Discrete Simulation. IEEE, pp. 74–81.

Perumalla, K.S., 2008. Efficient execution on GPUs of field-based vehicular mobility models. In: Proceedings of the ACM/IEEE/SCS Workshop on Principles of Advanced Discrete Simulation. IEEE, p. 154.

Perumalla, K.S., Brandon, G.A., Srikanth, B.Y., Sudip, K.S., 2009a. In: GPU-based real-time execution of vehicular mobility models in large-scale road network scenarios.International Workshop on Principles of Advanced and Distributed Simulationpp. 95–103.

Perumalla, K.S., Aaby, B.G., Yoginath, S.B., Seal, S.K., 2009b. In: GPU-based real-time execution of vehicular mobility models in large-scale road network scenarios. Proceedings of the ACM/IEEE/SCS Workshop on Principles of Advanced Discrete Simulation, IEEE, pp. 95–103.

Singapore-MIT Alliance for Research and Technology (SMART), DynaMIT, http://137.132.22.82:15014/w/index.php/Main_Page. [(Accessed 1 March 2014)].

Strippgen, D., Nagel, K., 2009. Multi-agent traffic simulation with CUDA.Proceedings of International Conference on High Performance Computing & Simulation, Leipzig, pp. 106–114.

Transportation Research Board, 2000. Highway Capacity Manual. Transportation Research Board, Washington, DC.

Wende, F., 2010. Simulation of Spin Models on NVIDIA Graphics Cards Using CUDA. .

Xu, Y., Tan, G., Li, X., Song, X., 2014. In: Mesoscopic traffic simulation on CPU/GPU,Proceedings of ACM SIGSIM Conference on Principles of Advanced Discrete Simulation, pp. 39–49.

Yan, X., Xiao, S., Zhiyong, W., Gary, T., 2014. An entry time based supply framework (ETSF) for mesoscopic traffic simulations. Simul. Model. Pract. Theory 47 (6), 182–195.

Further Reading

Auld, J., Hope, M., Ley, H., Sokolov, V., Xua, B., Zhang, K., 2015. POLARIS—general purpose agent-based modeling framework specialized for high-performance transportation simulations. Transp. Res. C 82, 653–672.

Aydt, H., Yadong, X., Michael, L., Alois, K., 2013. In: A multi-threaded execution model for the agent-based SEMSim traffic simulation.Proceedings of AsiaSim. vol. 40, pp. 1–12.

Brodtkorb, A.R., Trond, R.H., Martin, L.S., 2013. Graphics processing unit (GPU) programming strategies and trends in GPU computing. J. Parallel Distrib. Comput. 73, 4–13.

Buck, I., 2007. GPU computing with NVIDIA CUDA.SIGGRAPH '07, New York, NY.

Burghout, W., Koutsopoulos, H.N., Andreasson, I., 2006a. Hybrid Microscopic-Mesoscopic Traffic Simulation. Doctoral Thesis, Royal Institute of Technology, Stockholm, pp. 218–225.

Burghout, W., Koutsopoulos, H.N., Andreasson, I., 2006b. In: A discrete-event mesoscopic traffic simulation model for hybrid traffic simulation.Intelligent Transportation Systems Conference.

Çetin, N., 2005. Large-Scale Parallel Graph-Based Simulations. Ph.D. Thesis, ETH Zurich, Zurich.

Christensen, E., 1990. Integrating distributed simulation techniques: DEVS and time warp.Proceedings AI and Simulation Eastern Multiconference, April.

Christensen, E.R., Zeigler, B.P., Rozenblit, J.W., 1990. Reducing the validation bottleneck with a knowledge-based, distributed simulation environment. Expert Syst. Appl 3, 329–342.

Dell'Orco, M., Marinelli, M., Silgu, M.A., 2016. Bee colony optimization for innovative travel time estimation, based on a mesoscopic traffic assignment model. Transp. Res. C 66, 48–60.

Denis, G., Jose-Juan, T., Samuel, A., Roshan, M.D.S., 2012. Graphics processing unit based direct simulation Monte Carlo. Simulation 88 (6), 680–693.

Di Gangi, M., Cantarella, G.E., Di Pace, R., Memoli, S., 2016. Network traffic control based on a mesoscopic dynamic flow model. Transp. Res. C 66, 3–26.

Fan, Z., Qiu, F., Kaufman, A., Yoakum-Stover, S., 2004. In: GPU clusters for high performance computing.Proceedings of SC'04pp. 47–58.

Gordon, D.B.C., Gordon, I.D.D., 1996. Paramics—parallel microscopic simulation of road traffic. J. Supercomput. 10, 25–53.

INRO, Dynameq. http://www.inro.ca/en/products/dynameq/. [(Accessed 1 March 2015)].

Jayakrishnan, R., Mahmassani, H.S., Hu, T.Y., 1994. An evaluation tool for advanced traffic information and management systems in urban networks. Transp. Res. C 2 (3), 129–147.

Kallioras, N.A., Kepaptsoglou, K., Lagaros, N.D., 2015. Transit stop inspection and maintenance scheduling: a GPU accelerated metaheuristics approach. Transp. Res. C 55 (2), 246–260.

Liu, H.X., Ma, W., Jayakrishnan, R., Recker, W., 2004. Large-scale traffic simulation through distributed computing of PARAMICS, California PATH Research. Report UCI-ITS-TS-WP-04-14.

Liu, J., Zhang, L., Tao, F., 2014. In: Service-oriented model composition. Summer Simulation Multi-Conference, 2014, July 6–10. The Hyatt Regency Monterey, Monterey, CA.

Liu, Y., Zhang, L., Zhang, W., et al., 2016. In: An overview of simulation-oriented model reuse. Asian Simulation Conference. Springer, Singapore, pp. 48–56.

Michalakes, J., Vachharajani, M., 2008. In: GPU acceleration of numerical weather prediction.IPDPS 2008: IEEE Int'l Symp. Parallel and Distributed Processingvol. 18. p. 531.

Nutaro, J.J., Zeigler, B.P., 2017. How to apply Amdahl's law to multithreaded multicore processors. J. Parallel Distrib. Comput. https://dx.doi.org/10.1016/j.jpdc.2017.03.006 For general theory of speedup with multicore.

Passerat-Palmbach, J., Mazel, C., Hill, D.R.C., 2011. Pseudo-random number generation on GP-GPU.International Workshop on Principles of Advanced and Distributed Simulation, pp. 1–8.

Perumalla, K.S., 2006a. In: Discrete event execution alternatives on gen-eral purpose graphical processing units (GPGPUs).International Workshop on Principles of Advanced and Distributed Simulation, pp. 74–81.

Perumalla, K.S., Brandon, G.A., Srikanth, B.Y., Sudip, K.S., 2012. Interactive, graphical processing unit based evaluation of evacuation scenarios at the state scale. Simulation 88 (6), 746–761.

Rickert, M., Nagel, K., 2001. Dynamic traffic assignment on parallel computers in TRANSIMS. Futur. Gener. Comput. Syst. 17, 637–648.

Taylor, N.B., 2003. The CONTRAM dynamic traffic assignment model. Netw. Spat. Econ. 3, 297–322.

Tian, Y., Chiu, Y.-c., 2011. In: Anisotropic mesoscopic traffic simulation approach to support large-scale traffic and logistic modeling and analysis.Winter Simulation Conferencepp. 1500–1512.

Tian, Y., Chiu, Y.-C., 2014. In: A computational efficient approach to retaining zone pair travel time in dynamic traffic assignment for activity-based model integration.Proceedings of Transportation Research Board 93rd Annual Meeting 14-5716.

Waraich, R.A., Charypar, D., Balmer, M., Axhausen, K.W., 2014. Performance Improvements for Large Scale Traffic Simulation in MATSim, In: *Computational Approaches for Urban Environments*. vol. 13, Springer International Publishing, pp. 211–233.

Xiao, S., Lin, Z., Dongjing, H., Zhiyun, R., 2012. A DEVS based modelling and simulation methodology—COSIM. Int. J. Appl. Math. Inform. Sci. 6 (2S), 417S–423S.

Xiaosong, L., Wentong, C., Stephen, J.T., 2013. GPU accelerated three-stage execution model for event-parallel simulation.International Workshop on Principles of Advanced and Distributed Simulation.

Xu, Y., Tan, G., 2012. hMETIS-based offline road network partitioning.Proceedings of AsiaSim, Shanghaivol. 323. pp. 221–229.

Yang, W., 2009. Scalability of Dynamic Traffic Assignment. Ph.D. Thesis, Massachusetts Institute of Technology.

Zeigler, B.P., Sarjoughian, H.S., 2017. Guide to Modeling and Simulation of Systems of Systems, second ed. Springer.

Zeigler, B.P., Zhang, L., 2015. Service-oriented model engineering and simulation for system of systems engineering. In: Concepts and Methodologies for Modeling and Simulation. Springer International Publishing, Chapter 2.

Zeigler, B.P., Nutaro, J.J., Seo, C., 2015. What's the Best Possible Speedup Achievable in Distributed Simulation: Amdahl's Law Reconstructed, DEVS TMS. SpringSim.

Zeigler, B. P., H. Praehofer and T. G. Kim, *Theory of Modeling and Simulation: Integrating Discrete Event and Continuous Complex Dynamic Systems*, 2nd Ed. Academic Press, USA, pp. 25–32.

Zhang, L., 2011. Model engineering for complex system simulation. Proceedings of 58th Forum on New Academic Views, October 15. China Science and Technology Press, Lijiang.

Zhang, L., Shen, Y., Zhang, X., Song, X., Tao, F., Liu, Y., 2014. The model engineering for complex system simulation. Proceedings of the I3M Multiconference, September 10–12, Bordeaux, France.

Zhou, X., Tanvir, S., Lei, H., Taylor, J., Liu, B., Rouphail, N.M., Christopher Frey, H., 2015. Integrating a simplified emission estimation model and mesoscopic dynamic traffic simulator to efficiently evaluate emission impacts of traffic management strategies. Transp. Res. D 37, 123–136.

Ziliaskopoulos, A.K., Waller, S.T., Li, Y., Byram, M., 2004. Large-scale dynamic traffic assignment: implementation issues and computational analysis. J. Transp. Eng. 130 (5), 585–593.

Simulating Discrete-Event Models in the Classic Worldviews: A New Approach to Simultaneous Events and Parallel Execution[a]

James Nutaro*

*Computational Sciences and Engineering Division, Oak Ridge National Laboratory, Oak Ridge, TN, United States

A model built with the classical worldviews of discrete-event simulation is inseparable from its realization in a particular software package (Schriber et al., 2014). This fact can be troublesome when a long-lived model must behave consistently across versions of a simulation package or when a model outlives the simulation package with which it is constructed and must be rebuilt in a more modern venue. The classical worldviews offer a more immediate challenge when the modeler wants to use parallel computing to speed up a complex simulation. One of their intrinsic aspects is a foundation in the serial computing paradigm, which prohibits parallelization in any simple or automatic way (see, e.g., Fujimoto, 2000).

At the same time, modeling approaches for discrete-event systems that resolve these issues (see, e.g., Zeigler et al., 2000; Nutaro, 2010) are frequently neglected in favor of the classical worldviews. In part, this is because of the intuitively appealing approach to modeling and

[a]This manuscript has been authored by UT-Battelle, LLC, under Contract No. DE-AC0500OR22725 with the US Department of Energy. The US Government retains and the publisher, by accepting the article for publication, acknowledges that the US Government retains a nonexclusive, paid-up, irrevocable, world-wide license to publish or reproduce the published form of this manuscript, or allow others to do so, for the US Government purposes. The Department of Energy will provide public access to these results of federally sponsored research in accordance with the DOE Public Access Plan (http://energy.gov/downloads/doe-public-access-plan). Research sponsored by the Laboratory Directed Research and Development Program of Oak Ridge National Laboratory, managed by UT-Battelle, LLC, for the US Department of Energy.

simulation execution that these worldviews enable. Another influencing factor is that many (possibly most) commercially available and open source tools for discrete-event simulation are based on the classical worldviews, and the large number of models relying on these tools make impractical fundamental changes to their essential algorithms.

This chapter reviews the classical worldviews and then presents a new approach to discrete-event simulation that preserves their most attractive features. An important part of this new approach is a two-phase simulation procedure derived from the simulation algorithm for cellular automata, coupled difference equations, and other synchronous models. The two-phase procedure is motivated by asynchronous cellular automata, which incorporate important elements of synchronous models while being efficiently simulated with discrete events (Fatès, 2013; Nutaro, 2010). Following the approach by Zeigler (1984), we will conclude by showing how the proposed two-phase approach is realizable in the discrete-event system specification (DEVS). This points naturally towards two-phase approaches to the other worldviews, which are likewise realizable in DEVS.

The new simulation procedure is described via an abstract algorithm that has numerous, distinct realizations. This has the practical effect of enabling parallel execution of a discrete-event model while preserving familiar modeling constructs. It also serves to separate the model construction and simulator implementation. In particular, it is sufficient for the modeler to assume execution via the abstract algorithm while the builder of a simulation tool is free to choose any implementation that yields the required behavior. A tertiary effect of the two-phase approach is to enable modeling synchronous systems within the discrete-event framework.

1 CLASSICAL WORLDVIEWS

The three classical approaches to discrete-event simulation are event scheduling, activity scanning, and process-oriented (see, e.g., Zeigler et al., 2000; Mansharamani, 1997; Muzy et al., 2013). In each worldview, the model's state variables change only at specific points in time, and a discrete-event simulation is characterized by the irregular spacing of these time points. The location of events in time are specified by the model via mechanisms native to the each worldview. While there are numerous variations on these three worldviews, we will limit our discussion to their simplest incarnations. This facilitates detailed discussions of implementation choices and their consequences while providing a foundation on which the proposed revisions of these worldviews can be carried into a more sophisticated setting.

1.1 Event Scheduling

The conceptual simplicity of the event scheduling worldview makes it particularly appealing, and it can form the basis for the more sophisticated activity scanning and process-oriented worldviews. An event in this worldview comprises a time stamp indicating when the event will occur and an event handler that changes the state of the model at that time. Pending events are placed into a data structure called the future event list that stores events in time stamp order, with the earliest event (i.e., smallest time stamp) at the front and most distant event at the back.

A simulation proceeds by removing the event at the front of the future event list, updating the simulation clock to the time stamp of this event, and executing the event handler. These steps are repeated until no more events remain in the future event list or some other desired stopping condition is met. When the event handler is executed, it may modify the state of the model, insert events into the future event list, delete events from the future event list, or take any combination of these actions.

A primary contributor to the execution time of an event scheduling simulation is the computational complexity of the future event list. This list must allow for inserting an event and for removing the event with the least time stamp. It may also allow removal of (i.e., canceling) an arbitrary event. Data structures suitable for realizing the future event list and their relative merits have been discussed in several places (see, e.g., the brief survey in Mansharamani (1997)).

The simplest data structure is a list. An event is inserted by traversing the list in order, starting at the first event, until the time stamp of the event to be inserted is less than the time stamp of the most recently inspected event. If no such event is found then the new event is inserted at the end of the list. Otherwise, the new event is inserted into the list at the position immediately preceding the most recently inspected event. An event is removed by finding it via traversal from front to back and then deleting it from the list.

This implementation of the future event list imposes a first in, first out (FIFO) policy on the ordering of simultaneous events. For example, suppose the future event list contains the time stamps 1, 2, 4, 5 and we must insert a new event with time stamp $\underline{4}$. The underline distinguishes the new event from the existing event at time 4. Using the insertion procedure described previously, the event list becomes 1, 2, 4, $\underline{4}$, 5. If no more events are scheduled at time 4 then the existing event (i.e., the first one to be inserted) will be processed before the new event. Hence, the processing is FIFO.

In this example we compare the new time stamp with four others. We can avoid some of these comparisons by changing how an event is inserted. The new procedure inserts an event by traversing the list in order from front to back until the time stamp of the event to be inserted is less than or equal to the time stamp of the most recently inspected event. If no such event is found then the new event is inserted at the end of the list. Otherwise, the new event is inserted into the list at the position immediately preceding the most recently inspected event.

Using this procedure the event list becomes 1, 2, $\underline{4}$, 4, 5. This is in time stamp order, requires just three comparisons of time stamps, and is not the same list as before. In this case, the execution order of simultaneous events is last in, first out (LIFO). An execution of the model using this LIFO list may produce a different result than an otherwise identical simulation using the FIFO list.

The choice of future event list, and its impact on the ordering of simultaneous events, highlights one important reason why a model cannot be disentangled from its simulation tool. The previous examples also illustrate the difficulty of parallelizing a simulation built on this worldview. While it may appear natural to execute events with the same time stamp in parallel (after all, they occur simultaneously in the simulation), race conditions in the event handlers may produce unanticipated and incorrect results.

For instance, suppose the event at time 4 sets a variable v to $\max(v,0)$ and the adjacent event at time 4 increments v by one. In the FIFO simulation, the new state following this sequence of events is $\max(v+1,0)$. In the LIFO simulation, it is $\max(v,0)+1$. When the two events are executed in parallel we might obtain either answer.

If both answers are sensible then we should expect each to occur with a given frequency in the system being modeled, and the statistics of the outcomes should to be an invariant feature of the model. However, if we naively rely on the race condition to produce these statistics then our conclusions will depend solely on how the computational platform schedules parallel tasks. Specifically, the frequency of each outcome will be particular to the computer hardware, operating system, and workload, and no characterization of the simulator's statistics can give us useful information about the model's behavior.

1.2 Activity Scanning

The activity scanning worldview distinguishes two types of events: future events and conditional events. Both types can change the state of the model and modify the future event list. Future events are scheduled as in an event scheduling simulation. Conditional events are activated when a rule is satisfied by the model's state, and a distinct rule may be attached to each conditional event. Conditional events can simplify the construction of a model, and this is the primary appeal of the activity scanning approach.

The simulation procedure for an activity scanning model builds on the event scheduling approach by adding a step in which the conditional events are scanned for satisfaction of their rules. A step in this simulation begins by removing the first event from the future event list, updating the simulation clock to match its time stamp, and executing its event handler. Next the simulator scans the conditional events. Upon encountering the first conditional event whose rule is satisfied, its event handler is executed, and then the list is scanned again. This scan and execute step is repeated until no conditional event is ready to execute. When this happens, the simulator advances to the next event in the future event list.

As with the event scheduling approach, the manner in which events are inserted into the event lists plays an important role in determining the outcome of a simulation. For example, suppose there are conditional events a and b and the future event list is an FIFO list. Further suppose that a and b each schedule a future event with time stamp one. Indicate these future events by $1a$ and $1b$, respectively. If the rules for a and b are satisfied and they are scanned in the order a, b then the resulting sequence of events will be $a, b, 1a, 1b$. If the conditional events are scanned in the order b, a then the sequence of events is $b, a, 1b, 1a$.

If we also perform simulations with a future event list that is LIFO then there are two more possible sequences of events. These are $a, b, 1b, 1a$ and $b, a, 1a, 1b$. Moreover, it is possible that a and b negate the satisfaction of one anothers rules. If so then the set of possible outcomes grows again to include $a, 1a$ if the conditional events are order a, b or $b, 1b$ if the conditional events are ordered b, a. As with the event scheduling worldview, simultaneous events link a model's behavior to how the event lists are implemented and they require sequential execution to avoid spurious statistics.

1.3 Process Oriented

The process-oriented worldview dispenses with explicit events by replacing these with processes that operate on the state of the system. A process can be in one of two modes: executing or waiting. The simulation clock is fixed while a process is executing, and changes to the model state occur at the current simulation time. A process switches from executing to waiting upon issuing a wait statement. The argument to the wait statement is the condition under which the process will resume execution. When this condition is satisfied, execution begins at the instruction immediately following the wait statement.

Wait statements cause time to advance in a process-oriented simulation. For example, if the process invokes a statement *wait for 5 seconds* then it will enter the wait mode and remain suspended for 5 seconds of simulation time. After this interval, execution resumes with the simulation clock advanced by 5 seconds. A wait condition can depend on variables other than time. For instance, if there is a variable v and the process invokes a statement *wait until $v \leq 0$* then the process will be suspended until some other process causes $v \leq 0$ to be true.

The powerful illusion of this approach is that each process comprises a single, unbroken sequence of instructions. The wait statements appear as blocking calls, which are familiar to most programmers. This illusion is made possible with *coroutines*, within which the instructions for a process are programmed. To the programmer, a coroutine looks like any other procedure. To the simulation engine, the coroutine consists of at least two items. These are

1. variables created within the coroutine, that is, the procedure's local variables; and
2. an instruction pointer indicating which instruction is to be executed next.

Variables local to the coroutine are not accessible to other coroutines. Exchanges of data between coroutines occur via variables that are shared in the global address space.

In one implementation of the process-oriented approach, the simulation engine allows for a single executing coroutine. The other, suspended coroutines are kept in two lists. The first is a future event list for coroutines that have indicated an explicit activation time in their suspending wait statement. The second is a conditional event list, and a coroutine is placed into this list if its wait statement includes an activation condition depending on variables other than time. A suspended coroutine can be in both lists. For example, if the activation condition is *wait for 5 seconds or $v \leq 0$* then the coroutine enters the future event list with a time stamp 5 seconds in the future and enters the conditional event list with the activation rule $v \leq 0$.

Because processes are realized as coroutines, we may use the two terms interchangeably to describe the simulation procedure. Initially, each coroutine is placed onto the future event list with a time stamp equal to zero. The process at the front of the future event list executes until it calls *wait*. At that time, the process is suspended and placed into the future and conditional event lists as is appropriate to its activation condition. Next, the simulator scans the conditional event list for a coroutine that can be activated. If one is found then it is removed from the future and conditional event lists to become the new executing process. Otherwise the simulation clock is advanced to the time stamp of the first coroutine in the future event list and that process becomes the new executing process. This procedure is repeated until the future and conditional event lists are empty, which indicates that all processes are permanently suspended, or until some other stopping criteria are met.

The future and conditional event lists in this approach are clearly related to those in the activity scanning worldview. Indeed, if we required wait statements to indicate a future time or an activation condition, but not both, then the activity-oriented simulation procedure could be used to select the process that will execute. In the more general case, it is only necessary to modify the activity scanning approach such that it removes the executing coroutine from both lists. The implications of event ordering in an activity scanning simulation have direct analogs with the ordering of coroutine execution in a process-oriented simulation.

2 SIMULTANEOUS ACTIONS

As we have seen, a procedure for handling simultaneous events is not intrinsic to the classical worldviews. Moreover, this situation is unique to discrete-event models. For instance, when modeling the motion of two objects using differential equations the question "who moves first?" does not occur because simultaneous action is intrinsic to the mathematics of the model. Similarly, when we model with interacting automata or coupled difference equations, the question of which system should act first at a given time does not occur because simultaneity is intrinsic to the mathematical formulation.

To illustrate how simultaneity is resolved in these types of modeling approaches let us consider a one-dimensional cellular automaton in which each cell is influenced by its two neighbors. These are a particular instance of discrete time models, and their simulation procedure is identical to that used for simulating coupled difference equations and explicit numerical algorithms for continuous systems (see, e.g., Wolfram, 2002; Nutaro, 2010; Zeigler et al., 2000, 2018). We can view the cellular automata as a collection of connected models each in the form of a function

$$q' = \delta(q, u_l, u_r) \tag{1}$$

where q is the present state of the cell, q' is the new state, and the new state is a function of q and the input u_r and u_l supplied by the neighboring cells.

In the familiar synchronous approach to simulating this cellular automaton, the natural numbers are used for time and the state of each cell is changed at times 1, 2, 3, …. The simulation procedure has three steps.

1. For each cell, calculate its left input u_l and right input u_r as functions of the states of the cells to the left and right.
2. For each cell, calculate its new state q' to be $\delta(q, u_l, u_r)$.
3. Advance the simulation clock by one and go to Step 1.

To demonstrate this procedure, consider a cellular automaton for which the state of a cell is 0 or 1 and its left and right input are the states of the adjacent cells. Define

$$\delta(q, u_l, u_r) = q + u_l + u_r \tag{2}$$

where + is binary addition. That is, $1 + 1 = 0, 0 + 1 = 1 + 0 = 1$, and $0 + 0 = 0$. A simulation of this model with five cells having alternating initial states is shown in Table 1. The cells at the edges receive their left or right input from the cell at the opposite edge of the space.

If the real numbers are used for time and updates for each cell may occur at any point in time then this gives us the most general case of an asynchronous cellular automata. The synchronous cellular automata described earlier are a special instance of this general case (for a review, see Fatès, 2013). Now suppose for a moment that in a particular asynchronous cellular

TABLE 1 Simulation Using Eq. (2)

Time	State				
0	0	1	0	1	0
1	1	1	0	1	1
2	1	0	0	0	1
3	0	1	0	1	0
4	1	1	0	1	1
5	1	0	0	0	1

automaton no two cells update themselves simultaneously. It is trivial to define an event scheduling simulation for this model.

In this simulation there is a single type of event which we will call δ and a one-dimensional array of numbers that are the states of the cells. If there are $n > 1$ cells and we denote the array element at position $1 \leq k \leq n$ by q_k then the event δ calculates the new state q'_k to be

$$q'_k = \begin{cases} q_k + q_{k-1} + q_{k+1} & \text{if } 2 \leq k < n, \\ q_k + q_n + q_{k+1} & \text{if } 1 = k < n, \\ q_k + q_{k-1} + q_1 & \text{if } 2 \leq k = n. \end{cases} \tag{3}$$

The simulation begins by placing an event into the future event list for each cell k with a time stamp equal to the time of its first update. The event for cell k changes its state and schedules a new event for cell k at a future instant commensurate with its update internal.

To demonstrate this simulation, consider five cells that update at intervals 1 and the irrational intervals $\sqrt{2}$, e, π, and the golden ratio $\Phi = (1 + \sqrt{5})/2$, respectively. These update intervals ensure no two events will occur simultaneously. The first five events in this simulation are calculated in Table 2.

Can we construct a single, event-oriented simulation that handles the case in which some cells may experience simultaneous updates while others do not? For instance, suppose that the cell updating at intervals of length $\sqrt{2}$ is changed to update at intervals of length 2. In this case collisions will occur at even times 2, 4, ... The correct behavior in this situation is to

TABLE 2 Asynchronous Simulation Using Eq. (2)

Time	State				
0	0	1	0	1	0
1	1	1	0	1	0
$\sqrt{2}$	1	0	0	1	0
Φ	1	0	1	1	0
e	1	0	1	0	0
π	1	0	1	0	1

evaluate the new states for the cells at $k = 1$ and $k = 2$ as we did with the synchronous algorithm. That is, compute the input of both cells and then change their states. Our event-oriented simulation does not do this and so it will give an incorrect result.

The asynchronous cellular automaton illustrates the circumstance in which to update one variable we need information about another variable but events are scheduled that will modify both variables at the same simulation time. Consequently, the order of the events determines the future course of the model and, for the cellular automata, all possible orderings are incorrect. Consequently, we cannot solve this problem in general with the event-oriented simulation approach or the activity scanning and process-oriented simulations derived from it.

The other circumstance where the ordering of simultaneous events matters happens when a single variable will be modified by two or more events. We can illustrate this problem with an extension of the asynchronous cellular automaton such that functions δ_1 and δ_2 operate on each cell but at possibly distinct rates. In the case of a collision, there are four types of cells: those that apply δ_1 then δ_2, those that apply δ_2 then δ_1, those that apply only δ_1, and those that apply only δ_2. Again, there is no single, event-oriented simulation that correctly handles a model in which this circumstance arises.

3 REVISING THE CLASSICAL WORLDVIEWS

Simultaneous events as they occur in asynchronous cellular automata and all other types of synchronous systems can be properly resolved by augmenting the classical worldviews. We begin by resolving the problem in which to update one variable we need information about another variable but events are scheduled that will modify both variables at the same time. The simulation procedure for cellular automata manages this situation with two phases. In the first phase, we gather copies of the data that will be used to update the model's state. In the second phase, we use these copies when applying δ to update the relevant state variables. This two-phase approach makes the order in which the events are applied unimportant.

These two phases can be realized in the event-oriented worldview by adding to each event a function that gathers data needed by the event handler. The augmented event is defined by its scheduled time of execution, event handler, and the new input preparation function. The input preparation function is used to implement the two-phase approach to updating the model's state. In the first phase, we advance the simulation clock to the time of the first event in the future event list, extract all events from the future event list that have this time stamp, and execute their input preparation functions. These functions may be executed in any order, or even in parallel, because they do not modify state variables. In the second phase, we execute the event handlers, which may also be done in any order, or even in parallel, if no two events modify the same variable.

Now we turn to the problem in which a single variable is modified by two or more simultaneous events. To address this issue, let us identify a partitioning of the state variables such that each event acts on a single partition. Partitions are called logical processes in the field of parallel discrete-event simulation (Fujimoto, 2000) and atomic models in DEVS (Zeigler et al., 2000).

Practitioners using logical processes or DEVS typically construct these partitions by hand in the course of designing the simulation model, and this approach is natural if we intend a parallel execution from the outset. The question of how to identify partitions automatically offers an interesting topic for future research, with work by Zeigler et al. (2000, 2018) and Zeigler (1984) on the relationships between modular and nonmodular modeling methods offering an attractive point of departure. In what follows we assume the partitions to be given and do not further consider their origin.

Our augmented simulation procedure is aware of these partitions, and it associates with each partition a state transition function responsible for invoking the event handlers that act on it. The purpose of this function is to manage simultaneous events in a manner appropriate to the model. The simultaneous events acting on a given partition must execute sequentially, but the partitions may change state in parallel. The augmented simulation procedure is summarized as follows.

1. Advance the simulation clock to the time of the first event in the future event list.
2. Gather and remove from the future event list all events with time stamps equal to the current simulation time. These are the imminent events.
3. Execute the input preparation function of each imminent event.
4. Partition the imminent events into lists according to the partition each acts upon and pass each list to the state transition function of its partition.
5. Go to Step 1.

This algorithm makes possible a general-purpose, event-oriented simulation for cellular automata. In this simulation, the state variable for each cell occupies its own partition. The input preparation function for partition k makes copies of variables q_{k+1} and q_{k-1} (or q_n, q_{k+1} for the leftmost cell; q_{k-1}, q_1 for the rightmost) and stores these as $u_{r,\,k}$ and $u_{l,\,k}$. The event handler calculates δ. The ordering function for each partition makes the event selection described at the end of Section 2.

The augmented simulation procedure can mimic the classical event-oriented worldview with an FIFO scheduler. To do so, we define a single partition and omit implementations of the input preparation functions. The state transition function executes events in the order that they appear in its list. To include event cancelation it is necessary for the state transition function to be aware of canceled events and skip these if they appear in its list.

3.1 Activity Scanning

An augmented activity scanning worldview can be constructed by introducing conditional events into the preceding event-oriented simulation. Each conditional event is constrained to act on a single partition, but its activation rule may examine variables in many partitions. Conditional events are activated following Step 4 of the event-oriented simulation described in Section 3. As before, there are two phases. In the first, the conditional events are scanned to find those eligible for activation and to invoke their input preparation functions. In the second phase, the eligible events are passed to their partition's state transition function. These scan and execute steps are repeated until no conditional events may be activated. The simulation procedure is as follows.

1. Get the smallest time stamp from the front of the future event list and advance the simulation clock to this time.
2. Gather and remove from the future event list all events with time stamps equal to the current simulation time. These are the imminent events.

3. Execute the input preparation function of each imminent event.
4. Partition the imminent events into lists according to the partition each acts upon, and pass each list to the state transition function of its partition.
5. Scan the conditional events and select the events whose activation rules are satisfied.
6. If no events are selected, go to Step 1.
7. Execute the input preparation functions of the selected events.
8. Partition the selected events into lists according to the partition each acts upon, and pass each list to the state transition function of its partition.
9. Go to Step 5.

Steps 3–5, 7, and 8 all offer opportunities for parallel execution. The input preparation functions may be executed in parallel because these do not modify state variables, and the conditional events may be scanned in parallel for the same reason. The state transition functions of the partitions may be executed in parallel, even if the events within their lists must be executed sequentially.

3.2 Process Oriented

In the process-oriented worldview, the processes form a natural partitioning of the state variables if each process modifies only its own state variables and observes shared variables only as part of its activation condition. This process-oriented simulation may be implemented using the activity scanning procedure in Section 3.1 to decide when a waiting coroutine must resume execution. Because the events do not operate directly on the process state variables, it is sufficient for the event handlers to resume execution of the coroutine and then return when that routine is suspended by a wait.

4 REALIZING THE WORLDVIEWS IN A DEVS

We now show how the simulation approach constructed in Section 3.1 can be realized in a coupled DEVS. The process-oriented worldview may be realized on top of the activity scanning simulation as described in Section 3.2, and a simple restriction produces the event scheduling worldview. By basing an implementation on DEVS we set the stage for parallel execution of the classical worldviews by embedding them into a parallel simulation of a DEVS model (see Zeigler et al., 2000, 2018; Zeigler, 1984).

An unusual feature of this implementation will be that the input preparation functions, which we have described as gathering or pulling information from other partitions, are invoked within the output function of an atomic model. This would appear to reverse the proper role of the output function, which is intended to push information to other atomic models, although a precedence for this type of pull mechanism appears in Barro's Heterogeneous Flow System Specification (Barros, 2002). We choose to pull, rather than push, because it is convenient for building models that will run on a shared memory multiprocessor.

This convenience could be done away with by forcing the time advance of the atomic model to be zero following the execution of any event and requiring the output function to project state variable values that must be visible to other partitions. The input preparation functions would be invoked during the internal, external, and confluent transitions of the atomic model using a local copy of the relevant data from other partitions. This would occur after updating these copies using the projections supplied as input to the external and confluent transitions. In this way, the proposed implementation can be transformed into one suitable for a distributed memory parallel computer.

Each partition of the activity scanning model is embodied in a DEVS atomic model. The state of this atomic model comprises a future event list, a list of conditional events, a list of events to be scheduled for other partitions, the current simulation time, a mode which may be F or C, and the state variables of the partition. The mode indicates whether our next action will be to process future events (F) or conditional events (C). Upon creation, the atomic model sets its mode to F and the current time to zero.

The time advance function of this atomic model returns one of three values.

1. the difference between the next future event and current time if the mode is F;
2. infinity if the future event list is empty and the mode is F; or
3. zero, which occurs when the mode is C.

Upon expiration of the time advance, the output function of the atomic model is invoked. If the mode is F then the output function invokes the input preparation functions of the events at the front of the future event list. Otherwise, the mode is C and output function scans the list of conditional events, selects those that can be activated, and invokes their input preparation functions.

If there are events to be scheduled for other partitions or events that will act on this partition then the output function emits one or more messages. Events for other partitions are sent to their destinations, where these events will appear as input to the external or confluent transition function. A message is also sent to all partitions, which are interested in this partition's variables. This message notifies those partitions that a change of value is imminent.

If the coupling between atomic models that realize a partition are specified by the modeler then messages are directed to only the relevant partitions. If the modeler is uncertain about what this coupling should be, then the safest assumption is all to all coupling and each message is broadcast to every atomic model. Like the problem of automatic partitioning described earlier, the question of automatically deriving couplings is of considerable practical interest. As before, prior work on modular and nonmodular modeling formalisms is an attractive starting point for investigating this issue (Zeigler et al., 2000, 2018; Zeigler, 1984).

A partition changes state in one of three ways. An external transition occurs if the atomic model receives a message prior to the expiration of its time advance. An internal transition occurs if the time advance expires prior to receiving a message. A confluent transition occurs if a message arrives simultaneously with expiration of the time advance.

When an external transition occurs, the atomic model is supplied with time that has elapsed since the prior change of state. The current time variable of the atomic model is advanced by this amount. The mode is set to C to induce a scan of the conditional events that could be activated by the new variable values in the signaling partition. If events have arrived from other partitions then these are placed into the future or conditional event list as appropriate.

When an internal transition occurs, the atomic model advances its current time variable by the amount of the time advance. If the mode is F then the imminent events are removed from the future event list and passed to the partition's state transition function. If the mode is C then the conditional events suitable for activation are passed to the partition's state transition function. If any future or conditional events were found and processed then the mode becomes C. Otherwise the mode becomes F. A confluent transition performs the steps of an internal transition followed by those of an external transition. Regarding the confluent transition, this is not the only possible definition and some latitude for the modeler could be allowed regarding how the new values of the partition's variables are calculated given the collection of simultaneous events.

5 CONCLUDING REMARKS

This chapter has introduced a new approach to the classical worldviews of event scheduling, activity scanning, and process-oriented simulations. This new approach enables parallel implementations of these worldviews by employing distinct phases of the simulation for gathering input data and computing new values for state variables. The two phases mirror the simulation approach for synchronous models such as cellular automata and interacting difference equations, and they extend the scope of the classical worldviews to naturally include these types of models.

One approach to a parallel implementation embeds the augmented worldview in a DEVS model, and we have presented one such implementation in this chapter.[1] This particular implementation can leverage the DEVS simulation protocol to process events in parallel simultaneously active partitions (see Zeigler et al., 2000, 2018). Indeed, this is a generalization of the approach introduced in Zeigler (1984), where it was shown how the classical worldviews can be embedded in a sequentially executing (classical) DEVS model. The same approach of embedding the augmented worldview in a DEVS model can also enable parallel execution based on optimistic and conservative simulation algorithms (see Nutaro, 2010; Nutaro and Sarjoughian, 2004).

An attractive feature of the proposed approach is its compatibility with the fundamental modeling constructs of existing, sequentially executing simulation tools. This creates the possibility of extending existing, classically oriented simulation packages to include features for parallel execution while maintaining compatibility with legacy models. Such an extension would permit maintainers of legacy models to incrementally exploit these new features and provide new models with the opportunity to leverage parallel computing from the start. The widespread availability of parallel computing in the form of multicore computers has created a corresponding interest by modelers to use this computing power. At the same time, the difficulty of using algorithms developed chiefly for high performance has greatly hindered their application in practice (Fujimoto, 1993; Zeigler et al., 2015). The proposed approach offers a practical means of bridging the gap between modern parallel computers and widely used, but sequentially executing, packages for discrete-event simulation.

---

[1] Source code is available as part of the ADEVS simulation package, which is available online at http://web.ornl.gov/~nutarojj/adevs and http://sourceforge.net/projects/adevs.

References

Barros, F.J., 2002. Modeling and simulation of dynamic structure heterogeneous flow systems. Simulation 78 (1), 18–27.

Fatès, N., 2013. A guided tour of asynchronous cellular automata. In: Cellular Automata and Discrete Complex Systems. Springer, Berlin, Heidelberg, pp. 15–30.

Fujimoto, R.M., 1993. Parallel discrete event simulation: will the field survive? ORSA J. Comput. 5 (3), 213–230.

Fujimoto, R.M., 2000. Parallel and Distributed Simulation Systems. Wiley, New York, NY.

Mansharamani, R., 1997. An overview of discrete event simulation methodologies and implementation. Sadhana 22 (5), 611–627.

Muzy, A., Varenne, F., Zeigler, B.P., Caux, J., Coquillard, P., Touraille, L., Prunetti, D., Caillou, P., Michel, O., Hill, D.R.C., 2013. Refounding of activity concept? Towards a federative paradigm for modeling and simulation. Simulation 89 (2), 156–177.

Nutaro, J.J., 2010. Building Software for Simulation: Theory and Algorithms, With Applications in C++. Wiley, New York, NY.

Nutaro, J., Sarjoughian, H., 2004. Design of distributed simulation environments: a unified system-theoretic and logical processes approach. Simulation 80 (11), 577–589.

Schriber, T.J., Brunner, D.T., Smith, J.S., 2014. Inside discrete-event simulation software: how it works and why it matters. In: Proceedings of the Winter Simulation Conference 2014pp. 132–146.

Wolfram, S., 2002. A New Kind of Science. Wolfram Media, Champaign, IL.

Zeigler, B.P., 1984. Multifacetted Modelling and Discrete Event Simulation. Academic Press, London.

Zeigler, B.P., Praehofer, H., Kim, T.G., 2000. Theory of Modeling and Simu-lation, second ed. Academic Press, London.

Zeigler, B.P., Nutaro, J.J., Seo, C., 2015. What's the best possible speedup achievable in distributed simulation: Amdahl's law reconstructed. In: Proceedings of the Symposium on Theory of Modeling & Simulation: DEVS Integrative M&S Symposium, DEVS '15pp. 189–196.

Zeigler, B.P., Muzy, A., Kofman, E., 2018. Theory of Modeling and Simulation: Discrete Event & Iterative System Computational Foundations, third ed. Elsevier, Amsterdam.

17

DEVSim++ME: HLA-Compliant DEVS Modeling/Simulation Environment With DEVSim++

Tag Gon Kim*, Changbeom Choi†

*KAIST, Daejeon, South Korea †Handong University, Pohang, South Korea

1 INTRODUCTION

Because modern society is becoming more complex and evolving more quickly, a model engineer should be able to develop a proper simulation model quickly and correctly. Therefore, the model engineer should be equipped with a proper theoretical background on models as well as development skills. Although various model engineering tools are available, an engineer cannot build a simulation model alone. Experts from various domains should collaborate with each other to develop the model.

Among the various model engineering tools, the DEVSim++ME (hereafter referred to as ME) toolset has been actively utilized in various fields as a tool for collaboration, development, and experiments. The design philosophy of ME is to help a model engineer design, develop, and extend a simulation model easily and acquire data from the simulation model effectively.

The ME toolset adopts the DEVS formalism as a foundation of the modeling tool, and it provides the execution environment, so that model engineers can design models and implement them in the DEVSim++ simulation environment. After implementing the simulation model, the model engineer may want to ensure that the model meets the requirement specifications and that the implementation meets the model's design; in that case, the model engineer may utilize the MVali and MVeri tools. To extend the simulation model for interoperation, the model engineer may use KHLA Adaptor and FOM2CPPClass. For the model engineer, analyzing the simulation results from the developed simulation model is also important. In that case, the model engineer may use DEXSim from the ME toolset to conduct multiple experiments throughout distributed computing resources.

Model Engineering for Simulation
https://doi.org/10.1016/B978-0-12-813543-3.00017-2

This chapter introduces a system-theory-based model engineering environment to help model systems related to the simulation objective and gather simulation data from the simulation model. We employ the discrete event system specification formalism, known as the DEVS formalism, as a conceptual modeling tool and the ME toolset as the M&S environment. The DEVS formalism is fully explained in the companion volume to this book (Zeigler et al., 2018). Here, we only introduce the aspects needed for its use in this chapter.

2 BACKGROUND

2.1 DEVS Formalism

The DEVS formalism is used to capture the behaviors of discrete event systems (DESs) in a set-theoretic manner. Modelers may specify the behaviors of the core models and the system's hierarchical structures in a modular fashion. The atomic model is the core model that generates the model's behaviors, and a coupled model defines the model's hierarchical structure. The coupled model may comprise multiple atomic or coupled models to build a complicated model and may create complex behaviors by executing itself. The following sections introduce the notations of the DEVS formalism and its simulation algorithms (Kim et al., 1996).

2.1.1 Definitions of the DEVS Formalism

In DEVS-based model engineering, a modeler may capture a system's behavior using an atomic model and a coupled model. The atomic model has three sets and four functions with which to capture the atomic behavior of a system component. The three sets are the input event set X, the output event set Y, and the state set S. The four functions are the external transition function δ_{ext}, the internal transition function δ_{int}, the output function λ, and the time-advance function, ta. Fig. 1 shows the notation of the atomic model.

A modeler may model two types of system behaviors using the atomic model: (1) behaviors after external events and (2) behaviors without external events. The former can be modeled as an external transition function, which determines the next state based on the previous state and the external events. For the latter, the internal transition specifies the behavior. Also, the output function and time-advance function are used to specify the output events that can be generated by the given atomic model. Therefore, the atomic model's behavior can be viewed

$AM = < X, Y, S, \delta_{ext}, \delta_{int}, \lambda, ta >$

, where

X : a set of input events;

Y : a set of output events;

S : a set of sequential states;

δ_{ext}: $Q \times X \rightarrow S$, an external transition function,

   where $Q = \{(s,e) | s \in S$ and $0 \leq e \leq ta(s)\}$, total state set of M;

δ_{int} : $S \rightarrow S$, an internal transition function;

λ : $S \rightarrow S$, an output function;

ta : $S \rightarrow R_{0,\infty}$, time advance function

FIG. 1   Notation of the atomic model.

$CM = <X, Y, M, EIC, EOC, IC, SELECT>$
, where
X : a set of input events;
Y : a set of output events;
M : a set of all component models;
$EIC \subseteq CM.X \times UM.X$, external input coupling;
$EOC \subseteq UM.Y \times CM.Y$, external output coupling;
$IC \subseteq M.Y \times UM.X$, internal coupling;
$SELECT : 2^M - \Phi \to M$, tie-breaking function

FIG. 2   Notations of the coupled model.

as an event sequence influenced by external input events X that generates output events Y. Since the state set S represents the unique description of an atomic model, the model engineer may just specify the model's next state inside of the external transition function. If an external event arrives at the elapsed time e, which is less than or equal to $ta(s)$ as specified by the time-advance function ta, then a new state s' is computed by the external transition function δ_{ext}. Then, a new $ta(s')$ is calculated, and the elapsed time e is set to zero.

A coupled model CM consists of components that can be atomic models or coupled models. Similar to the atomic model, the coupled model uses three sets and four functions to model the system's structure and generates the behaviors of a composite system. The first two sets of the coupled model are the input event set X and the output event set Y. The last set of the coupled model, which is different from the atomic model, is the model set $\{M_i\}$. The four functions of the coupled model are the external input coupling function EIC, the external output coupling function EOC, the internal coupling function IC, and the tie-breaking function SELECT. Fig. 2 shows the notation of the coupled model.

The elements of the model set $\{M_i\}$ can be atomic models and another coupled model. Therefore, a modeler may compose the given model to build a complex simulation model using structure-defining functions. The EIC function connects the external input event to a model that exists inside of the coupled model. On the other hand, the EOC function connects the output event from the internal models of the coupled model to the outside of the coupled model. In addition, the IC function resolves the connections among the internal models. Finally, the SELECT function resolves the execution order among the simulation models. Mainly, the coupled model contains several models as components. Consequently, some models can be executed in the same time frame and may generate conflicting behaviors. In that case, the SELECT function solves the execution priority among the conflicting models. The SELECT function is used to order the processing of simultaneous internal events for sequential simulation. Thus, all of the events occurring at the same time in a system can be ordered by this function.

2.1.2 Simulation Algorithms of the DEVS Formalism

The simulation algorithms of the DEVS formalism use the abstract simulator concept for simulation, as proposed by Zeigler (1984). The concept defines a virtual processor that processes the model dynamics specified by the DEVS formalism. In other words, the simulation engine triggers the abstract simulators to be executed, and the simulators interpret the specifications of the DEVS model. Therefore, the DEVS model may not contain other computer code or may only contain the code for generating model dynamics.

There are three types of virtual processors: simulator, coordinator, and root-coordinator. The first processor is a simulator for an atomic model. The coordinator is used to interpret the coupled model. Finally, the root-coordinator is a special coordinator that controls the entire simulation and is not associated with any other models (Zeigler et al., 1994).

Fig. 3 shows a diagram of a simulator algorithm. The simulator algorithm processes four types of messages: the external event (x, t), star message $(*, t)$, output event (y, t), and done message (done, t_N). Since the simulator is associated with an atomic model, the model is always a leaf node of the simulation structure. Therefore, the model may generate an output message and a done message, but it cannot receive output messages or done messages. In other words, the coordinator algorithm converts the output message into an external input event message.

When a simulator receives the external event (x, t), the simulation algorithm processes the external event using the external transition function of the associated atomic model. As shown in Fig. 3, $M:\delta_{ext}$ denotes the external transition function of the associated atomic model. When the simulation algorithm finishes processing the external transition function, the simulation algorithm invokes the time-advance function to calculate the next deadline of the model. After calculating the next event time, the simulator algorithm sends the done message with the next event time, which is an amount of time to wait.

When the elapsed time of the simulation meets the next event time, the simulator algorithm will receive the star message. Then, the algorithm will execute the output function $(M: \lambda)$ and the internal transition function $(M: \delta_{int})$. After executing the output function, the simulator sends the output message (y, t), if one exists. Based on the DEVS formalism, the state of the atomic model may be changed. Therefore, the algorithm calculates the new deadline for the model and sends it to the coordinator.

Fig. 4 shows a diagram of a coordinator algorithm. The difference between the coordinator and root-coordinator is that the root-coordinator is not associated with any other models, and it decides the simulation components to be executed. Therefore, the root-coordinator does not route the message (y, t) to the parent but it routes the message (y, t) to the children. The simulation flow of the coordinator algorithm starts with the wait state. When other processors send a message (x, t), it routes that message based on the EIC function and waits until every

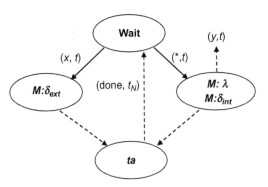

FIG. 3   Simulator algorithm for the atomic model.

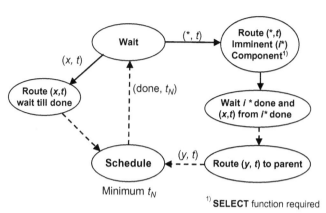

FIG. 4   Coordinator algorithm for coupled model.

child model sends the (done, t_N) message. When the coordinator receives every (done, t_N) message from the child models, the coordinator algorithm calculates the minimum next event time and sets the time of the (done, t_N) message. Then, the coordinator sends its (done, t_N) message to its parent model.

When a coordinator receives the star message (*, t), it identifies the imminent components i^*. Note that the coordinator may contain multiple components, and the components may request the same next event time; so, the coordinator should resolve the execution order. In this case, the coordinator algorithm will utilize the SELECT function. Since a model engineer provides the SELECT function, that model engineer is responsible for resolving the execution priority in the simulation model. Also, the coordinator waits until the imminent component sends the (done, t_N) message. During the execution, a child component model may send an output message (y, t). In this case, the coordinator should consider two cases. The first is a message to the component inside of the coupled model. Then, the coordinator algorithm uses the IC function to route the message. For the second case, the coordinator should use the EOC function to deliver the output message outside. After receiving the done message from every imminent component model, the coordinator may reschedule the minimum next event time.

2.2 IEEE1516 Standards for Simulation Interoperation

IEEE1516 standards are well-known standards for a distributed architecture and its implementation for interoperation between heterogeneous simulations. Mainly, the IEEE1516 standards consist of the distributed architecture, the high-level architecture (HLA), its implementation, the runtime infrastructure (RTI), and the federate development process and execution process (FEDEP) for developing a simulation with interoperable simulators.

2.2.1 High-Level Architecture

The HLA provides a common framework for the modeler to compose various simulators concerning the overall simulation objective in an extensible and scalable manner (IEEE Standards Association, 2012). To achieve the simulation objective, a space known as the federation can be created for simulators to collaborate and achieve given simulation objectives. Each member of the federation is referred to as a federate. Therefore, the HLA defines the services for the federates of a given federation to interact with each other.

The HLA of the IEEE1516 standards contains the HLA rules, the federate interface specification, an object model template (OMT), and the FEDEP, which are a set of basic rules that define the interoperability and the responsibilities of the federates and federation for HLA-compliant simulations. Also, the federate interface specification (HLA Working Group, 2000) provides a specification that a simulation may perform, or be asked to perform, during the simulation. It includes RTI services that are available to each simulation and the callback functions that each federate must provide to the RTI.

The OMT (Simulation Interoperability Standards Committee, 2000) is a standard description for a data model in the federation. The OMT is a common template for specifying the simulation information regarding a hierarchy of object classes, their attributes, and so on. The OMT is used to describe the simulation object model (SOM), which describes the information that the federate can produce or consume, and the federation object model (FOM), which is the conceptual data model for the information exchanged between federates within the federation.

The FEDEP mainly defines a process framework for federation developers and model engineers. The standard utilizes the HLA for interoperating heterogeneous simulators by adopting a systems engineering approach. Accordingly, the modelers may easily construct a federation by following FEDEP.

2.2.2 Runtime Infrastructure

The HLA defines the rules and interface among the heterogeneous simulators, and model engineering researchers need to implement the middleware for interoperation. The middleware of the HLA is called RTI and is software that supports the federation execution and conforms to the IEEE1516 standards. The RTI provides several services to the federates for them to exchange data and synchronizes the simulation time during the simulation. The RTI has a local RTI component (LRC) to provide interoperation services to each federate in the federation. Therefore, the RTI may utilize distributed computing resources to execute distributed simulations.

For multiple simulators to interoperate, a federation developer integrates an RTI library and embeds the RTI service code within the federate. The RTI library contains the RTI executive, the RTI ambassador, and the federate ambassador. RTI providers provide the library as software components, so that the developer may choose appropriate software components based on their development environment. The RTI ambassador and the federate ambassador control the information exchange among the federates. In general, the RTI ambassador component contains the services API provided by the RTI. A federate may invoke services through the given interface to request a service. The federate ambassador component is, in general, a pure virtual class that defines the callback functions for the federate developer to implement processing the data from other federates through the RTI. The callback functions provide a mechanism for the RTI to invoke operations and communicate back to the federate. While the RTI ambassadors are built into the RTI, federate ambassadors are written by the simulation developer for each federate.

The RTI improves the scalability of distributed simulations by controlling the amount of data exchanged among the federates. The data distributed in an HLA federation can be broken down into attributes and interactions. Attributes are properties of objects, while interactions are events that involve two or more objects. The FOM is an essential model of federation. The FOM defines the data elements of the entire federates of a given federation. The developer defines the data types, units of measurement, and attribute names. These data are used to provide information about the simulation to the RTI during runtime. The RTI does not transfer data that is not declared in the FOM. Before initiating a simulation, each federate registers itself to the RTI using the declaration management service and declares its publish and subscribe information. As a result, a federate can receive the data, which are subscribed before the simulation, from the RTI. Moreover, a federate can publish data that the federate declared to publish.

3 DEVSim++ME TOOLSET

In general, the purpose of model engineering is to model a system easily and correctly, so model engineers can gather data from the model quickly and analyze the data effectively. For example, another chapter of this book, (Chapter 14) introduces model management and execution methods based on scalable M&S architectures. Modelers can use several tools to help them model a system and acquire data from the model. For example, modelers may utilize popular programming languages, such as C++ or Java, to build a model from scratch. While a modeler can build a simulation model without using an M&S environment, it may require significantly more time and funding to develop the model. On the other hand, a modeler may employ M&S environments, such as MS4 Me (Seo et al., 2013) or VLE (Quesnel et al., 2009), which offer many supporting features, to build DEVS models more quickly and reliably. This chapter describes the advantages that the DEVSim++ME toolset offers over other existing environments, especially when targeting DEVS models for deployment in HLA-distributed simulations.

This section introduces DEVSim++ME. The ME toolset is a total solution for modeling engineers to capture a system's behavior effectively, verify and validate their model easily, and manage the execution of the models to acquire simulation data efficiently. In addition, model engineers may use the ME to build and execute their models efficiently or compose existing simulators to construct a union of simulators to build a simulation with one simulation objective utilizing HLA/RTI. This section introduces the tools of the ME and its design philosophy.

3.1 Organization of the DEVSim++ME Toolset

The ME toolset is designed to help model engineers build and construct their model effectively. The ME tools can be placed into four categories: (1) modeling, (2) simulation and interoperation, (3) validation and verification, and (4) data acquisition.

For modeling, the ME supports the DEVS formalism. Since the DEVS formalism is set-theoretic and may model a system in hierarchical and modular form, the ME utilizes object-oriented programming languages, especially the C++ programming language, to represent a model. Therefore, the ME provides a template for essential models, so model engineers can easily extend their model by composing various simulation models or inherit useful features from other simulation models (Roberts and Dessouky, 1998; Zeigler et al., 2000).

Further, the ME environment manages the execution of the simulation and supports the interoperation of simulations. After a model engineer designs a model using model templates from the ME, the ME simulation environment executes the model using simulation algorithms from the DEVS formalism. *Moreover, the ME provides model engineers with interoperation support features.* When model engineers want their models to interoperate with other models, the model engineers should understand the interoperation standards, integrate interoperation services into their model, and control the simulation to interoperate with other simulation models. Furthermore, the model engineer may utilize the adaptor for interoperation in the DEVSim ++ toolset.

There are several things to consider for the validation and verification (V&V) of a simulation model. First, to build a new simulation model, engineers should express their ideas in a design document using formal representations. Afterward, the model engineer may implement the model using a programming language. The implementation can then be checked to prove whether the computer simulation code has been developed correctly against the design documents. The ME provides V&V tools to help model engineers build models correctly, including popular design document tools, PowerPoint and Visio.

Finally, the ME provides a distributed simulation execution framework to acquire data effectively. As the efficiency of the computing environment increases, a model engineer can not only utilize local computing resources but also any computing resources, especially multi-core supercomputers, at the remote site. Approaches to improving the simulator's performance can be placed into two categories: those that improve the simulator itself and those that reduce the data-acquisition time. The former approaches consider the parallelism of the simulation model. Therefore, the model engineer should consider modifying existing simulators to support multicore environments. Unfortunately, modifying the existing simulator to support a multi-core environment is not an easy task. The engineer should consider adopting other formalisms, such as parallel DEVS (Chow and Zeigler, 1994), to support the multicore environment as well as develop models to enable parallel execution in the simulators. Chapters 15 and 16 in this book discuss methods for simulating discrete event models on high-performance simulators.

On the other hand, reducing data-acquisition time is a simple method of utilizing multicore environments. In general, the model engineer may analyze data after conducting experiments involving massive simulation scenarios. Since the objectives in M&S include analyzing the data quickly and providing alternative solutions over time, model engineers may take advantage of reduced data-acquisition times when analyzing the data. Among the various methods that utilize multi-core environments, the parallel execution of a simulation is the simplest. Note that parallel simulation and parallel execution of a simulation are different methods. The former method is used to execute multiple simulation models simultaneously with a given scenario, while the latter method involves executing one simulator with multiple scenarios. Chapter 13 of the companion volume to this book discusses a theory for understanding the relative costs in terms of time and effort between parallel simulation of a model and parallel execution of multiple models (Zeigler et al., 2018).

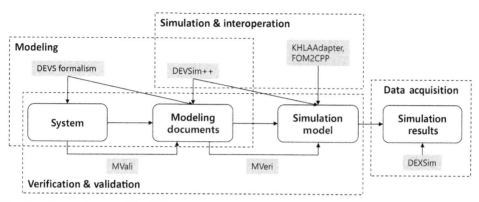

FIG. 5 Overall organization of the ME toolset.

Fig. 5 shows the model-development process and the corresponding ME tools. When a model engineer wants to model a system, the engineer may utilize the DEVS formalism to capture the behavior and characteristics of the system. Since the DEVS formalism is hierarchical and modular in nature, the model engineer may easily build a model or construct a complex model using existing models. After designing a model, the model engineer may write design documents. Therefore, the model engineer can use the DEVS formalism to describe the model rigorously. After writing the design documents, the model engineer can validate the model documents against the model's requirements using MVali, which checks the consistency between Unified Modeling Language (UML) diagrams and the developed DEVS models. After the validation, the model engineer may proceed to develop a computer simulation model using DEVSim++, which is a simulation environment based on the C++ programming language (Kim and Park, 1992) that supports the DEVS formalism. This enables the model engineer to translate descriptions of the atomic model and coupled model into the atomic model and coupled model classes. In addition, DEVSim++ controls the execution of the DEVS models. When the model engineer finishes developing the simulation model, he or she may utilize the MVeri tool to check the computer simulation code against the modeling document. The MVeri tool automatically embeds the probe and verification code into the simulation code using an aspect-oriented programming language. Additionally, the engineer may utilize the KHLA Adaptor and FOM2CPP tools to transform the simulation code into an interoperable form (Kim et al., 2006). Finally, DEXSim supports experiments with the developed simulation model by gathering simulation data from distributed computing resources. We will provide more details about these tools later.

3.2 Model Development Tool: DEVSim++

As indicated above, DEVSim++ is a simulation environment for the DEVS formalism based on the C++ programming language. It was developed in and has evolved from 1992 and has been widely utilized to implement various models in C++. Since the DEVS formalism and object-oriented concepts have many things in common, DEVSim++ has adopted C++ as a model description language. In addition, DEVSim++ was implemented using C++ so that a model engineer may utilize various third-party libraries to extend his or her simulation models. For instance, the simulation model may be given visualization capability by connecting it to the SIMDIS tool using the network library.

DEVSim ++ realizes the DEVS formalism and executes the virtual processor of the simulation model to implement abstract simulator concepts. For a dynamic structure, DEVSim ++ supports various extensions of the DEVS formalism, such as dynamic DEVS (Barros, 1997), variable-structure DEVS (Hu et al., 2005), and multi-resolution-model DEVS formalisms. In addition, DEVSim ++ can be used for modeling and simulating real-time and interactive systems. During such development, model engineers must implement the atomic model and coupled model to complete their simulators. To help them, DEVSim ++ provides template codes and the CModel, CAtomic, and CCoupled classes.

Fig. 6 shows the class diagram of the template models in DEVSim ++. The CModel class is the parent class of the CAtomic and CCoupled classes. In addition, the CModel class may be utilized as an interface for the DEVSim ++ simulation engine. The CModel class provides the basic structure to the model engineers to implement a DEVS model. The CModel class defines the common attributes of the atomic and coupled models. In particular, the CModel class contains attributes such as the name of the model, information about the input/output ports, the state of the model, and time. In addition, the CModel class has common functions for both atomic and coupled models. For example, the CModel class provides AddInPorts and AddOutPorts methods, which the CAtomic and CCoupled models inherit. Furthermore,

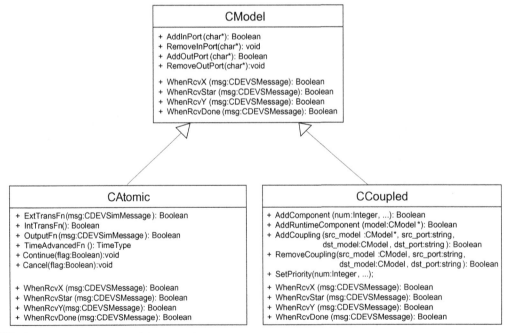

FIG. 6   Class diagrams of models in DEVSim ++.

the CModel contains interface functions for the abstract simulation algorithm such as WhenRcvX, WhenRcvY, WhenRcvDone, and WhenRcvStar, which are implemented by CAtomic and CCoupled.

These class designs display the important design philosophy of DEVSim++: the separation of concerns. The objective of model engineers is to model a system's behaviors. Therefore, model engineers concentrate on modeling and implementing a model's behavior rather than on implementing simulation algorithms. When provided with model template code, the model engineers may focus on expressing the behaviors and implementing the simulation models. In addition, the simulation engine has an advantage under the philosophy. As model engineers can develop their simulation model freely, the simulation engine of the DEVSim++ should have preliminary information of the simulation model to execute it. However, by utilizing the polymorphism features of the object-oriented programming paradigm, the simulation engine may consider the CAtomic and CCoupled models as CModel class instances and execute the simulation based on the CModel class.[1]

3.3 Model Verification and Validation Tools: MVeri and MVali

Model engineering is a multidisciplinary process, and various experts participate in developing a model over long periods. Many faults may occur during the research and development phases due to human error. As a result, verification and validation (V&V) of a simulation model are important in the model engineering domain, and the model engineer should check that the models are well designed in relation to the simulation's objectives and environments.

In general, as indicated above, V&V involves successive steps of checking the consistency among design documents and software code. The first step is to rigorously define the requirements specification using UML or System Modeling Language (SysML) (Fowler, 2004; Friedenthal et al., 2014). Therefore, a model engineer may check the UML or SysML diagrams to make sure the diagrams are well described and embrace the requirements. Then, a model

---

[1] The CAtomic class is the structural template of the atomic model in DEVSim++. The abstract simulation algorithm for the atomic model is implemented within the CAtomic class. If a model engineer wants to make a new atomic model, he or she may define three sets and complete four functions. The model engineer can utilize the AddInPorts and AddOutPorts functions to define three sets.

The CCoupled class is the structural template of the coupled model in DEVSim++. Similar to the CAtomic class, the CCoupled class has three types of attributes and four functions that implement the simulation algorithm. The model engineers may inherit from the CCoupled class to implement the coupled model. In particular, a model engineer adds input and output ports using functions from the CModel class to express the input and output event set. The coupled model has another set with which to express the system components. Accordingly, the model engineer can use the AddComponent function to add models to the coupled model. In addition, the CCoupled model has the AddCoupling function to connect the components. The function may connect the external input event ports to internal components' input event ports. In addition, the function may be used to connect the internal components' output ports to the coupled model's output ports. Similarly, the function may be used to connect the output ports and input ports among internal components. Finally, the CCoupled class has the SetPriority function, which can resolve the priority among component models.

engineer may compare the modeling results and the requirements specification. During this step, the model engineer checks the consistency of event sequences that are generated from the UML/SysML diagrams and DEVS model. Finally, the model engineer checks the event sequences of the DEVS models and the generated event sequences of the computer simulation model.

In the ME toolset, MVali and MVeri support the V&V of a model (Byun et al., 2009; Kim, 2008). MVali is a model-validation tool that examines the model's specifications against the requirements. An important feature of MVali is that the tool adopts commercial software for its input/output data format. To appeal to a wide variety of disciplines, MVali utilizes Microsoft Visio and PowerPoint. MVali processes UML and DEVS models, which are modeled using Visio stencils, and generates the verification results in PowerPoint files. In this way, the model engineer can directly present MVali results at review meetings.

Fig. 7 shows the requirement template in the ME environment. The requirement template of ME contains necessary information based on the user's requirements. In particular, the template is used to specify a model's static structure, dynamic behavior, and timing information. A model's static structure contains the class type, attributes, and operations of each object in the simulation. The dynamic behavior specifies the interactions among messages and operation sequences in ascending order. Finally, the timing information denotes the time consumption for each operation. In addition, if a model's behavior cannot be denoted in a single table, the requirement template of the ME will include ID, name, precedents, and successor fields to organize the model's behaviors. Then, the model engineer draws the class diagrams, sequence diagram, and DEVS diagrams using Visio. Fig. 8 shows the validation process using MVali.

As shown in Fig. 8, a model engineer utilizes Visio to capture the structural and event sequences of a system using class diagrams and sequence diagrams from the requirements. In addition, as indicated, the system's behavior characteristics may be specified using DEVS diagrams in Visio. Then, Visio converts each diagram into XML documents. Finally, MVali accepts the generated XML documents and produces the verification results in PowerPoint format.

ID	Document ID	Name	Scenario Name
Goal	Goal of a scenario		
Precedents	Document ID of a precedent scenario		
Successors	Document ID of a next scenario		
Participated object			
Trigger event	An event to initiate a scenario		
Main scenario	Step	Action	Time consumption
Sub - scenario	Step	Action	Time consumption
Notes			

FIG. 7   Requirement template for V&V in ME toolset.

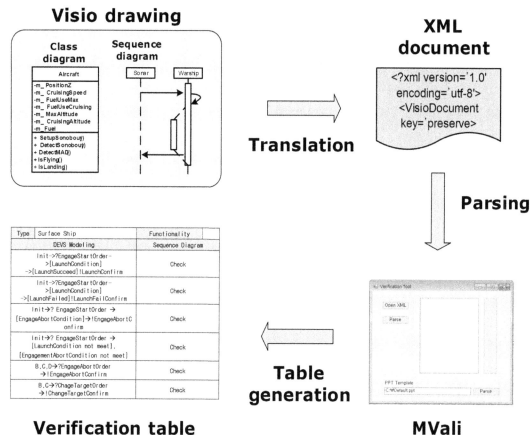

FIG. 8   Validation process using MVali.

The MVeri tool checks the DEVS model against the computer simulation code. The design philosophy of the MVeri tool is to resolve the traditional testing problems and code-tangling and code-scattering problems during verification.

The code-tangling problem in model verification is that probe code should be injected to extract desired event sequences during the simulation. Since the probe code is not part of the model's behaviors, injecting and removing the probe code are likely to be a source of much human error. Similarly, code-scattering errors arise when removing code that specifies the components that generate the simulation logs. Such pernicious and accidental effects can be resolved with the MVeri tool.

The MVeri tool utilizes aspect-oriented programming (AOP) to do away with manually injecting the probe code into the simulation code, as shown in the Fig. 9. In addition, the MVeri tool utilizes duplicated simulation code, so that the model engineer can leave the simulation code unaltered when the MVeri tool cannot find a problem. On the other hand, the engineer can modify the simulation code when the MVeri tool finds a problem.

To utilize the MVeri tool, a model engineer may simply define the probe points and describe the verification code. Then, the MVeri tool combines the probe code with the given

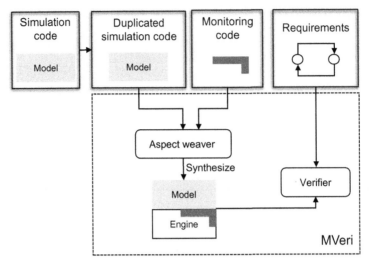

FIG. 9 Architecture of MVeri.

simulation code. After the synthesis, the MVeri tool verifies the event sequences against the DEVS diagram, which has been verified by the MVali tool.

3.4 Distributed Simulation Replications Tool: DEXSim

With the theoretical basis of the SSMS concept, the ME has a distributed simulation replication tool called the Distributed Execution Simulation (DEXSim) framework (Choi et al., 2014). The DEXSim framework can accelerate the simulation experiments by replicating the simulation application to local computing resources or distributed computing resources. The design philosophy of the DEXSim framework is to model a model engineer to support automated execution of simulation executions while maximizing the computing utilization. In particular, the tool simulates the behavior of a model engineer, which means it finds the idle computing resource and allocates the simulation scenarios to the resource.

To support a heterogeneous execution environment, DEXSim was designed to utilize the HLA/RTI. The key components of the DEXSim are the federation DEXSim and the federate DEXSim, as shown in Fig. 10.

Federation management (FM) is a service for federate DEXSim management. In the HLA/RTI, the FM service defines how federates create, join, and resign from the federation. In the DEXSim framework, the federation and the federate DEXSim are the federates in the HLA/RTI, which are involved in exchanging node information and simulation results. Therefore, a model engineer may easily append an additional computing resource by executing federate DEXSim on the computer. Then, federate DEXSim may join the experiment environment and receive the necessary data for the experiment.

In addition, DEXSim utilizes declaration management (DM) and object management (OM) services to control and deploy simulation data to each federate DEXSim. As the federation DEXSim controls the experiment and distributes the data for the experiment centrally, it

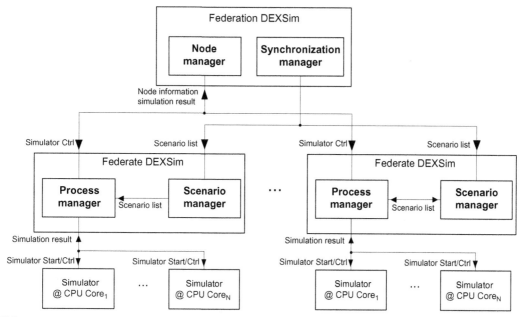

FIG. 10   Architecture of the DEXSim.

publishes the process control message and the necessary data, while the federate DEXSim subscribes the messages using the DM services. The federation DEXSim utilizes the object of the OM services to distribute the simulator, scenario, and scenario list files. Therefore, when an additional computing resource joins the DEXSim framework, necessity data can be synchronized by OM services. Moreover, as the federate DEXSim subscribes the process control messages, the federation DEXSim can control the local execution environment through federate DEXSim by sending a process control message as an interaction.

Finally, the DEXSim uses the time management (TM) services for abnormal situation-handling schemes for the DEXSim framework. In the HLA/RTI, each federate manages the local time and coordinates data exchange with other members of a federation based on the local time. To implement the abnormal situation-handling scheme for the DEXSim framework, the federation DEXSim does not proceed its local time until it receives each experiment status. When the federation DEXSim receives the experiment status of the federate DEXSim, the federation DEXSim decides on the abnormal terminations or abnormal occupancy of replicated simulators under certain conditions, which are given by the engineer.

3.5 RTI Interfacing Tool: KHLA Adaptor and FOM2CPPClass

When model engineers want to make an interoperable simulation model, they must select the appropriate RTI vendor and implement the simulation model using an RTI software development kit provided by an RTI vendor. In addition, the model engineers must understand the time synchronization concepts and implement the time synchronization using HLA/RTI. This process is an entry barrier for the software engineers to develop an interoperable simulation model. In general, the software engineers are used to working with typical network protocols, such as TCP/IP and UDP, rather than using IEEE 1516 standards. The design philosophy of the ME's RTI interfacing tool bridges this gap. The RTI interfacing tool, the KHLA Adaptor, and FOM2CPPClass help software engineers to build an interoperable simulation model without any knowledge of IEEE1516 standards. The KHLA Adaptor is an interoperation manager that controls the interoperation of a simulation using HLA/RTI protocols and converts simulation messages into HLA-compliant messages.

Another problem of the RTI interfacing is the halting simulation problem. The halting simulation problem originates from the conservative time synchronization method. In HLA/RTI, the federation gathers all time requests from the federates and broadcast the time grant message to the federates. Therefore, if a federate does not send the time request message, the federation cannot determine whether the federate is normal but is taking some time to complete the message or is not normal and has entered an infinite loop situation.

The key idea of the KHLA Adaptor is to provide HLA-compliant interoperation features to a simulation model that does not consider the interoperation among simulators at the designing phase. In other words, the simulation model and the KHLA Adaptor simulate an interoperable simulation model of the given model. To make the simulation model and the adaptor mimic the federate, the KHLA Adaptor becomes the interoperation engine of the simulation model. The KHLA Adaptor utilizes two types of network libraries, which are the TCP/IP network library and the RTI library, at the same time. As the KHLA Adaptor should mimic every service of the RTI, the KHLA Adaptor provides the templates of the services. Therefore, when a model engineer wants to use the special services of the HLA/RTI, the engineer may choose to override the necessary interfaces. Fig. 11 shows the interoperation concept of the KHLA Adaptor.

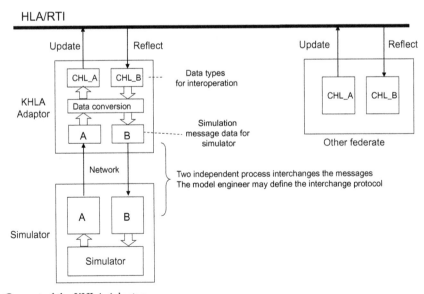

FIG. 11   Concept of the KHLA Adaptor.

As shown in Fig. 11, the KHLA Adaptor is an interoperation engine for a simulation model that participates in the HLA/RTI network. When the simulation model crashes due to unusual reasons, the KHLA Adaptor may send a time request and receive a time grant message for the crashed simulation model. In addition, the KHLA Adaptor may receive simulation data and keep them during the simulation. Then, the simulation model may restore the simulation progress based on the received simulation data.

FOM2CPPClass is designed to help a model engineer reduce the number of tedious jobs involved in developing interoperable simulators. The FOM2CPPClass translates the FOM data into C++ class code. During the federation's development, the most time consuming and most tedious task is developing code for encoding and decoding the simulation messages, which may be exchanged during the simulation. As all the simulation messages during the simulation are described in the FOM, the model engineers should refer to the FOM to build a code for data marshaling. The FOM2CPPClass generates the code skeletons for model engineers to enhance their productivity. The FOM2CPPClass extracts information from two files: the Object Model Definition file and a template file, which contains the information about the user-defined types. First, the FOM2CPPClass processes objects and interactions. Each object and interaction structure contains attributes and parameters. Therefore, the FOM2CPPClass generates the data marshaling code for each attribute and parameter using the template code. When a model engineer wants to customize the marshaling code, the model engineer should insert additional code into the code skeletons.

Additionally, when a model engineer wants to use other RTI interfacing tools, the model engineer may use the FOM2CPPClass tool by changing the template class. As the FOM2CPPClass is separated from the KHLA Adaptor, any RTI interfacing tools that utilize C++ as a programming language may use the FOM2CPPClass to generate a code skeleton for the interoperation of a simulation.

4 M&S PROCESS USING DEVSim++ME

In general, model engineering is a multidisciplinary field in which various technologies are involved to build a domain-specific model. For example, assume a stakeholder wants to build a weather forecast model. It is probable that the stakeholder will execute the model in a supercomputing environment. Consequently, software engineers will participate to develop the weather forecast model. To develop the weather forecast model, the software engineer should process the weather data from the past few years. However, the engineers cannot complete the weather forecast model without the help of a meteorologist because of their lack of knowledge of meteorology. Therefore, developing a complex domain-specific model is tough work, and model engineers should consider defining a development process for the model. In particular, the model engineers should consider how to proceed with collaborative research among various experts, such as domain experts, M&S experts, and platform experts.

The domain experts are those who have special knowledge of a given domain and who understand the system that requires modeling. The domain knowledge includes use case scenarios, workflows, and mathematical formulas. The domain experts usually write the system's requirement specification and check the model against this specification. The M&S

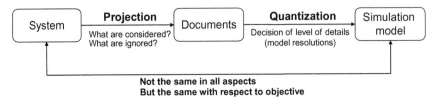

FIG. 12   Model development process and outputs.

experts are the experts who model a system with a stakeholder's objectives in mind. The M&S experts quantize the system to be executed in the high-performance computing system. Furthermore, the M&S experts may help the domain experts to analyze the simulation results from the model (Sung and Kim, 2012). The platform experts are the engineers who implement the model into the computer simulation software and deploy the software to the real systems.

Fig. 12 shows the process of modeling a system to develop the corresponding simulation model. This process is an objective-driven abstraction of a given system. To develop a simulation model, modelers, the domain experts, and the M&S experts develop design documents. In this projection phase, modelers must ask themselves what needs to be considered and what needs to be ignored to design the simulation model. The simulation objective drives these questions. Based on the simulation objective, the modeler may choose the essential part of the system to express the system's behavior. The results of the phase are the design documents. The design documents contain the necessary information to develop the simulation model. After the projection phase, the modelers decide the level of details of the simulation model. As the simulation models are executed on the computer system, each model should be quantized into a computer simulation model. The platform expert may implement the simulation model using various simulation tools. The most important aspect of the process is that the system and the developed simulation model are not the same in all aspects. However, both the system and the simulation model behave identically with regard to the given objective.

The following subsections introduce the cooperative model development process using ME in the defense domain.

4.1 Characteristics of M&S for a Domain-Specific Defense System

In general, the M&S for a domain-specific defense system can be categorized into three types of defense M&Ss: training, analysis, and acquisition. The training model is used to verify the system functionalities and operations of soldiers in a real-world situation. The training model should provide real-time simulation and human-in-the-loop simulation. As the operator may intervene in the training, the training simulation model should support an external event handling mechanism that may change the simulation results. In addition, the model should support a recovery method to respond to an unexpected situation during the training. Finally, the model should support the replay functionality to aid the after-action review (Page and Smith, 1998).

The analysis model is a simulation model to analyze the effectiveness and performance of a system. To examine the performance and the effectiveness of a system, the simulation model should provide a Monte Carlo simulation method. In addition, the model should be fast enough to analyze the simulation results, as the model needs to be able to examine massive scenarios. To enhance the simulation speed, the analysis model adopts the logical time simulation method. The difference between real-time simulation and logical time is the waiting time handling mechanism in the simulation engine. During the simulation, the simulation engine will collect the next event time from the models; it will then select the minimum event time from among the submitted next event times and will send the grant message to the model that submits the minimum event time. During this process, the real-time simulation waits for the difference between the elapsed time and the minimum event time. For example, when a minimum next event time is 1.5s and the unit of real-time simulation is 1s, the real-time simulation method will wait for 1.5s to proceed with the simulation. Therefore, the real-time simulation is called a wall-clock time simulation. Unlike the real-time simulation, in the logical time simulation method the simulation engine does not wait and just processes the next event. Consequently, when a model engineer utilizes the logical time simulation, the simulation engine may utilize high-performance computing resources to maximize its performance as much as possible and may finish the simulation as quick as possible.

The final model of the defense M&S is the acquisition model. The objective of the acquisition model is to acquire various pieces of information to decide whether the real equipment has correct functionality and meets the performance specification. The acquisition model is essential in the defense of M&S, as the deployment of a military system requires an astronomical budget. In addition, the simulation results should provide the theoretical foundation for the deployment of the military system. The simulation must examine various combinations of scenarios to provide theoretical confidence. Therefore, the model should be equipped with a Monte Carlo simulation method and hardware-in-the-loop simulation to test the equipment's hardware and software.

4.2 Expert Cooperation in Defense System Modeling

As various experts collaborate during the development of an M&S application, the model engineer should distribute the appropriate tasks to the experts and should manage the overall development process. Interestingly, collaboration among various experts in system development is not a new concept. The collaborative process has been proposed in the hardware/software codesign and system integration fields (Adams and Thomas, 1996; Chittister and Haimes, 1996; Weiming et al., 2008). The collaborative process focuses on allocating the right research and development resources to the right places.

Similarly, the collaborative development process of the defense M&S requires three experts—the military experts, the M&S experts, and the platform experts—who collaborate to build a defense M&S model (Sung and Kim, 2012). The primary role of the military experts is to analyze the given system and specify the requirements concerning the simulation objectives based on the military knowledge. The M&S experts may model the discrete behaviors of the system and develop the discrete event model concerning the requirements. Finally, the platform experts realize the conceptual model as a computer simulation model. In addition, the platform experts implement the experimental environments to help the domain experts and M&S experts conduct the experiments.

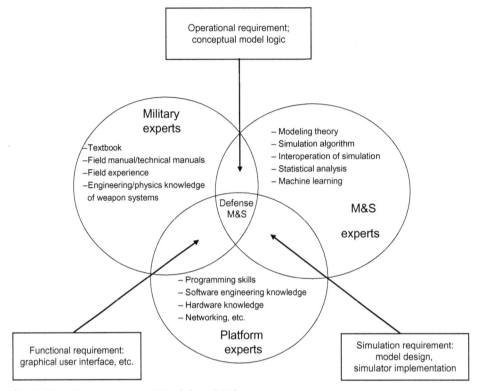

FIG. 13   Collaboration among experts in defense M&S.

As shown in Fig. 13, the military experts should understand the content of the textbook, the field manual, and the technical manuals. As the defense M&S model should be executed according to the military doctrine, the military experts should develop the requirements rigorously and completely. In addition, the military experts may be requested to share their field experiences to make the defense M&S model resemble the real-world system. Finally, the military experts should decide on the level of sharing information of the military system. This role is one of the essential reasons as to why experts should collaborate in defense M&S. In particular, the model of the defense M&S cannot guarantee the correctness or validity of the simulation results when other experts do not have domain knowledge. However, such information may be confidential so that other experts may not access the information. Therefore, the military experts should participate deeply in the development of the defense M&S model.

The M&S experts must have confidence in the M&S theories, such as modeling know-how, simulation algorithms, interoperation of simulation, and knowledge of simulation data analysis. Finally, the platform experts must deal with the software. They should have programming skills, knowledge of software and hardware development, and knowledge of how to utilize network libraries.

By experts cooperating, the defense M&S system may be developed. To develop the system successfully, the model engineer must consider three types of requirements of the development of the M&S application. The first requirement specification is the operational requirement. The operational requirement contains the conceptual model logic. Therefore, the military experts and the M&S expert collaborate to decide the model logics concerning the simulation objectives. The second requirement specification is the simulation requirements. For the simulation requirements, the M&S experts and the platform experts collaborate to decide the structure design of a model from the perspective of software engineering. In addition, they cooperate to specify other simulation requirements, such as the interoperation of the simulation features and the hardware-in-the-loop simulation.

The final requirements specification is the functional requirements. As the end-users of the M&S application are the military experts, the military expert and the platform expert cooperate to decide the functional requirements of the simulation model. For instance, the military expert may want to use 3D visualization tools to visualize the simulation status or use separate windows that display the simulation results in graph form. Therefore, the military expert and the platform expert must collaborate to develop the functional requirements concerning the development period and the budgets.

4.3 Cooperative Model Development Process Using DEVSim++ME

As the previous subsections showed, the development of the defense M&S software is a cooperative process with various experts. During the development phases, each expert group has its role and shares some information to model a defense system, transfer the phase outcomes to other groups, and refine their outcomes based on the phase outcomes. However, cooperation among expert groups, which have different characteristics and cultures, is very difficult work. For instance, most of the requirement specifications are written in natural language; so, each expert group may misunderstand the documents and may produce unwanted outcomes. Moreover, when an expert is involved in the development process, the expert may produce human errors and thereby delay the entire process. To resolve the problems, the model engineer may utilize the tools in the ME toolset.

The ME toolset provides a tool to describe the model rigorously, and the tool effectively manages the models to conduct simulation experiments. Furthermore, the ME toolset provides semi- or fully automatic supports for the development process so that the ME tool may prevent human error in advance (Kim et al., 2011b).

Fig. 14 shows the relationship between the development process of the defense M&S and the ME tool. As shown in Fig. 14, the domain expert, the M&S expert, and the platform expert cooperate to develop the model requirements and specifications. They then decide on the object-oriented modeling architecture for the simulation model. The next phase is the modeling phase. In this phase, the M&S expert and military expert separate the modeling tasks. The M&S expert models the DES modeling, and the military expert models the domain model. During this phase, the military expert and the M&S expert may utilize the MVali tool to verify the DEVS model against the requirements. After modeling the DES and the domain model, the M&S expert and the platform expert will implement the model based on the modeling results. During the implementation phase, the experts may utilize

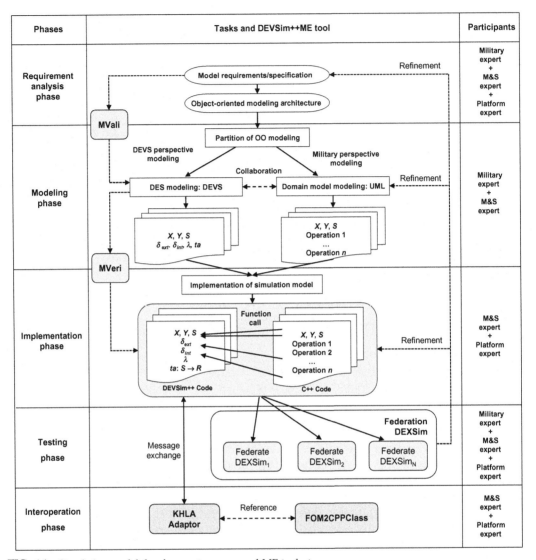

FIG. 14 Simulation model development process and ME toolset.

the DEVSim++ to implement the model efficiently. To verify the implementation against the requirement, the experts may also utilize the MVeri tool. Furthermore, DEXSim can be utilized to test the developed model. By executing multiple scenarios simultaneously, the experts may find the glitches easily and quickly. When they find the glitches, they can refine the requirement specification, the model, and the implementation based on the testing results. Finally, the experts may utilize KHLA Adaptor and FOM2CPPClass to make the simulation model into an interoperable simulation model. By utilizing the KHLA Adaptor as an interoperation engine, the experts may easily add interoperation functionalities to the simulation model.

5 EXAMPLE CASE STUDIES

This section introduces empirical studies using the ME toolset. The ME toolset is a total solution to build a simulator; however, a model engineer may use an individual tool to develop a model, construct interoperable simulators, apply the V&V process, and utilize massive computation resources to acquire simulation data.

In particular, ME toolsets have been applied and verified to support the military branches of the Republic of Korea since 2002. They have been used to model various military systems, such as submarines, torpedo, aircraft, etc., and to support simulation experiments, such as Monte Carlo simulation. In addition, ME toolsets have been used to develop a hardware-in-the-loop simulation, virtual-constructive simulation interoperation, and hybrid simulations. During the development, the model engineers have used ME tools to validate and verify the requirements, test the implementation, gather simulation results efficiently, and interoperate the model with other simulation models. Although the ME has been applied to various military domains, this section introduces the Naval Warfare Simulator for anti-torpedo (NWSimAT) warfare and anti-air-to-surface ship missile warfare simulation.

Naval warfare is combat involving various maritime systems, such as surface ships, warships, torpedos, and decoys (Kim et al., 2011a; Seo et al., 2014). In addition, various combat systems may participate in the battle with marine systems. For example, an aircraft may fire an air-to-surface missile (ASM) at a surface ship. The surface ship may then utilize a defense system to intercept the missile. Although there are many simulation objectives in naval warfare, this section introduces a naval warfare simulator that gives the survival rates for a surface ship. Therefore, the main simulation objective is to maximize the survival ratio of an underwater-to-surface warfare situation and an air-to-surface warfare situation.

In the underwater-to-surface warfare situation, a surface ship detects an underwater threat and may evade the threat using various means. In particular, the NWSimAT detects a torpedo and utilizes various decoys to evade the threat. In the air-to-surface warfare situation, a surface ship detects a threat from the sky. As this threat has high speed, the surface ship should utilize effective countermeasures to evade the threat. Therefore, the naval warfare simulator analyzes the survival rate of the countermeasure tactics based on the distance between the threat and the surface ship.

Fig. 15 shows the developed naval warfare simulations and the relationship among ME tools and the simulations.

As shown in Fig. 15, the various model engineers utilized the ME toolset throughout the development of the naval warfare simulator (the NWSimAT). The initial design of the NWSimAT is to analyze the countermeasure tactics against the torpedo, the development of which is supported by the ME. If the requirement of the NWSimAT evolved later, the interoperation features of the ME would support the evolution.

Details of the development process of the NWSimAT and use of the ME toolset are presented in Appendix A.

6 CONCLUSION

This chapter introduced the DEVSim++ME, a development environment for model engineering. Although the ME toolset supports the entire process of a model's development, there

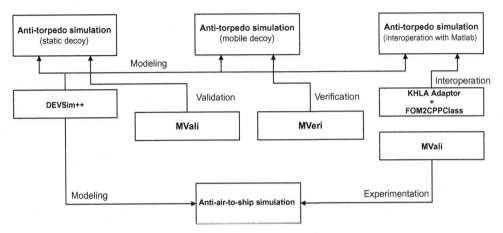

FIG. 15   Overview of the NWSimAT.

is still room to contribute to model engineering tools. As the demand for the modeling and simulation field grows, new model engineering tools will be needed to enable various domain experts to participate in developing models and conducting simulation experiments in various scenario settings. In summary, the DEVSim++ ME toolset offers advantages over other existing environments, especially when targeting DEVS models to be deployed in the HLA-based distributed simulation. These include ease of modeling, automatic model verification and validation, distributed experimentation of a single simulation with multiple scenarios, and extending non-interoperable DEVS models to HLA-compliant ones. The toolset utilizes the full features of the C++ programing language, which allows modelers to easily utilize algorithms and equations in existing C++ codes in the full development phase. Automatic model verification and validation can be done by MVeri and MVali tools, respectively. DEXSim utilizes computing resources efficiently to run a single model with multiple scenarios in a parallel/distributed manner. Finally, KHLA Adaptor implements a full set of services and callbacks defined in HLA, by which non-interoperable DEVS models are easily extended to HLA-compliant ones.

APPENDIX A: EXAMPLE CASE STUDY: NAVAL WARFARE SIMULATOR FOR ANTI-TORPEDO SIMULATION

NWSim for Anti-Torpedo (NWSimAT) analyzes the effectiveness of anti-torpedo warfare tactics and its countermeasure system, the decoy system. In addition, NWSimAT may be utilized to determine the performance indices of the underwater weapon system. To build NWSimAT, various model engineers collaborated to model the anti-torpedo system. First, the model engineers identified the systems involved in the anti-torpedo warfare situation.

Fig. A.1 shows the conceptual process of the anti-torpedo warfare situation.

As shown in Fig. A.1, the surface ship, the submarine, the torpedo, and the decoy are the system components of anti-torpedo warfare. When a hostile submarine detects a friendly

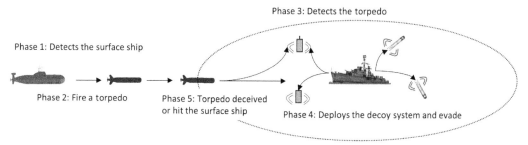

FIG. A.1 Operations of naval warfare for anti-torpedo tactics.

surface ship in the operating area, the hostile submarine launches a torpedo at the surface ship. When the surface ship detects the torpedo and identifies it as a threat, the torpedo launches the decoy system and makes a detour to evade the threat. The torpedo system may be deceived by the decoys. When the torpedo is deceived, the torpedo may utilize the sensor and apply various search patterns to find the surface ship. If the torpedo is not deceived, the torpedo will shoot down the surface ship.

A.1 Initial Modeling of Anti-Torpedo Simulation

To model the combat between a surface ship and a torpedo, the model engineer models the surface ship, submarine, torpedo, and decoy. Each surface ship, torpedo, and submarine model contains a maneuver model and a sensor model. Therefore, the models may move and detect the other simulation models. Every model of the NWSimAT is modeled using the DEVS formalism. Coupled Model was used to model the surface ship, the submarine, and the torpedo, as the models are a platform for subsystems, such as an engine, a sensor, and a C2. In addition, Atomic Model was used to model the behaviors of the subsystem. Fig. A.2 illustrates the Coupled Model and the Atomic Model of the surface ship. As shown in Fig. A.2, a surface ship may have several types of equipment, such as sensors, a command and fire control system model, and a launcher. Therefore, the model engineer may compose different types of surface ships based on the simulation objectives. The left side of Fig. A.2 illustrates the Damage Atomic Model. The Damage Atomic Model receives the damage from the other platform and calculates it. It then computes the damage status and broadcasts the message to the other models. The Damage Atomic Model may then be reused by other platforms such as the submarine, the torpedo, and the aircraft.

A.2 Extension of Anti-Torpedo Simulation

When the first version of the anti-torpedo simulation was developed, the model engineers used the simulator to analyze the survival rate of the surface ship. Then, the model engineers wanted to test a new type of decoy system, the mobile decoy system. As the existing decoy system is a static system, it mimics the sound patterns of the surface ship at the landing

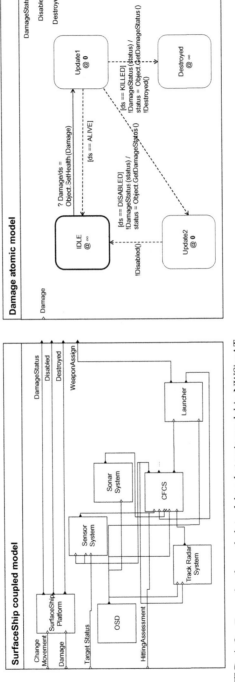

FIG. A.2   Example of coupled model and atomic model in NWSimAT.

FIG. A.3 Extended NWSimAT.

location. However, the mobile decoy system has a maneuver system so that the decoy may move in various directions. Therefore, the decoy system may deceive the sensor system of the torpedo easily.

As there was no mobile decoy system existing at that time, the model engineers designed and implemented the mobile decoy system based on the static decoy system. Therefore, the engineers designed the mobile decoy system as a coupled model, and they inserted the static decoy model and the maneuver model into it. As the DEVS formalism has modular characteristics and the DEVSim++ supports the modularity of the DEVS formalism, the engineers could extend the existing anti-torpedo simulation by simply composing the mobile decoy system and reusing the existing simulation models. Fig. A.3 shows the execution status of the extended version of the anti-torpedo simulation.

A.3 V&V of the Naval Warfare Simulator

A model engineer may utilize MVali and MVeri to apply the V&V process to the models of NWSimAT. To build a correct model, the model engineers developed the requirements specification tables. Fig. A.4 shows the portion of the requirement specification for a submarine and its requirement table.

As shown in Fig. A.4, a model engineer may identify the behaviors of the participated simulation object and its time constraints easily. The dotted line of Fig. A.4 displays the mapping between the underlined statements in the requirement specifications and the statements in the requirement table.

A.3.1 Validation Using MVali

After defining the requirements, the model engineer should draw the class diagram, sequence diagram, and DEVS diagram. Fig. A.5 shows each diagram based on the requirements.

Finally, the model engineer may use MVali to generate the verification table. Fig. A.6 shows the verification table of an engagement scenario involving a submarine.

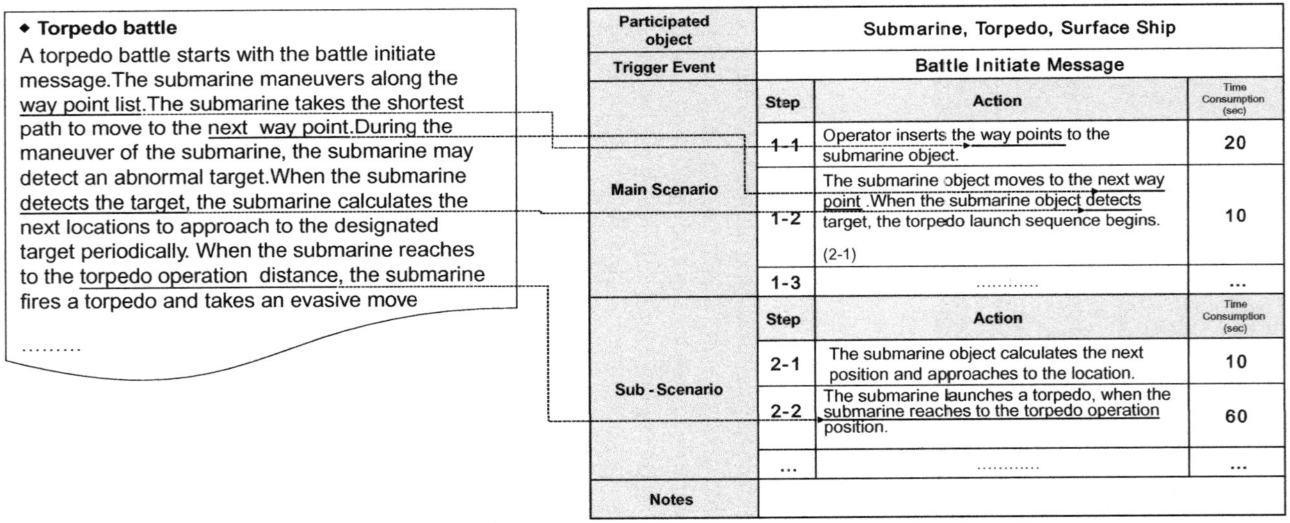

Participated object	Submarine, Torpedo, Surface Ship		
Trigger Event	Battle Initiate Message		
	Step	**Action**	**Time Consumption (sec)**
Main Scenario	1-1	Operator inserts the way points to the submarine object.	20
	1-2	The submarine object moves to the next way point. When the submarine object detects target, the torpedo launch sequence begins.	10
	1-3	(2-1)	...
	Step	**Action**	**Time Consumption (sec)**
Sub-Scenario	2-1	The submarine object calculates the next position and approaches to the location.	10
	2-2	The submarine launches a torpedo, when the submarine reaches to the torpedo operation position.	60
	...	....	...
Notes			

♦ **Torpedo battle**

A torpedo battle starts with the battle initiate message. The submarine maneuvers along the way point list. The submarine takes the shortest path to move to the next way point. During the maneuver of the submarine, the submarine may detect an abnormal target. When the submarine detects the target, the submarine calculates the next locations to approach to the designated target periodically. When the submarine reaches to the torpedo operation distance, the submarine fires a torpedo and takes an evasive move

....

FIG. A.4 Natural requirements specification and requirement specification template.

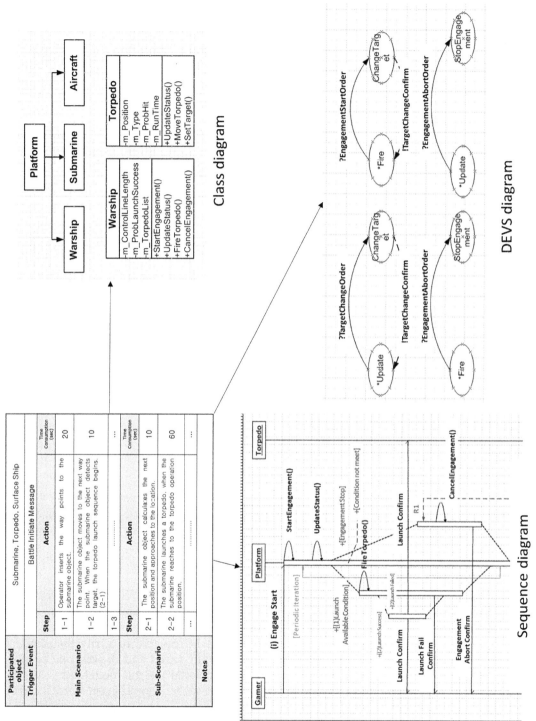

FIG. A.5 Requirement diagrams from requirement specification.

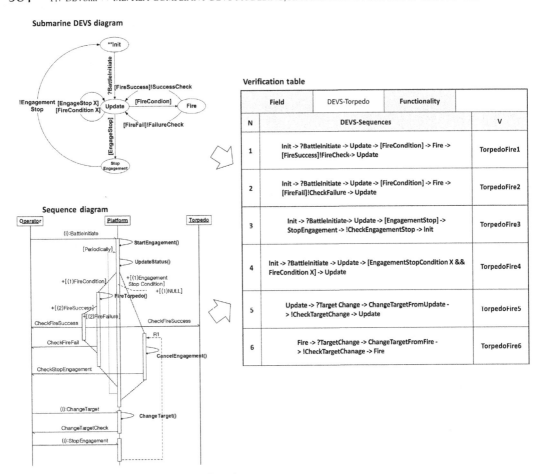

FIG. A.6 Verification results of the MVali tool.

As shown in Fig. A.6, the MVali tool generates possible event sequences from the UML diagrams and checks the execution sequence against the DEVS diagrams. In the verification table, the column "V" denotes the verified event sequences that can be checked against the event sequences in the DEVS graphs.

A.3.2 Verification Using MVeri

Once a model engineer confirms that the DEVS diagrams match the UML diagrams using the MVali tool, they may develop the simulation model based on the DEVS diagrams. The model engineers may use various tools to develop their models; however, if they use the DEVSim++ as a development tool, they may utilize the MVeri tool to verify their code. In general, a simulation model contains the domain-specific logics and the discrete event-specific behaviors simultaneously, which means that verification of the model code can be a difficult job. To check the model's behavior, the model engineer should inject a probe code and check the execution sequence against the DEVS diagram. Consequently, the model engineer may make a mistake when he or she checks the model manually.

```
01:aspect CMveri
02:{
03:    Stateinfo* LUTable;
04:
05:    CMveri() {
06:        LUTabl = new Stateinfo(); e
07:    }
08:    pointcut ext_trans () = execution("% MobileDecoy::ExtTransFn(...)");
09:    pointcut int_trans () = execution("% MobileDecoy::IntTransFn(...)");
10:    pointcut output() = execution("% MobileDecoy::OutputFn(...)");
11:    pointcut timeadv () = execution("% MobileDecoy::TimeAdvanceFn(...)");
12:    pointcut close () = destruction("VerificationModel");
13: public:
14:    advice close() :before ()
15:    {
16:        tjp->target()->VeriTable ->PrintOn();
17:        std::cin.get();
18:    }
19:};
```

FIG. A.7   Example of aspect in MVeri.

Fig. A.7 shows a code snippet from the MVeri tool. As introduced in Fig. A.7, the MVeri tool utilizes the aspect-oriented programming features. The CMveri aspect has several pointcuts: ext_trans, int_trans, output, timeadv, and close. Each pointcut denotes the probe point. This probe point is the location at which the model engineer should inject the probe code. However, when the model engineer utilizes the MVeri tool, the MVeri tool automatically injects the probe code into the pointcut location.

Fig. A.8 shows an example of the timeadv pointcut. When the DEVSim++ simulation engine executes the simulation algorithm, the engine may invoke the time advanced function. Then, the timeadv advice will be activated and will execute the code. Line 07 in Fig. A.8 records the previous state, and the MVeri tool proceeds with the simulation. After the execution, the model may change its state and return the time advanced value. Then, the MVeri tool

```
01: advice timeadv () : around ()
02:{
03:    std::string mname = tjp->target()->GetName();
04:
05:    if(strcmp(mname.c_str(),LUTable->ModelStateList.at(0)->modelname.c_str())==0)
06:    {
07:        int previous Status = tjp->target() ->state;
08:        tjp->proceed ();
09:
10:        CTraceEle* pLOG = new CTraceEle("TA");
11:        pLOG->Present_State= LUTable->GetStateName(mname, previousStatus);
12:        pLOG->TimeAdvance= *(tjp->result());
13:
14:        CCoupled* pRoot = DEVSINTERFACE_PTR->GetModel();
15:        CModel* pModel  = pRoot->GetComponent ("VerificationModel ");
16:        DEVSINTERFACE_PTR->SendTSOMessage( pModel, mname.c_str(), pLOG, tjp->target()->GetTime());
17:    }
18:    else
19:        tjp->proceed ();
20:}
```

FIG. A.8   Example of probe code in MVeri.

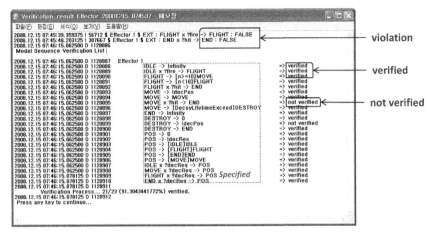

FIG. A.9   Verification results of MVeri.

gathers the information from the simulation model and sends it to the verification model, which checks the event sequence. Fig. A.9 shows the verification results of the MVeri tool.

As illustrated in Fig. A.9, the MVeri tool generates verification reports from the simulation model code. The first region shows the violated event sequences. The violated event sequences may occur because a model engineer may make a mistake or refine the model during the development. Therefore, the violated sequence may be the evidence to correct the code or the model specification.

The dotted line region shows the specified model behavior by a model engineer. The MVeri tool continuously monitors the execution sequence of the simulation model and checks whether it is verified or not. The verified sequences are the desired sequences for which the model should be generated, and the not-verified sequence means that the MVeri tool cannot confirm the behavior because it did not appear during the simulation. Therefore, the model engineer may change the initial condition and execute the simulation again to monitor the behavior.

A.4 Interoperation of the Naval Warfare Simulator

To increase the fidelity of a simulation model, model engineers may consider two methods: rebuilding the model from scratch or enhancing the model based on the existing model. There are pros and cons to each method. For instance, rebuilding the simulation model from scratch may involve redesigning the entire simulation based on the previous development experience so that the new simulation model may have high fidelity and other useful functionalities that had not been considered before. However, building a simulation from scratch requires lots of time and a high budget. Besides, enhancing the model from the existing model may save time and funding by reusing the existing simulation model, but it may be difficult to improve the model, as the model engineers should understand the behavior of the simulation model and the model's source code. To tackle the problem, the ME gives the interoperation tools to the model engineers.

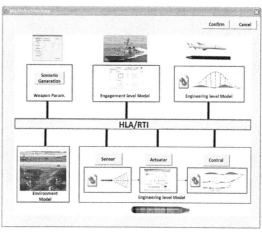

FIG. A.10 Extension of the NWSimAT using KHLA Adaptor.

Fig. A.10 shows the interoperation version of the anti-torpedo simulation. The left side of Fig. A.10 shows the RTI and the pairs of the KHLA Adaptor and the simulation model. The right side of Fig. A.10 shows the control user interface of the anti-torpedo simulation. As shown in Fig. A.10, the model engineers may control the multiple simulators and the KHLA Adaptors. Therefore, model engineers may develop a control program using a graphical user interface to control each KHLA Adaptor and the simulation model. Fig. A.11 shows the execution status of the interoperable simulation. The left side of Fig. A.11 shows the simulation visualization tool, SIMDIS (Bigelow and Kowalchuk, 2012). By utilizing the KHLA Adaptor, the model engineers may easily attach the 3D visualization tool.

A.5 Data Acquisition of the Naval Warfare Simulator

A model engineer may develop a naval air defense simulator to validate an anti-air defense doctrine for the sea. The main objective of the simulator is to assess the survival ratio of the surface ship in response to various anti-air defense doctrines. In general, many warships adopt defense strategies based on the distance of the threats. The strategies divide the defense stages based on the distance. To lessen the threats from a long-distance target to a short-distance target, the surface ship tries to lessen the threat level by utilizing three types of weapon systems.

The three types of weapon systems are the Ship to Air Missile (SAM), Electrical Warfare (EW), and Closed In Weapon System (CIWS). When the surface ship detects the threat, the Command & Control (C2) system commands and operates the weapon systems. Fig. A.12 gives an example of scenarios involving naval air defense.

When multiple missiles approach a surface ship, the C2 system of the surface ship may detect the missiles that exist within the detection radius of the surface ship. Then, the C2

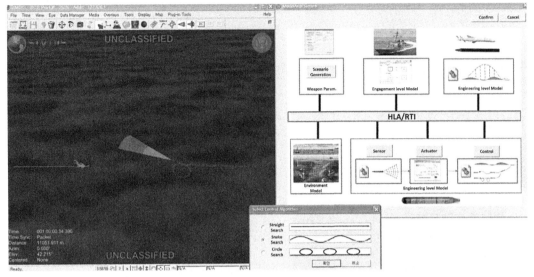

FIG. A.11   Execution screen of the NWSimAT.

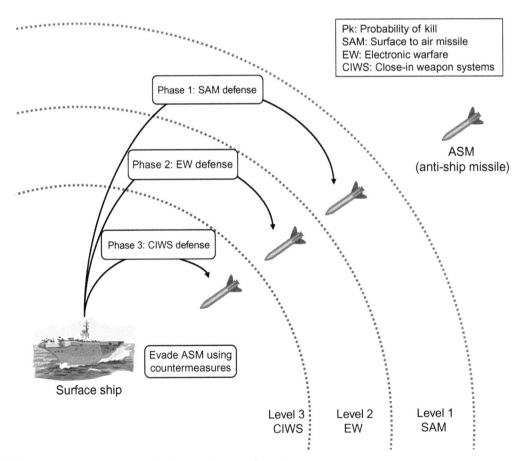

FIG. A.12   Scenario example of anti-air-to-ship missile tactics.

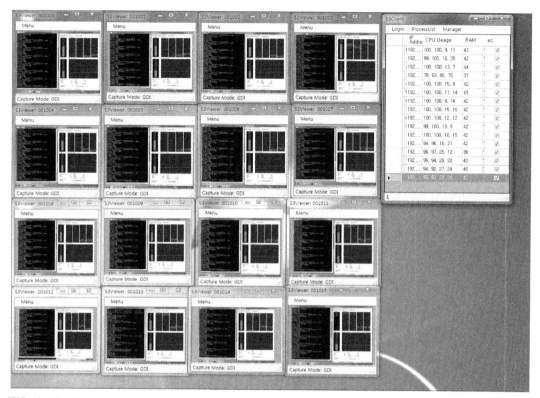

FIG. A.13   Control software of the DEXSim.

system of the surface ship should decide to utilize the surface ship's weapon systems. When the missiles are within the firing range of SAM, the C2 may allocate SAM to eliminate the threat. If the SAM fails, the surface ship may utilize the EW method. Again, if the EW method fails, the surface ship will use the last countermeasure system, the CIWS. When the CIWS cannot intercept the missile, the surface ship will be destroyed by the missile. During the experiments, the simulation procedure will continue to execute all possible combinations of the input parameters. In addition, if a model engineer decides to replicate the scenario 100 times to achieve confidence in the statistical evaluation, he or she should conduct the experiments 7200 times. Fig. A.13 shows the federation DEXSim and the status of the federate DEXSim.

As model engineers may share the computing environment, several experiments may be executed on the DEXSim environment. Fig. A.14 shows the data acquisition time of the NWSimAT with 7200 scenarios in a homogeneous environment. In addition, Fig. A.15 shows the data acquisition time in a heterogeneous environment.

As Figs. A.14 and A.15 show the data acquisition time of the NWSim decreases linearly as the computing resources increase linearly. However, when the number of the process increases, the DEXSim may occupy more computing resources so that the NWSim cannot utilize some computing resources.

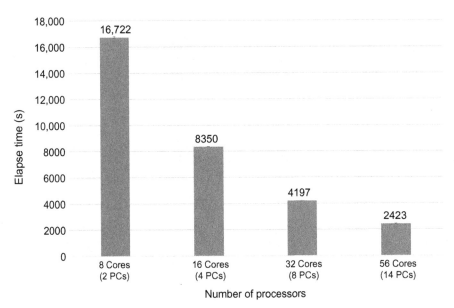

FIG. A.14   Data acquisition time of NWSim in a homogeneous environment.

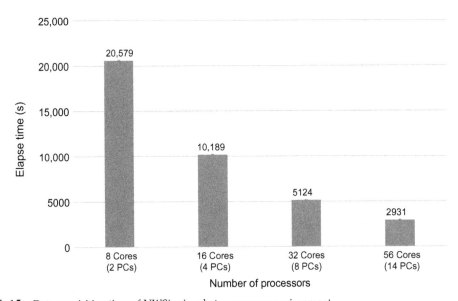

FIG. A.15   Data acquisition time of NWSim in a heterogeneous environment.

References

Adams, J.K., Thomas, D.E., 1996. In: The design of mixed hardware/software systems.33rd Design Automation Conference Proceedings, 1996, 3–7 June 1996, pp. 515–520.

Barros, F.J., 1997. Modeling formalisms for dynamic structure systems. ACM Trans. Model. Comput. Simul. 7, 501–515.

Bigelow, M.S., Kowalchuk, S.A., 2012. SIMDIS for real-time hardware-in-the-loop simulation visualization of rocket systems.AIAA Modeling and Simulation Technologies Conference, Minneapolis, MN, pp. 13–16.

Byun, J.H., Choi, C.B., Kim, T.G., 2009. Verification of the DEVS model implementation using aspect embedded DEVS. Proceedings of the 2009 Spring Simulation Multiconference. Society for Computer Simulation International, p. 151.

Chittister, C.G., Haimes, Y.Y., 1996. Systems integration via software risk management. IEEE Trans. Syst. Man Cybern. Syst. Hum. 26, 521–532.

Choi, C., Seo, K.-M., Kim, T.G., 2014. DEXSim: an experimental environment for distributed execution of replicated simulators using a concept of single simulation multiple scenarios. Simulation 90, 355–376.

Chow, A.C.H., Zeigler, B.P., 1994. In: Parallel DEVS: a parallel, hierarchical, modular modeling formalism. Simulation Conference Proceedings, 1994. Winter. IEEE, pp. 716–722.

Fowler, M., 2004. UML Distilled: A Brief Guide to the Standard Object Modeling Language. Addison-Wesley Professional, Boston.

Friedenthal, S., Moore, A., Steiner, R., 2014. A Practical Guide to SysML: The Systems Modeling Language. Morgan Kaufmann, Boston.

HLA Working Group, 2000. IEEE Standard for Modeling and Simulation (M&S) High Level Architecture (HLA)-Framework and Rules. IEEE Standard, pp. 1516–2000.

Hu, X., Hu, X., Zeigler, B.P., Mittal, S., 2005. Variable structure in DEVS component-based modeling and simulation. Simulation 81, 91–102.

IEEE Standards Association, 2012. 1516–2010-IEEE Standard for Modeling and Simulation (M&S) High Level Architecture (HLA). IEEE Standards Association, Piscataway, NJ.

Kim, D.H., 2008. In: Method and implementation for consistency verification of devs model against user requirement. ICACT 2008. 10th International Conference on Advanced Communication Technology. IEEE, pp. 400–404.

Kim, T.G., Park, S.B., 1992. In: The DEVS formalism: hierarchical modular systems specification in C++.Proc of 1992 European Simulation Multiconference, pp. 152–156.

Kim, K.H., Seong, Y.R., Kim, T.G., Park, K.H., 1996. Distributed simulation of hierarchical DEVS models: hierarchical scheduling locally and time warp globally. Transactions 13, 3.

Kim, J.-H., Hong, S.-Y., Kim, T.G., 2006. In: Design and implementation of simulators interoperation layer for DEVS simulator.Proceedings of M&S-MTSA'06, pp. 195–199.

Kim, J.H., Choi, C.B., Kim, T.G., 2011a. Battle experiments of naval air defense with discrete event system-based mission-level modeling and simulations. J. Def. Model. Simul. 8, 173–187.

Kim, T.G., Sung, C.H., Hong, S.-Y., Hong, J.H., Choi, C.B., Kim, J.H., Seo, K.M., Bae, J.W., 2011b. DEVSim++ toolset for defense modeling and simulation and interoperation. J. Def. Model. Simul. 8, 129–142.

Page, E.H., Smith, R., 1998. In: Introduction to military training simulation: a guide for discrete event simulationists. Simulation Conference Proceedings, 1998. Winter. IEEE, pp. 53–60.

Quesnel, G., Duboz, R., Ramat, E., 2009. The virtual laboratory ennvironment—an operational framework for multi-modelling, simulation and analysis of complex dynamical systems. Simul. Model. Pract. Theory 17, 641–653.

Roberts, C.A., Dessouky, Y.M., 1998. An overview of object-oriented simulation. Simulation 70, 359–368.

Seo, C., Zeigler, B.P., Coop, R., Kim, D., 2013. DEVS modeling and simulation methodology with MS4 Me software tool.Proceedings of the Symposium on Theory of Modeling & Simulation—DEVS Integrative M&S Symposium, 2013. Article No. 33.

Seo, K.-M., Choi, C., Kim, T.G., Kim, J.H., 2014. DEVS-based combat modeling for engagement-level simulation. Simulation 90, 759–781.

Simulation Interoperability Standards Committee, 2000. IEEE Standard for Modeling and Simulation (M&S) High Level Architecture (HLA)–Object Model Template (OMT) Specification. IEEE Std. 1516.2-2000.

Sung, C., Kim, T.G., 2012. Collaborative modeling process for development of domain-specific discrete event simulation systems. IEEE Trans. Syst. Man Cybern. Part C Appl. Rev. 42, 532–546.

Weiming, S., Qi, H., Mak, H., Neelamkavil, J., Xie, H., Dickinson, J., 2008. In: Systems integration and collaboration in construction: a review.2008 12th International Conference on Computer Supported Cooperative Work in Design, 16–18 April 2008, pp. 11–22.

Zeigler, B.P., 1984. Multifacetted Modelling and Discrete Event Simulation. Academic Press Professional, Inc., San Diego, CA.

Zeigler, B.P., Song, H.S., Kim, T.G., Praehofer, H., 1994. In: DEVS framework for modelling, simulation, analysis, and design of hybrid systems. International Hybrid Systems Workshop. Springer, pp. 529–551.

Zeigler, B.P., Praehofer, H., Kim, T.G., 2000. Theory of Modeling and Simulation: Integrating Discrete Event and Continuous Complex Dynamic Systems. Academic Press, San Diego, CA.

Zeigler, B.P., Muzy, A., Kofman, E., 2018. Theory of Modeling and Simulation: Discrete Event & Iterative System Computational Foundations. Academic Press.

Further Reading

Eriksson, L., Johansson, E., Kettaneh-Wold, N., Wikström, C., Wold, S., 2000. Design of Experiments. *Principles and Applications*, Learnways AB, Stockholm, pp. 172–174.

Rubinstein, R.Y., Melamed, B., 1998. Modern Simulation and Modeling. Wiley, New York.

Modeling and Simulation of Versatile Technical Systems Using an Extended System Entity Structure/Model Base Infrastructure

T. Pawletta*, U. Durak†, A. Schmidt*
*Wismar University of Applied Sciences, Wismar, Germany
†German Aerospace Center (DLR), Brunswick, Germany

1 INTRODUCTION

The development of versatile technical systems requires enhancements of methods and tools for effective modeling, simulation, and control. Aerospace systems (Sampigethaya and Poovendran, 2013) and smart factories according to the "Industrie 4.0" trend (Hermann et al., 2016) are typical examples. They are modular structures aggregated by smart components, which communicate and cooperate with each other and with humans, and make decentralized decisions. Following the definition in Lee (2008), such systems are cyber-physical systems (CPS). The term versatile here emphasizes that these systems are flexible, reconfigurable, and reactive. Flexibility means the ability to adapt quickly and with little effort, within the limits of a given range, to changing conditions. Reconfigurability describes the ease with which a system can be employed for different purposes or objectives by quickly changing or adapting its physical units or controls. Flexibility and reconfiguration of a system in this sense implies a reactive behavior according to changing conditions. Such characteristics have been of interest in control engineering for many years and are part of industrial automation problems, as shown in Hummer et al. (2006).

In software engineering, versatility is considered under the term variability and often in the context of Software Product Line (SPL) engineering, although not necessarily in combination with reactivity. Bosch (2013) considers SPL engineering as an approach for managing functionality for a family of products, in Apel et al. (2016) variability has been defined as the ability to derive different products from a common set of core assets, and in Jezequel (2012) variability is described as a characteristic of a software system that can be configured, customized, or extended for employment in a particular context. Moreover, Jezequel (2012) emphasizes that the basic idea of the SPL engineering process is the same as in traditional engineering disciplines, which use modeling techniques. According to Pohl (2006), SPL engineering is subdivided into domain engineering and application engineering. Domain engineering focuses on core assets development and subsequently, application engineering addresses the usage of the core assets to develop a final product based on customer requirements. In addition, Jezequel mentions, that the variability itself has to be modeled to handle complex variability problems and that the whole process needs some form of automation. Several concepts of this kind of automation are known as model-driven engineering (MDE) and are discussed in Cretu and Dumitriu (2015). MDE methods are well accepted in several engineering disciplines, particularly in industrial automation, automotive, aerospace, or robotics. MDE methods that have been developed in software engineering are often inspired by and used for solving software-intensive technical applications, such as in Hummer et al. (2006), Zander (2008), and Haber et al. (2013).

Many of the problems discussed in software engineering or other engineering disciplines regarding versatility or reactivity have been studied—if not before, then simultaneously—by the modeling and simulation (M&S) community focusing on system theory. In this context, we would like to point out the work on multifaceted and variable structure modeling (Zeigler, 1984; Zeigler and Praehofer, 1989; Rozenblit and Zeigler, 1993; Zeigler et al., 2000; Couretas, 2006; Zeigler and Hammonds, 2007; Zeigler and Sarjoughian (2013)Seo et al., 2014; Pawletta et al., 2016a,b; Santucci et al., 2016), variable and dynamic structure systems simulation (Pawletta and Pawletta, 1995; Barros, 1997; Uhrmacher, 2001; Kim and Kim, 2001; Pawletta et al., 2002; Hu et al., 2003; Park and Hunt, 2006; Mittal et al., 2006), or dynamic model updating with multimodels (Ören, 1987; Yilmaz and Ören, 2004). Last but not the least, Chapter 2 presents structure variations for achieving a desired behavioral approximation.

To model the variability of a product, an SPL model defines features and variation points, where different variants of a product can be derived for varying requirements. Features represent abstract system characteristics and variation points denote selection possibilities for deriving a distinct variant. According to the layers of the meta-object-facility (MOF) hierarchy in OMG (2006), variability mechanisms can be defined at different levels of abstraction. They can be specified using a variant or variability model at the metamodel layer. This type of a variability model is combined with reusable core assets, organized in a library or implemented as a 150% model, such as in Dziobek et al. (2008), Steiner et al. (2013), and Lackner et al. (2014). Then, using MDE-based transformation methods, executable system models or target code for real devices are generated. This approach is basically comparable with the System Entity Structure and Model Base (SES/MB) approach in M&S, as described in Zeigler (1984), Zeigler and Praehofer (1989), and Zeigler et al. (2000). Particularly, in software for technical systems, variability and core assets are often modeled directly on the system level in combination with specific configuration methods, such as in Hummer et al. (2006), Zander (2008), Kliemannel et al. (2010), and The Math Works (2017). These approaches are partly comparable with methods developed for variable and dynamic structure systems simulation in M&S, such as in Barros (1997), Uhrmacher (2001), Kim and Kim (2001), Pawletta et al. (2002), Hu et al. (2003), Park and Hunt (2006), and Mittal et al. (2006). However, there is often a gap between the theoretical research in the M&S stage and the development of appropriate tools and infrastructures for engineers.

This chapter introduces a novel infrastructure for the M&S of versatile and reactive technical systems, based on an extended concept of the SES/MB. It briefly introduces the basic theory and discusses some extensions to make it more pragmatic for implementation in an engineering tool or infrastructure such as MATLAB/Simulink. According to the previously introduced concepts, an SES specifies a set of structures of system designs, including variation points. A system design is defined by its system structure and parameter configuration. The core assets are specified as a set of configurable basic models, which are organized in an MB. Moreover, the SES/MB framework defines a set of transformation methods for generating executable simulation models (SMs). Besides an extended concept of the SES/MB framework, the presented infrastructure consists of an execution unit (EU) and an overall experiment control (EC) to enable automated, reactive processing of SES models in combination with different model bases during the experiment.

The chapter is organized as follows. Section 2 gives an overview to the new infrastructure and summarizes essential basics of the underlying SES/MB framework. In Section 3, an advanced industrial robot (IR) control is introduced as a typical example of a versatile technical system and the basics of a model-based control design using the simulation-based control (SBC) approach following Pawletta et al. (2009) are described. This robotic example is then used to demonstrate the basic steps of domain and application engineering in Section 4, using the new infrastructure. After that, Section 5 provides a short insight into a toolbox implementation for MATLAB/Simulink. The chapter concludes with our results and provides an outlook to some future works.

2 INFRASTRUCTURE BASED ON AN EXTENDED SES/MB APPROACH

The infrastructure is based on the SES/MB framework and extends it by an EU and EC unit. Starting with a general overview to the several components and their complex interactions, we explain how the infrastructure can be used for the M&S of versatile dynamic systems. In this context, we discuss the problem of state consistency and suggest an experimental frame (EF) for such systems corresponding with the other elements of the infrastructure. Then, we summarize basic concepts for modeling system variability using system entity structures and introduce some extensions to that approach.

2.1 General Overview of the Infrastructure

In this context, the M&S of versatile systems comprises of: (i) modeling of a set of alternative system configurations, also called multifaceted modeling, and (ii) automated, reactive generation of system configurations, which are executable within a simulation environment.

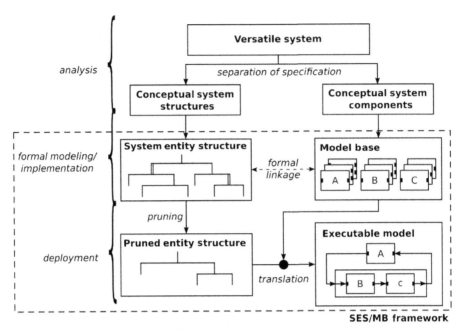

FIG. 1   Procedure model and classic SES/MB framework.

The first part can be tackled using the classic SES/MB framework as introduced in Zeigler et al. (2000); for the second, the framework has to be extended.

Fig. 1 illustrates the procedure model for the M&S of a versatile system based on the concept of the classic SES/MB approach and the corresponding framework presented in Zeigler et al. (2000). As part of the *domain engineering phase*, the versatile system has to be analyzed regarding its different system configurations. That means, possible system structures, parameter configurations, and core assets have to be identified. The result is a set of *conceptual system structures*, including possible parameter configurations and a set of *conceptual basic system components*, which can be composed in a modular, hierarchical manner. As illustrated in Fig. 1, the next step in the domain engineering phase is called *formal modeling and implementation*. The dynamic behavior of the conceptual systems has to be modeled. This results in formal *basic systems or models*, which have to be implemented as reusable software or model components with defined input and output interfaces and organized in an MB. On the other hand, the conceptual system structures and parameter configurations have to be modeled with an SES, which defines simulator-independent links to basic models in the MB. That means, the resulting *SES is a formal model of all possible system configurations*. In the *application engineering phase*, executable models are generated with the transformation methods called *pruning* and *build*. This step is summarized in Fig. 1 under the term *deployment*. Based on user requirements, the pruning operation derives a unique formal system structure with parameter configurations from the SES. The result is a decision-free tree structure, called *pruned entity structure (PES)*. Then, an explicit executable model can be generated using a simulator-dependent build method based on the information of PES and basic software components from the MB.

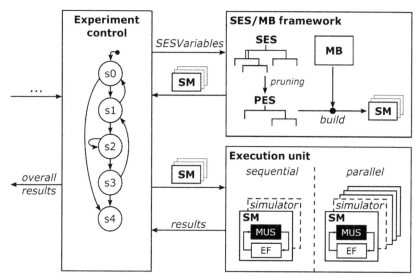

FIG. 2   Extended SES/MB-based infrastructure.

The classic SES/MB framework does not define techniques for an automated, reactive generation of explicit executable models. For such a task of deployment, the framework has to be extended, as illustrated in the *infrastructure* in Fig. 2. The interactions between individual components of the infrastructure are described in more detail in Fig. 3.

Besides the SES/MB framework, the infrastructure consists of an *EU* and an overall *EC unit* to enable automated, reactive processing of an SES in combination with an MB during experimentation runtime. Therefore, the classic SES/MB framework is extended by an *input and output interface* using *SES variables (SESvar)*. The selection of a particular system configuration depends on the current settings of the SESvar.

The EC manages the order of explicit system configurations that have to be generated. Based on an arbitrary application-dependent algorithm, the EC computes current settings for the SESvar as inputs for the SES/MB framework to generate an executable *SM or a set of SMs*. However, in this paper, we will stick to one SM. The EC transmits the SM to an EU. Additionally, the EC configures and starts the EU. It provides the necessary settings of simulation execution parameters, and initiates the execution. The EU performs three major tasks: (i) linking an SM with a simulation engine (simulator), (ii) executing a simulation run, and (iii) collecting the results. Once the execution is complete, it returns the results to the superior EC. The EC collects all intermediate results from the EU. Depending on its internal algorithm and current dataset, it starts a new cycle or finishes the experiment. Finally, the EC provides the overall results to the user or another software component.

A first application of the introduced infrastructure was published in Schmidt et al. (2016) for solving reactive model-based testing problems in the domain of aeronautics. Moreover, the authors show that it is useful to extend the SM by an *EF* according to Zeigler (1984). The concept of EF supports more generalized system models, called *model under study (MUS)* in Fig. 2. Thus, the SM consists of the MUS and a corresponding EF, as illustrated in Fig. 2.

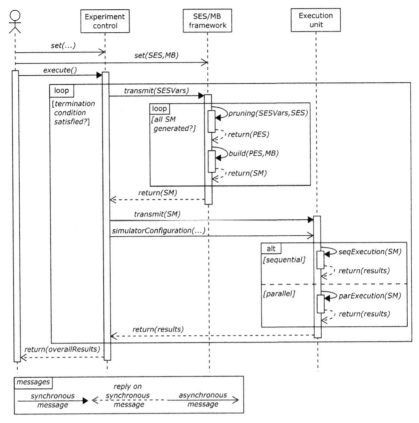

FIG. 3   Sequence diagram describing interactions among infrastructure components.

At this point, it should be emphasized that the dynamic structure and parameter configurations are not modeled on the system model level, but on the higher abstraction level of SES. Thus, the *problem of state consistency between reconfigurations* cannot be handled inside the system model alone. This problem is solved in the infrastructure of the EC unit using the introduced SESvar and the concept of EF. The basic structure and function of an appropriate EF is discussed in the next subsection.

2.2 Structure of a Corresponding EF

The general concept of the EF was introduced by Zeigler (1976), further elaborated in Zeigler (1984) and has been continuously advanced and applied, such as in Rozenblit (1991), Zeigler et al. (2000), Traorè and Muzy (2006), and Ponnusamy et al. (2014). Basically, according to Zeigler (1984), an EF specifies "a limited set of circumstances under which a system or model is to be observed or subject to experimentation." In the case explored by this paper, this means the circumstances of the simulation experiment with the MUS. Hence, the EF has to generate admissible input values X for the MUS and has to observe its output

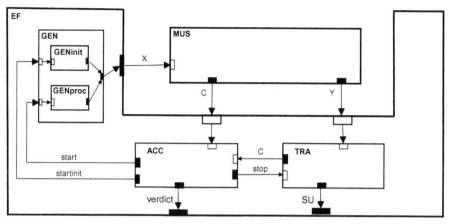

FIG. 4   Basic EF-MUS structure for the introduced infrastructure.

values Y. In addition, the EF has to control the initialization, continuation, and termination of an experiment. Therefore, Zeigler introduces a set of run control variables, in short C. While the concept of EF can be implemented in various ways, Zeigler recommends a modular and structured model specification for a coupled EF structure—consisting of a generator (GEN), acceptor (ACC), and transducer (TRA)—which can be connected via external inputs and outputs to a set of corresponding MUSs. Based on this theory, we derived the basic EF structure in Fig. 4, which is compatible with the infrastructure described in Fig. 2.

The basic EF consists of at least two GEN subtypes, an ACC and a TRA. At the start time of an experiment, the ACC sends a startinit event to the generator GENinit, which sends the initial conditions to the MUS. After a successful initialization of the MUS, determined by evaluating the control variables C, the ACC sends a start event to the generator GENproc. During the continuation phase of a simulation run, the GENproc delivers the MUS input values X and the TRA receives the MUS output values Y and computes control variables C for the ACC. Based on its current input values, the ACC determines whether a simulation run will be continued or has to be terminated. In case of termination, the ACC sends a stop event to the TRA, which reacts by computing and sending the set of summary mappings SU as output values of EF. Simultaneously, the ACC sends an output value verdict to EF, which indicates whether the simulation run has been valid or invalid. Of course, this is a basic EF structure and, depending on the application, it can define additional coupling relations.

The EF outputs verdict and SU correspond to the results of the EU in Fig. 3. Thus, the set SU can contain state variables, which have to be saved for the initialization of a subsequent simulation run. The EC unit can select, save, and transmit these values as the new values of SESvar to the SES/MB framework to start configuration for the next experiment. Our basic EF structure specifies no inputs from outside because the total set of admissible configurations to build an SM, consisting of an MUS and an EF, are specified in the SES. The configuration of an SM is only determined by the pruning method, depending on the current settings of SESvars, and the build method using basic models from the MB. Hence, the MB has to also organize basic models to build an appropriate EF.

2.3 Basic Concepts of the System Entity Structure and Some Extensions

Following Zeigler and Hammonds (2007), the SES itself is an ontology that reflects system engineering concepts of hierarchical decomposition and specialization. We will focus on concepts of the SES, which are important for modeling a set of modular, hierarchical SM configurations and for building an executable SM using basic models from an MB. This consideration is based on the baseline definitions in Zeigler and Hammonds (2007). Additionally, we briefly introduce the concept of SES Variables (SESvars) and SES Functions (SESfcn) from Pawletta et al. (2014). Fig. 5 part one (Fig. 5p1) and Fig. 5 part two (Fig. 5p2) illustrate essential concepts that shall be discussed. To start with, an example SES and MB (Fig. 5a) is analyzed, with individual sections examined in Fig. 5b–f. These figures show a section of the SES on the left side and the specified model structures of the respective section on the right.

The modeling of an SES is based on six axioms: (i) *uniformity*, (ii) *strict hierarchy*, (iii) *alternating mode*, (iv) *valid brothers*, (v) *attached variables* or more general attributes, and (vi) *inheritance*. An SES can be represented as a directed labeled tree (Fig. 5a). The tree nodes are divided into two fundamental types, entity and descriptive nodes, which alternate. Entity nodes describe system elements and the system itself (root node). Descriptive nodes express relationships between entities and are subdivided into: *aspect, specialization*, and *multiaspect* nodes. The different types of descriptive nodes are marked with different edge types (one, double, or triple vertical lines) and usually with a specific name suffix; in this case: Dec for aspect node, Masp for multiaspect node, and Spec for specialization node.

The entities specify basic, coupled, or abstracted models. Basic models are always represented by leaf nodes. They define a specific attribute *mb* that specifies a link to a corresponding MB (Fig. 5a). They can define more attributes (*para*) for configuring the referenced basic model.

Coupled models are always described by a subsequent aspect node (Fig. 5b). An aspect describes a decomposition relationship between its parent entity and its children entity nodes, whereby the coupling relations—the external input (EIC), internal (IC) and external output (EOC) couplings—are defined as attribute. Each coupling relation is defined as a four-tuple(`from entity, port name, to entity, port name`).

Alternative system compositions can be described using aspect siblings, as at entity B with its successors B1Dec and B2Dec (Fig. 5c). The selection is specified by an additional aspect node attribute called an aspect rule. In the example given, the aspect rule, named arule, uses an SESvar, named SESvar1.

The multiaspect node is a special kind of aspect node (Fig. 5d). It stands for a multiplicity relationship that specifies that the parent entity is a composition of multiple entities of the same type. Hence, it defines an additional attribute, called numRep, to specify the multiplicity. The multiplicity can be defined by a fixed value, an interval, or variable depending on an SESvar or SESfcn. Note that depending on the multiplicity, several coupling relationships have to be defined and that during the pruning process, the replications are numbered. Different parameter configurations of replications can be defined by multisets.

A specialization describes the taxonomy of an entity (Fig. 5e). The parent entity node represents an abstracted model and the children entity nodes are possible specializations. In contrast to the AND relationship of aspect and multiaspect, a specialization describes an XOR relationship regarding its children. The selection is defined by an attribute called the selection rule—named *srule* in Fig. 5e—analogous to the aspect rule for aspects. During the pruning

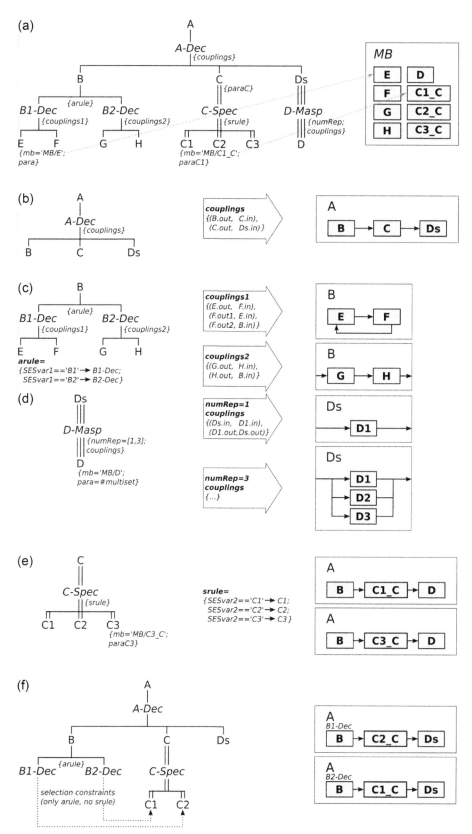

FIG. 5 Fig. 5 p1: An example SES and MB (a) and the basic concept of aspect (b) and aspect siblings (c). p2: The basic concept of multi-aspect (d), specialization (e) and selection constraints (f).

operation, the parent entity and the selected child entity will be aggregated according to the inheritance axiom. Fig. 5e illustrates only the simplest case for a specialization with name and attribute aggregation. More complex cases are described in Zeigler and Hammonds (2007).

Moreover, the SES supports the specification of cross-tree relations, called selection constraints, which are defined by directed edges (Fig. 5f). A selection constraint determines that the selection of a certain node causes the selection of other nodes. Selection constraints can be combined with aspect and selection rules.

The concept of SESvar has been introduced as an input interface for an SES and for general cross-tree information exchange between node attributes. SESvars, which represent input variables for an SES, require a value assignment before the pruning operation. In addition, semantic conditions can be defined using SESvar. Such conditions can limit the value range of SESvar and can exclude specific system configurations. In the example, in Fig. 5c–e, SESvar1 was used in the aspect rule *arule* to select the aspects *B1Dec* or *B2Dec* and SESvar2 in the selection rule *srule* of the specialization *CSpec*. If, for example, we want to exclude the system configuration B1Dec with the specialization C3, this can be specified with the following semantic condition: $SESvar1=='B1' \wedge \neg(SESvar2=='C3')$. Semantic conditions are always checked before pruning.

The concept of SESfcn has been introduced to specify complex variability within node attributes with minimal effort and to keep a lean SES tree. Typical examples include the definition of varying coupling relations, varying port numbers of systems, or the definition of variable parameter configurations in attributes. The multiaspect node in Fig. 5d is considered as an example. Note that depending on the multiplicity attribute *numRep*, several coupling relationships have to be defined. This can be specified using an SESvar, named NUM, for the multiplicity attribute *numRep* and the subsequent SESfcn, named *cplg*.

```
cplg(children, parent, num)
for i=1 to num
cplg(i)={(parent, 'in', children(i), 'in'), -EIC
(children(i),'out', parent, 'out')} -EOC
end
return(cplg)
```

Then, the *DMasp* node in Fig. 5d can simply define the attributes:

```
{numRep=NUM; couplings=cplg(children,parent,NUM)}
```

The SESfcn *cplg* is evaluated during the pruning operation. For effective coding of SESfcn, the implicit attributes *parent* and *children* are introduced for each SES node. They encode the parent and children node names, respectively. Our practical experience showed that an SESfcn can often be used for a variable configuration of attributes at several nodes. The above SESfcn *cplg* specifies a general case of a parallel coupling structure.

3 A ROBOT CONTROL AS EXAMPLE FOR A VERSATILE SYSTEM

IR applications are typical examples of versatile technical systems because modern IRs are the most flexible and reconfigurable units in manufacturing. We introduce a classical advanced robotic control problem. Then, we briefly discuss the SBC approach as basic for the subsequent problem solution. The robotic control example will be used as a basis for explanations in the subsequent section.

3.1 Problem Statement

Fig. 6 shows the layout of the robot application that will be studied. The overall goal is to identify different part types at an input buffer (IB) and to transport them with an IR to a part type depending on the output buffer ($OB_1...OB_n$). The part types are randomly distributed in the IB. We assume that every time the next part for handling is located at a fixed pick position for the IR. Before handling by the IR, the part type has to be identified using an electronic scale (SCA). Depending on this first identification, only some exclusive part types need a second identification by an image sensor (CAM). After successful identification of the current part, the IR picks it up, moves it to the corresponding part type dependent OB, and places it at a fixed location.

The control software (CS) needs to be designed for a variable number of different part types. Thus, the problem is versatile, because the number of OBs depends on the total number of part types and the necessary identification operations depend on the current part type.

3.2 The Simulation-Based Control Approach

The control development should be carried out stepwise and should be model based, using reusable basic models following (e.g., the SBC) approach introduced in Pawletta et al. (2009). Fig. 7 illustrates schematically the SBC approach. This approach uses an SM and refines it stepwise across the development phases: planning, automation, and operational use of the CS. During transition from planning to the automation phase, the SM is separated into a control model (CM) and a process model (PM) for a clear arrangement of the control design.

The transition from the automation to the operational phase is usually known as code generation. Depending on the real-time requirements of the control application, the SBC approach distinguishes between explicit and implicit code generation. The first type is the classical method for high real-time requirements in conjunction with mostly embedded controller hardware using a compiler. For applications with rather slow timing, like our robot example, implicit code generation is suitable. Then, the SM from the automation phase can be extended by an interface model (IM) and directly used as a CS (Freymann et al., 2016).

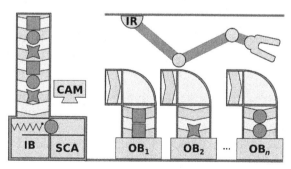

FIG. 6   Layout of the robotic application.

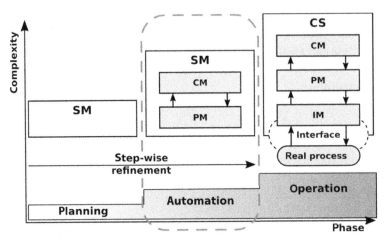

FIG. 7   Simulation-based control (SBC) approach.

The IM provides a process interface enabling a software-in-the-loop (SIL) control according to the definition in Abel and Bollig (2006). Following the SBC approach, the CS always includes a PM. Thus, observer concepts can easily be realized.

In Schwatinski et al. (2012), Maletzki (2014), and Freymann et al. (2016), the SBC approach has been successfully applied for task-oriented control (TOC) applications. According to Siciliano (2008), the principle of TOC is to divide a control problem into a set of tasks and their couplings. Tasks are logical, mostly independent, abstract operations. Once identified, tasks are coupled together to map the control problem. However, a TOC specification is not executable directly because tasks are an abstract description of operations. To perform tasks, a transformation method is required. All tasks have to be transformed into control commands in order to execute them. In the SBC approach, the CM contains the TOC specification. The PM still has a component-oriented structure according to the elements of the real process and is also the place of generalized task transformation. That means the CM and PM are independent of vendor-specific robots. Vendor-specific task mapping is the domain of the IM. Subsequently, we will focus our considerations only relating to the SM requirements in the automation phase. Details for developing a corresponding IM can be found in Freymann et al. (2016).

4 DOMAIN AND APPLICATION ENGINEERING OF THE ROBOT CONTROL

This section presents the principal steps of domain and application engineering using the introduced infrastructure through the example of the robot control. We start with a system analysis and the conceptual modeling of core assets regarding basic system components and system structures. Then the basic steps for formal modeling and for implementing a suitable MB are explained, with a particular focus on the modeling of system designs with an SES.

Finally, the deployment of the SES/MB-based robot control is shown by discussing the basics of an appropriate algorithm for an EC.

4.1 Analysis and Conceptual Modeling

Taking into account the problem constraints defined in the robot control example, the requirements of SBC, and the previous considerations relating to the EF mentioned in Section 2.2, we can derive the general conceptual system structure for an SM of our robot application as illustrated in Fig. 8.

The SM (sm) is made up of an MUS, comprising of a control model (cm) and a process model (pm), and a corresponding EF (ef). Note that the original SM from Fig. 7 is now the *mus*. The conceptual structure of the EF has already been analyzed and illustrated in Fig. 4. The entity *gen* in Fig. 8 represents a general GEN component that can be further decomposed or organized as a complex, configurable basic model in an MB. The same applies to the acceptor *acc* and the transducer *tra*.

The further decomposition of *pm* is given by the real system layout in Fig. 6. The decomposition of control model (cm) results from the task-based control design. Each component can be implemented as a configurable basic model and stored in an MB:

- scale (sca), image sensor (cam), input buffer (ib), robot (ir), and a generalized output buffer (ob).
  The decomposition of *cm* results from the task-based control design. Based on the problem description, the following four tasks can be derived and implemented as configurable basic models for storing in an MB.
- identification using the SCA (isca); identification using the CAM (icam); picking a part with the IR at the IB (pick); and moving the part with the IR to an OB and placing it (place)

The task *place* is a composite task, which involves moving and placing a part. As the two single tasks themselves are simple, combining them in one basic model reduces the total complexity. The tasks are executed in the order:

- identification of each new part using SCA at IB,
- second identification of only exclusive parts using CAM at IB, depending on the previous identification result with the SCA,

FIG. 8 SES tree representing the conceptual system structure of an SM for the robot control.

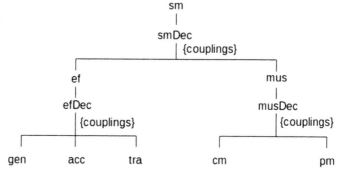

- picking the part with the IR at IB and moving/placing it at the corresponding OB.

In addition to the identification and robot control, the concurrent arrival of randomly distributed part types at the IB has to be modeled in the *pm*, but this operation has no direct effect on the *cm*. Moreover, it can be assumed that both identification tasks require a negligible period of time. Hence, the concurrent part arrival at the IB has to be considered only during the pick-and-place task of the IR. Based on these considerations, the entire problem can be modeled using three system structures, illustrated as conceptual block models in Figs. 9 and 10. The system structures for both identification operations differ only in the basic model for the identification device (sca | cam) in the *pm* and for the control task (isca | icam) in the *cm*. In contrast, the pick-and-place operation with the IR requires quite a different system structure. Note that this system structure depends on the number of part types, because they influence the number of OBs in the system. The system structure in Fig. 10 contains only the coupling relations for two OBs.

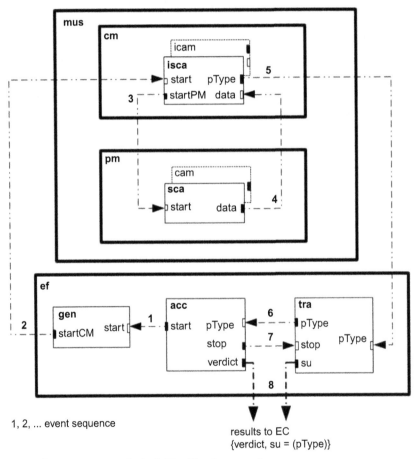

FIG. 9 Conceptual system structure for both identification operations.

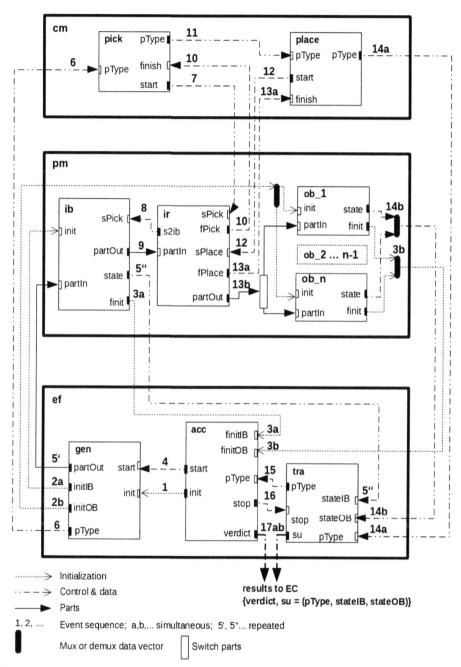

FIG. 10 Conceptual system structure for the pick-and-move/-place operation.

Subsequently, we will summarize some essential characteristics regarding the system structures. As discussed in Section 2.2, the initialization, continuation, and termination of the simulation is controlled by an EF. Both identification operations (Fig. 9) require no specific initialization. The current part identification is started by the two consecutive events of *acc* (msg#1) and *gen* (msg#2). Then, the identification task (*isca* | *icam*) in *cm* sends an event (msg#3) to the identification device (*sca* | *cam*) in *pm*, which returns the identification data as an event (msg#4) to the task model in *cm*. Based on these data, the identification task (*isca* | *icam*) determines the part type (*pType*) and sends this information (msg#5) to the *tra* in *ef*, which relays the *pType* (msg#6) to the *acc*. The *acc* evaluates the identification result and stops the continuation phase with an event (msg#7) to the *tra*, whereupon the *tra* and the *acc* produce an EOC event of *ef* (msg#8)—the *acc* yields a *verdict* regarding the experiment and the *tra* a summary mapping *su = (pType)*. These output events are essential for the further process. In the case where *verdict == false*, the experiment was wrong; otherwise, the result in *tra.su* is valid and the information in *pType* determines the next step. This can be a second identification using the image sensor (cam) or a pick-and-place operation with the IR. In the second case, the part type information in *pType* is necessary to initialize the system structure.

The system structure for performing a pick-and-place task (mus in Fig. 10) with the IR needs three pieces of information for initialization: (i) the stocking in IB, (ii) the stocking in each OB, and (iii) the previously identified type (*pType*) of the part to be actually handled by the IR. The initialization phase is started by an event from *acc* (msg#1) to *gen*, which sends initialization events to the *ib* and to each *ob* in *pm* (msg#2a, #2b). They send an initialization finish event to the *acc* (msg#3a, #3b). In case of successful initialization, the *acc* starts the actual execution phase ("continuation") with an event to the *gen* (msg#4), which randomly generates new parts (msg#5') for the *ib* in *pm* that reports each state change (msg#5") to the *tra* in *ef*. Additionally, the *gen* sends the previously identified *pType* as event to the *pick* task in *cm* (msg#6), whereupon the pick operation using the IR (*ir*) is executed (msg#7, #8, #9, #10). We assume that the basic model *ir* of IR is configured with the trajectories to the location points at the IB and at each OB. After receiving the finish event (msg#10), the *pick* task starts the *place* task in *cm* (msg#11). Based on the information in *pType* (#msg11), the *place* task basic model determines the appropriate OB (*ob1...ob_n*) and sends the corresponding buffer address with an event (msg#12) to the *ir* in *pm* to execute the handling operation, which finishes with the simultaneous events (msg#13a, #13b). Then, the *place* task sends the *pType* to *tra* in *ef* (msg#14a). Analogous to the *ib* in *pm*, each *ob* reports its state to *tra* (msg#14b). The termination phase by the *acc* is principally identical to the system structure for the identification operations, but the output *su* of *tra* contains as additional information the last states of *ib* and each *ob* in *pm*.

4.2 Formal Modeling

According to the procedure model set out in Fig. 1, the next steps are:

1. the formal modeling of basic models, their implementation and organization in an MB and
2. the formal modeling of all system configurations regarding their structure and parameter configuration with an SES.

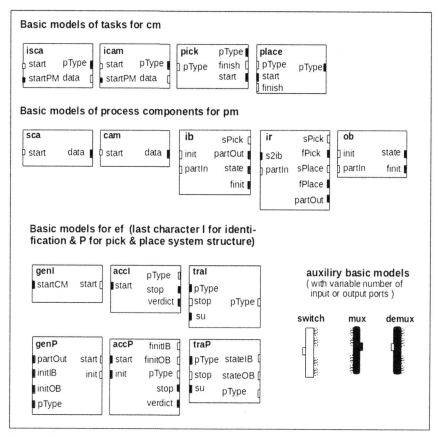

FIG. 11 MB with basic models for the robotic application.

Since the first set of steps is commonly known, we will only consider the basic models that have to be implemented. The set of appropriate basic models for the MUS (*mus*) have already been discussed in the previous subsection. Regarding the basic models for the EF (*ef*), it is suitable to develop different basic models for the structures in Figs. 9 and 10. Accordingly, Fig. 11 shows the resulting MB.

Using the infrastructure introduced in Section 2, the different system configurations of the versatile system can be mapped to a set of executable SMs, which are linked in the deployment phase using the execution unit (EU) of the infrastructure. Subsequently, the main steps for developing a corresponding SES are explained.

The SES tree is illustrated in Figs. 12 and 13. Before discussing the SES tree, the following set of SESvar is defined through some semantic relations.

```
SETofPartTypes ...input parameter for SES
NUMob = |SETofPartTypes| // || for cardinality
```

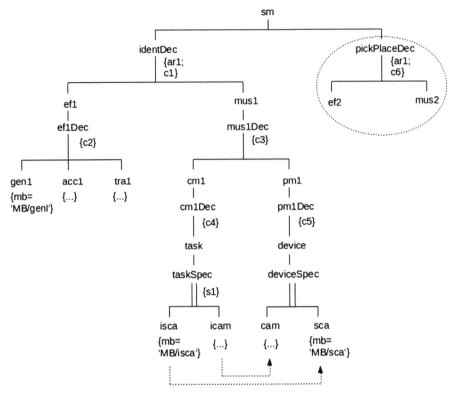

FIG. 12 SES main tree for the robotic control example.

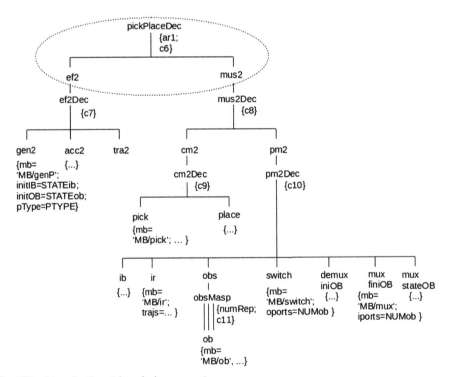

FIG. 13 SES subtree for the pick-and-place operation.

```
PTYPE ∈ {SETofPartTypes∪NIL}
STATEib ...vector with state of input buffer occupancy
STATEob ...matrix with state of each output buffer occupancy
PHASE ∈ {TASKisca, TASKicam, TASKpickplace}
```

The SES has to model three different configurations for an SM according to Figs. 9 and 10. For this reason, the decomposition of the SM (*sm*) is specified with the two aspect siblings *identDec* and *pickPlaceDec*. The subtree of *identDec* specifies the system structures of the two different identification operations and the other one, depicted in detail in Fig. 13, the system structure of the part handling operations using the IR. The selection is specified as an aspect rule with the subsequent conditions in the respective attribute *{ar1}* of the nodes *identDec* and *pickPlaceDec* node. It depends on the current setting of the SESvar *PHASE*.

```
identDec:ar1={PHASE==TASKisca | PHASE==TASKicam}
pickPlaceDec:ar1={PHASE==TASKpickplace}
```

Additionally, each aspect node specifies the decomposition of *sm* into an MUS (*mus*) and EF (*ef*) and defines the coupling relationships as attributes (*{c1},{c6}*) analogous to the theory in Fig. 5c. The node names for the two MUS and EF are different because of the SES *uniformity axiom*.

The SES in Fig. 12 specifies the same EF (*ef1*) for both identification operations. The aspect node *ef1Dec* defines its decomposition in a generator (*gen1*), acceptor (*acc1*), and transducer (*tra1*) and the coupling relationships as attribute *{c2}*. The leaf nodes specify the link to a basic system in the MB in their attributes (Fig. 11), as shown at node *gen1*. The decomposition of the *mus1* in a control model (*cm1*) and a process model (*pm1*) is modeled with the aspect node *musDec* and its corresponding attribute *{c3}*. The subsequent aspect nodes *cm1Dec* and *pm1Dec* describe the decomposition of *cm1* and *pm1*, respectively. The entities *task* and *device* represent abstract systems, which are specialized by their subsequent specialization nodes *taskSpec* or *deviceSpec*. The selection is described by the selection rule in attribute *{s1}* and depends on the value in the SESvar *PHASE*. The selection of the corresponding identification device is defined with the *Selection Constraints* edges.

```
taskSpec:s1={PHASE==TASKisca → isca;
PHASE==TASKicam → icam}
```

The SES part of the pick-and-place handling operation using the IR, the subtree of the aspect *pickPlaceDec*, is depicted in Fig. 13. Besides the actual MUS (*mus2*), it specifies a specific EF (*ef2*) with its coupling relations in attribute *{c7}*. As described in Section 4.1 (Fig. 10), the generator (*gen2*) has to provide initialization data for the IB (*ib*) and for each OB (*ob*) and information about the part type to be handled by the IR (*ir*). All these data are defined as attributes of node *gen1* using SESvar.

Regarding the MUS (*mus2*), the control model (*cm2*) is specified as a simple static composition of the *pick*-and-*place* task (*cm2Dec*). The structure of the process model (*pm2*) depends on the number of part types, because it influences the number of output buffers in the system, as defined by the SESvar setting `NUMob=|SETofPartTypes|` in the beginning. Based on this fact and the conceptual model structure in Fig. 10, the aspect *pm2Dec* defines the decomposition of *pm2* in an IB (*ib*), IR (*ir*), an entity *obs*—which is further refined—and auxiliary components such as the part switch (*switch*) and components for decomposing and composing vector-based data (*demuxiniOB, muxfiniOB, muxstateOB*). Of course, the coupling relationships of *pm2*, defined in the attribute *{c10}*, are also influenced by the number of OBs in the system, which means they depend on the current setting of the SESvar *NUMob*. This variability in the coupling relations can be specified using an SESfcn, as shown in the subsequent code excerpt and the specification in *{c10}* is the corresponding SESfcn call: `{couplings=cplg_c10 (children,parent,NUMob)}`.

```
cplg_c10(children,parent,num)
// static couplings
...
// variable couplings between entity switch and entity obs
for i=1 to num
cplg(i)={(children.switch,out(i), children.obs,in.partin)}
end
...
return(cplg)
```

The decomposition of the entity *obs* in a variable number of type identical entities (*ob*), which represent the output buffers, is specified by the multiaspect node *obsMasp*. It has the following attribute definitions:

```
{numRep=NUMob;
couplings=cplg_c11(children,parent,NUMob)}
```

The specification of SESfcn *cplg_11* is similar to the one above. Besides the coupling relations, the input/output port configurations of some basic systems—such as those specified at the entities *switch* and *muxfiniOB*—depend on the number of OBs in the system, which is specified using the SESvar *NUMob*.

4.3 Deployment Using the Infrastructure

Subsequently, the deployment of the SES/MB-based robot CM, using the infrastructure shown in Fig. 2, is presented. For this, it is necessary to implement the application-dependent algorithm for the EC, which links the classic SES/MB framework with the EU. The following algorithm shows the major steps for the robot control.

For complex problems, it is advisable to develop the algorithm using formal approaches, such as finite state machines as presented for a model-based testing application in aviation in Schmidt et al. (2016).

```
//initialization
1. SES=      load(SES of robot control)  //Fig.12&13
2. MB =      load(MB of robot control)   //Fig.11
//set input SESvars
3. SETofPartTypes          ={p1,p2,p3)      //e.g. 3 diff. types

// NUMob = |SESofPartTypes| explicitly defined in SES
4. STATEib      =[p1,p1,p3,p1,p2]    //initial part of IB
5. STATEob      =[NIL;NIL;NIL]  //all OB are empty
6. PTYPE=NIL
//start pick & place, but only for initialization ib & each ob
//(Fig.10)
7. PHASE=       TASKpickplace    //set next PHASE
8. SESvars={SETofPartTypes,STATEib,STATEob,PTYPE}
9. SM=SESMBframework(SES,MB,SESvars)    //pruning & build
 //config. execution unit & execute only initialization
10. EU.configure(simulator,[t0,t0],sequential,…)
11. results=EU.execute(SM)
12. if results.verdict==FALSE then Stop Experiment
 //continue exp. also when results.verdict==Only_pType_NIL
13. STATEib=results.stateIB
14. STATEob=results.stateOB
15. tnext=t0                    //set next start time

//do 1st identification with scale (Fig.9)
16. PHASE=           TASKisca  //set next phase
17. SESvars={SETofPartTypes,STATEib,STATEob,PTYPE}
18. SM=SESMBframework(SES,MB,SESvars) //pruning & build
 //config. execution unit & execute simulation
19. EU.configure(simulator,[tnext,inf],sequential,…)
20. results=EU.execute(SM)
21.             tnext=EU.tfinal //save next start time
22. if results.verdict==FALSE then Stop Exp. (ERROR)
 //continue exp. also when results.verdict==Only_pType_NIL
23. PTYPE=results.pType

//is 2nd identification with camera necessary?
24. if PTYPE==NIL ? then
 //do 2nd identification with camera (Fig.9)
 24.1 PHASE=              TASKicam   --set next phase
 24.2 execute steps 17-22
 24.3 if results.verdict==Only_pType_NIL then
 Stop Exp.(ERROR)
 else
 PTYPE=    results.pType   //PTYPE≠NIL

//do pick & place handling with robot (Fig.10)
25. PHASE=           TASKpickplace  //set next phase
26. execute steps 17-22
27. if results.verdict==Only_pType_is_NIL then Stop Exp.(ERROR)
28. PTYPE=results.pType;
29. STATEib=results.stateIB
30. STATEob=results.stateOB

//check, if overall experiment break-off condition is reached
31. if overall_break-off==TRUE then Stop Exp.(Finished)
30. repeat execution starting with step 16 (operating next part)
```

LISTING 1   Specification of the experiment control for the robot application.

5 PROTOTYPE IMPLEMENTATION IN MATLAB/SIMULINK

The suggested infrastructure has been prototyped using the SES toolbox for MATLAB/ Simulink (Pawletta et al., 2014). The SES toolbox provides a graphical environment for SES-based modeling within MATLAB. It comprises a graphical editor and different methods like (i) *pruning* for derivation of a PES from an SES, (ii) *flattening* for hierarchy reduction of an PES, (iii) *merging* to synthesize various SES into a large one, and (iv) *validity checking* of the SES and PES, for example, to see whether it satisfies the general SES or the PES axioms and the specific *semantic conditions*. All the methods can be called from the GUI of the SES toolbox. Fig. 14 shows a view of the GUI with the SES parts tailored for the robotic control example. Furthermore, an MB analogous to Fig. 11 was implemented using MATLAB/Simulink and MATLAB/SimEvents.

The GUI consists of a menu bar, tab bar, and the three subwindows: *Node Properties* (left), *SES Hierarchy* (middle), and general *Global Settings* (right). In the middle subwindow, an SES tree can be edited in a manner similar to a data manager. Node attributes like coupling relationships or selection rules are edited and displayed in the left subwindow, *Node Properties*. In the view displayed above, the aspect rule for choosing the *identDec* or *pickplaceDec* aspect is displayed. Global properties and cross-tree relations of an SES, such as SES Variables, SES Functions, and Semantic Conditions, are managed in the subwindow *Global Settings*. After the SES is specified using the GUI, it can be saved as a MATLAB specific *mat* file or an XML file.

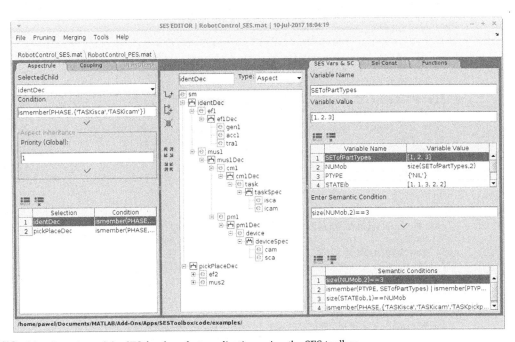

FIG. 14   Overview of the SES for the robot application using the SES toolbox.

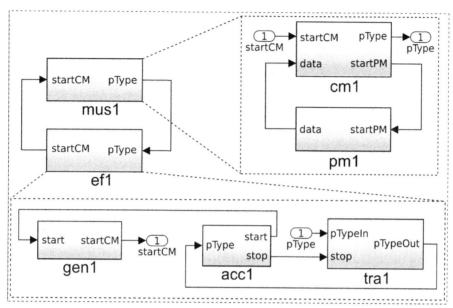

FIG. 15   Generated simulation model for the identification task in Fig. 9.

Moreover, the toolbox implements an API in order to enable access from other computational components, such as an EC unit, which can be, for instance, a MATLAB program. Additionally, the API provides a generic *build* method for generating SMs for different EUs, using basic models from a corresponding MB.

Fig. 15 shows the generated SM using the infrastructure for the identification task according to Fig. 9 and the SES in Fig. 12. It should be pointed out that the output ports *verdict* of *acc1* and *su* of *tra1* in Fig. 15 are different from those in Fig. 9. The reason is that these ports are modeled using the *ToWorkspace* blocks from the MATLAB/Simulink library within the corresponding basic models.

The build method can be configured via MATLAB scripts depending on the specific EU. Currently, the toolbox supports the generation of Simulink models, including SimEvents, Simscape, and Stateflow components (Pawletta et al., 2016b) and MATLAB DEVS models (Deatcu et al., 2017). Moreover, first experiments for generating Dymola and OpenModelica models have been carried out. The toolbox can be used for free and downloaded after a registration, according to the information in Pawletta et al. (2015).

6 CONCLUSION AND FUTURE WORK

Technical systems are evolving to be smarter and more connected. Networks of technical systems are showing versatile and reactive characteristics, which map to enhanced and adaptive feature sets that have never been achieved. Yet, the advancement is not without constraints. All fields of study in technical systems development now face various new challenges. One of the important questions in the field of model engineering is how to model and simulate versatile systems where reactive dynamic variability in the system structure is indispensable.

This chapter presented an approach based on SES/MB. SES/MB was extended to tackle the characteristic challenges of versatile systems. SES was introduced as a means of conceptual system structure modeling in order to tackle variability. Thereafter, the SES/MB framework was enriched and expanded with various components such as EC or EU, and various features such as SES Functions. It was proposed as a mechanism to simulate reactive systems with dynamic reconfiguration. The proposed infrastructure was explained in detail, using an example from the domain of industrial robotics. We also presented a prototype implementation of the infrastructure.

Future technical systems are expected to be open, autonomous, decentralized, networked, adaptive, and emergent. Effective model engineering methodologies are necessary to support M&S-based development.

Research in the future is expected to explore two different fields. One is to advance the proposed methodology in order to achieve a well-accepted engineering technique supported by proper tools. This requires further application to cases from various domains, such as from aeronautics for multiple drone scenarios or from the automotive industry for autonomous and connected cars.

The second involves investigating machine learning for dynamically evolving conceptual system structures that is reflected to run time expending SES trees. Then, we can begin with a baseline of all possible system configurations, while the system will uncover more configuration possibilities by itself.

Acknowledgments

The authors Thorsten Pawletta and Artur Schmidt gratefully acknowledge the grant from German Science Foundation, DFG (PA 631/2). Moreover, the authors would like to thank Peter Junglas, who contributed valuable work to the development of a generic model builder for the SES toolbox; Daniel Pascheka, who implemented a first version of the graphical editor within MATLAB; Birger Freymann, who redesigned the editor and implemented a model builder for the MatlabDEVS toolbox; Hendrik Folkerts, who has been developing the new SES toolbox with Python/Qt; Christina Deatcu and Sven Pawletta for their valuable cooperation in the project; and our former colleague, Tobias Schwatinski, who provided preliminary work. Last but not least, we like to thank Bernie Zeigler for motivating and supporting our work.

References

Abel, D., Bollig, A., 2006. Rapid Control Prototyping. Springer Pub, Berlin & Heidelberg, Germany (in German).

Apel, S., Batory, D., Kästner, C., Saake, G., 2016. Feature-Oriented Software Product Lines. Springer-Verlag, Heidelberg, Germany.

Barros, F., 1997. Modeling formalisms for dynamic structure systems. ACM Trans. Model. Comput. Simul. 7 (4), 501–515.

Bosch, J., 2013. Software product line engineering. In: Capilla, R., Bosch, J., Kang, K.-C. (Eds.), Systems and Software Variability Management. Springer Pub, Heidelberg, Germany, pp. 3–24.

Couretas, J.M., 2006. System architectures: legacy tools/methods, DoDAF descriptions and design through system alternative enumaretion. J. Diagn. Med. Sonogr. 3 (4), 227–237.

Cretu, L.G., Dumitriu, F., 2015. Model-Driven Engineering of Information Systems: Principles, Techniques, and Practice. Apple Academic Press, New Jersey, USA.

Deatcu, C., Freymann, B., Pawletta, T., 2017. In: PDEVS-based hybrid system simulation toolbox for MATLAB.Proc. SpringSim-TMS/DEVS, Virginia Beach, VA, USA, pp. 989–1000.

Dziobek, C., Loew, J., Pryzstas, W., Weiland, J., 2008. Model Diversity and Variability—Handling Functional Variants in Simulink Models. vol. 2. Elektronik Automotive, Germany, pp. 33–37.

Freymann, B., Pawletta, S., Schmidt, A., Pawletta, T., 2016. Design, simulation and operation of task-oriented multi-robot applications with MATLAB/Stateflow. Simul. Notes Eur. 26 (2), 83–90.

Haber, A., Kolassa, C., Manhart, P., Nazari, P.M.S., Rumpe, B., Schaefer, I., 2013. In: First-class variability modeling in MATLAB/Simulink. Proc. 7th Int. Workshop on Variability Modelling of Software-Intensive Systems. ACM Press, Pisa, pp. 11–18.

Hermann, M., Pentek, T., Otto, B., 2016. Design principles for industrie 4.0 Scenarios. 49th Hawaii International Conference on System Sciences (HICSS) 5–8 January 2016. IEEE Computer Soc. Pub., Kauai, Hawaii, pp. 3928–3937.

Hu, X., Zeigler, B.P., Mittal, S., 2003. Variable structure in DEVS component-based modeling and simulation. Simulation 81 (2), 91–102.

Hummer, O., Sunder, C., Zoitl, A., Strasser, T., Rooker, M.N., Ebenhofer, G., 2006. Towards zero-downtime evolution of distributed control applications via control based on IEC 61499.IEEE Conf. on Emerging Technologies & Factory Automation, Prague, Czech. Rep, pp. 517–524.

Jezequel, J.-M., 2012. Model-driven engineering for software product lines. ISRN Softw. Eng. 2012: 670803.

Kim, J.-H., Kim, T.G., 2001. DEVS-based framework for modeling/simulation of mobile agent systems. Simulation 76 (6), 345–357.

Kliemannel, F., Rock, G., Mann, S., 2010. In: A custom approach for variability management in automotive applications.Proc. Int. Workshop on Software-intensive Systems (VaMoS), Linz, Austria, pp. 155–158.

Lackner, H., Thomas, M., Wartenberg, F., Weißleder, S., 2014. In: Model-based test design of product lines: raising test design to product line level. Proceedings of 7th International Conference on Software Testing, Verification and Validation (ICST). IEEE Press, pp. 51–60.

Lee, E.A., 2008. Cyber physical systems: design challenges. Electrical Engineering and Computer Sciences Dep., Univ. of California at Berkeley. Technical Report No: UCB/EECS-2008-8. Available from: http://www.eecs.berkeley.edu/Pubs/TechRpts/2008/EECS-2008-8.html.

Maletzki, G., 2014. Rapid Control Prototyping of Complex and Flexible Robot Controls Based on the SBC Approach. (Ph.D. thesis)Wismar Univ. of Applied Sc. & Univ. of Rostock, Germany (in German).

Mittal, S., Mark, E., Nutaro, J., 2006. DEVS-based dynamic model reconfiguration and simulation control in the enhanced DoDAF design process. J. Diagn. Med. Sonogr. 3 (4), 239–267.

OMG, 2006. Unified Modeling Language: Infrastructure version 2.0. Available from: http://www.omg.org/spec/UML/2.0/.

Ören, T.I., 1987. In: Model update: a model specification formalism with a generalized view of discontinuity.Proceedings of Summer Computer Simulation Conference, Montreal, QC, Canada, pp. 689–694.

Park, S., Hunt, A., 2006. In: Coupling permutation and model migration based on dynamic and adaptive coupling mechanisms.Proc. SpringSim-TMS/DEVS, Huntsville, AL, USA, pp. 46–53.

Pawletta, T., Pawletta, S., 1995. In: Design of simulator for structure variable systems.Proceedings 5th International IMACS-Symposium on System Analysis and Simulation, Berlin, SAMS, 1995. vol. 18–19. pp. 471–474.

Pawletta, T., Lampe, B., Pawletta, S., Drewelow, W., 2002. A DEVS-based approach for modeling and simulation of hybrid variable structure systems. In: Engel, S., Frehse, G., Schnieder, E. (Eds.), Modelling, Analysis, and Design of Hybrid Systems. In: Lecture Notes in Control and Information Science, vol. 279. Springer Pub., Berlin & Heidelberg, Germany, pp. 107–129.

Pawletta, T., Pawletta, S., Maletzki, G., 2009. Integrated modeling, simulation and operation of high flexible discrete event controls.Proc. Mathematical Modelling MATHMOD 2009, Vienna, Austria. Argesim Report No.: 35.

Pawletta, T., Pascheka, D., Schmidt, A., Pawletta, S., 2014. Ontology-assisted system modeling and simulation within MATLAB/Simulink. Simul. Notes Eur. 24 (2), 59–68. 8 pp.

Pawletta, T., Schmidt, A., Pascheka, D., Freymann, B., 2015. SES Toolbox for MATLAB/Simulink. Available from: https://www.mb.hs-wismar.de/cea/SES_Tbx/.

Pawletta, T., Schmidt, A., Zeigler, B.P., Durak, U., 2016a. Extended variability modeling using system entity structure ontology within MATLAB/Simulink.Proc. SpringSim-ANSS, Pasadena, CA, USA, pp. 62–69.

Pawletta, T., Schmidt, A., Junglas, P., 2016b. A multimodeling approach for the simulation of energy consumption in manufacturing. Simul. Notes Eur 27 (2), 115–124.

Pohl, K., 2006. Software Product Line Engineering: Foundations, Principles and Techniques. Springer-Verlag, Berlin.

Ponnusamy, S.S., Albert, V., Thebault, P., 2014. In: Modeling & simulation framework for the inclusion of simulation objectives by abstraction.Proceedings SIMULTECH 2014: 4th Int. Conf. on Simulation and Modeling Methodologies, Technologies and Applications, Vienna, Austria, pp. 385–394.

Rozenblit, J.W., 1991. Experimental frame specification methodology for hierarchical simulation modeling. Int. J. Gen. Syst. 19 (3), 317–336.

Rozenblit, J.W., Zeigler, B.P., 1993. In: Representing and constructing system specifications using the system entity structure concepts.Proceedings Winter Simulation Conference, Los Angeles, CA, USA, pp. 604–611.

Sampigethaya, K., Poovendran, R., 2013. Aviation cyber-physical systems: foundations for future aircraft and air transport. Proc. IEEE 101 (8), 1834–1855.

Santucci, J.-F., Capocchi, L., Zeigler, B.P., 2016. System entity structure extension to integrate abstraction hierarchies and time granularity into DEVS modeling and simulation. Simulation 92 (8), 747–794.

Schmidt, A., Durak, U., Pawletta, T., 2016. Model-based testing methodology using system entity structures for MATLAB/Simulink models. Simulation 92 (8), 729–746.

Schwatinski, T., Pawletta, T., Pawletta, S., 2012. Flexible task oriented robot controls using the system entity structure and model base approach. Simul. Notes Eur. 22 (2), 107–114.

Seo, C., Kang, W., Zeigler, B.P., Kim, D., 2014. In: Expanding DEVS and SES applicability: using M&S kernels within IT systems.Proc. SpringSim-TMS/DEVS, Tampa, FL, USA, pp. 46–53.

Siciliano, B., 2008. Springer Handbook of Robotics. Springer Pub, Berlin, Germany.

Steiner, E., Masiero, P., Bonifacio, R., 2013. Managing SPL variabilities in UAV simulink models with pure::variants and hephaestus. CLEI Electron. J. 16 (1), 6.

The Math Works, 2017. Variant Configuration Management in Simulink. [Video] Available from: https://ch.mathworks.com/videos/variant-configuration-management-in-simulink-80677.html.

Traorè, M.K., Muzy, A., 2006. Capturing the dual relationship between simulation models and their context. Simul. Model. Pract. Theory 14 (2), 126–142.

Uhrmacher, A.M., 2001. Dynamic structures in modeling and simulation—a reflective approach. ACM Trans. Model. Comput. Simul. 11 (2), 206–232.

Yilmaz, L., Ören, T.I., 2004. In: Dynamic model updating in simulation with multimodels: a taxonomy and a generic agent-based architecture.Proceedings Summer Computer Simulation Conference, San Jose, CA, USA, pp. 3–8.

Zander, J., 2008. Model-based Testing of Real-Time Embedded Systems in the Automotive Domain. (Ph.D. thesis). Technical Univ. Berlin, Berlin.

Zeigler, B.P., 1976. Theory of Modeling and Simulation. Wiley Interscience, New York.

Zeigler, B.P., 1984. Multifaceted Modelling and Discrete Event Simulation. Academic Press, London.

Zeigler, B.P., Hammonds, P.E., 2007. Modeling and Simulation-Based Data Engineering. Elsevier Academic Press, London.

Zeigler, B.P., Praehofer, H., 1989. In: Systems theory challenges in the simulation of variable structure and intelligent systems.International Conference on Computer Aided Systems Theory, EUROCAST '89, pp. 41–51.

Zeigler, B.P., Sarjoughian, H.S., 2013. Guide to Modeling and Simulation of Systems of Systems. Springer International Publishing AG, Cham, Switzerland.

Zeigler, B.P., Praehofer, H., Kim, T.G., 2000. Theory of Modeling and Simulation, second ed. Academic Press, London.

Model Validation of Control Systems With an Application in Abnormal Driving State Detection

Weicun Zhang*, Ying Liu†

*School of Automation and Electrical Engineering, University of Science and Technology Beijing, Beijing, China †School of Automation Science & Electrical Engineering, Beihang University, Beijing, China

1 INTRODUCTION

Model is always built for some certain application purpose (Zhang, 2011). Thus, model validation can be conducted through application effectiveness. From the view point of methodology, practice is the sole criterion for testing the truth. Consequently, to judge whether the model is good or bad, the sole evaluation criterion should be practiced, that is, the effectiveness of model in application. To be specific in a closed-loop control system, for a given input signal, the effectiveness of the model should be evaluated according to some certain difference index between the model output and the real plant output. If the difference index cannot meet the requirement of engineering application, or if the requirement needs to be increased with the progress of practice, then we need to further improve (change) the model.

We may use a logic diagram as shown in Fig. 1 to express the relationship between the model and the practice, which is a reciprocal feedback progress.

In the above-described process of modeling and practice, modeling methods, model types, and model validation methods are all in a dynamical progress. Thus, related research and innovation are always necessary.

Model validation refers to all stages from model construction to model utilization. We will focus on the model validation in application stage; meanwhile we also pay attention to model validation in modeling stage. For model validation in modeling stage, we need to consider the

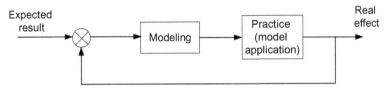

FIG. 1 Modeling and practice (including model validation).

complexity of the model, that is, model should be as simple as possible. Thus, the model order information is usually considered as penalty item in performance index, such as, Akaike information criterion (AIC), and final prediction error (FPE) (Yang and Gu, 1986; Akaike, 1969).

2 MODEL VALIDATION BASED ON PERFORMANCE EVALUATION OF CONTROL SYSTEM

Applying the above-mentioned observations to control system application, we have the logic diagram for control system performance evaluation in control system. In Fig. 2, model M can be obtained by any possible means (first-principle modeling, system identification, etc.). Based on model M, controller C is designed according to a given control strategy (robust control, adaptive, predictive control, etc.), P is the real plant (including the environment of application).

As well known, the basic requirements of a control system are: stability, accuracy, and quickness. In detail, the closed-loop system is stable; the system output signal tracks the reference signal (steady-state error is zero or small enough), and the transient process is quite short.

Classical performance evaluation of control systems (Harris et al., 1999; Huang and Shah Sirish, 1999) relies on the difference between the reference signal and the real plant output, that is, $e(k) = y_r(k) - y(k)$. There are some problems with the classical evaluation method:

(1) When reference signal changes, so will the difference signal $e(k)$, which will disturb the evaluation index.
(2) The basis for performance evaluation (comparison) is the best performance of minimum variance control of the plant. The correctness of the basis also depends on the accuracy of the model.

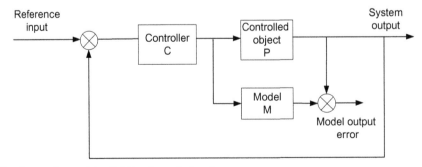

FIG. 2 Sketch map for model validation in control system.

To improve the above-mentioned situation, based on virtual equivalent system (VES) theory (Zhang, 2010), we propose in this research an alternative performance index $e_m(k) = y_m(k) - y(k)$ to replace the classical one, that is, $e(k) = y_r(k) - y(k)$. With this new performance index, hopefully the two problems listed above can be solved or improved.

Initially VES concept and methodology was put forward to give a unified analysis of the stability and convergence of self-tuning control systems. Later on, it was also successfully adopted to analyze the stability of multiple model adaptive control (MMAC), active disturbance rejection control (ADRC), and U-model-based nonlinear control, etc.

Actually, VES could be used for stability analysis of all kinds of model-based control systems. Along this line of thinking, it also paves a new way for performance evaluation of control systems that are designed based on some certain models.

Fig. 3 shows the concept of VES, which is equivalent, in the input-output sense, to its original system as shown in Fig. 2.

Owing to the equivalence in the input-output sense between the VES and its corresponding real control system, the stability and performance of these two systems are also equal. Thus, some difficult analysis problems of complex control systems become easy to deal with. The main merit of VES is that with the help of VES, original nonlinear dominant (nonlinear in structure) problem can be converted into a linear dominant (linear in structure) problem with a nonlinear compensation signal.

According to VES theory, system as shown in Fig. 3 can be decomposed into two subsystems, as shown in Figs. 4 and 5, respectively.

Then we have the following conclusions.

If a control system as shown in Fig. 2 satisfies the following condition:

In design stage, controller C and M formulate a stable and tracking closed-loop system with prescribed performance index.

Model output error is quite small (detailed error index will be given in the later sections). Then the real control system as shown in Fig. 2 is stable with prescribed performance index, that is to say that the performance index of real control system (Fig. 2) will converge to that of the "ideal" control system (Fig. 4).

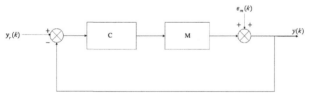

FIG. 3   Sketch map of the VES for original system (Fig. 2).

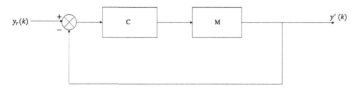

FIG. 4   Decomposed subsystem 1.

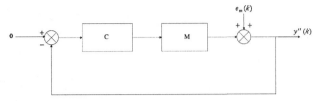

FIG. 5   Decomposed subsystem 2.

3 DRIVER'S ABNORMAL STATE DETECTION

As an application of VES-based control system model validation, this section provides an engineering application, that is, driver's abnormal state detection based on system performance evaluation.

It is worth pointing out that in this application, the model is a controller model, that is, a model for driving behavior of a human driver. The input signal of the controller/driver model is the lane departure signal (vehicle lateral displacement), which can be calculated in accordance with the detection of the two white lines of the lane. The output signal of the controller/driver model is the steering angle.

Considering that normal driving behavior corresponds to normal driving data, abnormal driving behavior corresponds to abnormal driving data, and vice versa, we may regard driver-vehicle-road as a closed-loop control system (Chen and Ulsoy, 1997, 2001; Pilutti and Ulsoy, 1999).

Our basic idea is to use system identification (modeling) method to build a normal driving behavior model with normal driving data. The normal model is then used to fit the real driving data of unknown driving state, then the fitting error is used to evaluate and monitor the driver's state with statistical process control (SPC) approach, that is, the control charts tools (Zhang and Yang, 2000; Liang and Qian, 2003). If the fitting error is evidently different from that of normal driving state, then abnormal driving state is recognized.

It should be noted that Pilutti and Ulsoy (1999) have also adopted system identification method to detect abnormal driving state, but in a different way, that is, the variation in model parameters, which is difficult to be realized for online detection.

3.1 Experiment Design and Data Acquisition

The data in this research were collected from the driving simulator VR-4, which was developed in University of Science and Technology Beijing. Five regular drivers, who have driver licenses with different driving experiences, and different ages, were arranged to drive the simulator in some typical highway scenario. The fitting error analysis indicates that the ARMAX model can yield residuals close to white noise. This implies the ARMAX model structure is a better choice than other parametric model structure. The FPE criterion was used to determine the model structure and time delay. The effectiveness of the proposed methodology was verified by comparing five driver's eye closure time with the alarm points on the SPC charts of model fitting error.

3.2 Modeling (System Identification)

3.2.1 Pretreatment of the Data

The time period from sobriety to drowsiness differs from one driver to another. So the experiment data lengths are also chosen differently. The original data sampling frequency is 60 Hz, to make the modeling data more efficient, the secondary sampling frequency set at 10 Hz.

Although generally speaking, every driver has his own time period from sobriety to drossiness, during the beginning 15 min they are all in normal driving state. During the beginning 2 min, the driving behavior may be unstable. Having taken into account all these considerations, we selected the data period from 3 to 12 min as the normal data for each driver modeling. To avoid "data saturation" in least-squares (LS) identification algorithm, the 10-min modeling data were divided into 10 equal groups. Each group of data was used to obtain corresponding model parameters. And then the final model parameters are obtained by making average of the 10 group parameters.

3.2.2 Identification of the Model Structure

Model structure is the first thing we need to decide in model identification/modeling. Model structures include auto-regressive with eXtra inputs (ARX) model, auto-regressive moving average with eXtra inputs (ARMAX) model, and Box Jenkins (BJ) model.

Through comparison and analysis, we found that the parameters of ARMAX are fewer than that of BJ, the residual error index of ARMAX is less than that of ARX. Then we finally choose the following ARMAX model as driver behavior model:

$$A\left(q^{-1}\right)\delta(k) = B\left(q^{-1}\right)S(k) + C\left(q^{-1}\right)e(k) \tag{1}$$

where $S(k)$ is the system input signal, that is, vehicle lateral displacement, $\delta(k)$ the system output signal, that is, steering wheel angle, and $e(k)$ is white noise. $A(q^{-1})$, $B(q^{-1})$, and $C(q^{-1})$ are polynomials. After model structure is determined, FPE criterion (Zhang and Yang, 2000; Liang and Qian, 2003) is adopted to select the order and time delay of the ARMAX model.

$$FPE = \frac{1 + \dfrac{n}{N}}{1 - \dfrac{n}{N}} \times LS \tag{2}$$

where N is the data length; n is the model order; and L is the lost function.

Lost function L is defined as follows:

$$L = \frac{1}{N}\sum_{k=1}^{N} e^2(k) \tag{3}$$

Through calculations based on MATLAB, the final model structure for five experiment driver is determined as follows:

$$\begin{cases} A\left(q^{-1}\right) = 1 + a_1 q^{-1} + a_2 q^{-2} \\ B\left(q^{-1}\right) = b_1 q^{-1} + b_2 q^{-2} \\ C\left(q^{-1}\right) = 1 + c_1 q^{-1} + c_2 q^{-2} \end{cases} \tag{4}$$

Then Eq. (1) can be rewritten as

$$\delta(k) + a_1\delta(k-1) + a_2\delta(k-2) = b_1 S(k-1)$$
$$+ b_2 S(k-2) + e(k) + c_1 e(k-1) + c_2 e(k-2)$$

(5)

3.2.3 Parameter Estimation

There are many parameter estimation algorithms, among them LS algorithm is the most widely used one. Considering recursive LS with forgetting factor (Li and Hu, 2006) has advantages of light calculation and memory burden, and suitable for tracking time varying parameters, we adopted it in this research.

In detail, from Eq. (5), we define

$$\begin{cases} \theta = [a_1, a_2, b_1, b_2, c_1, c_2]^T \\ h(k) = [-\alpha(k-1), \ -\alpha(k-2), \ S(k-1), \ S(k-2), \ e(k-1), \ e(k-2)]^T \end{cases}$$
$$\hat{e}(k) = z(k) - h^T(k)\hat{\theta}(k-1)$$
$$h(k) = [-\alpha(k-1), \ -\alpha(k-2), \ S(k-1), \ S(k-2), \ \hat{e}(k-1), \ \hat{e}(k-2)]^T$$
$$\begin{cases} \hat{\theta}(k) = \hat{\theta}(k-1) + K(k)\big(z(k) - h^T(k)\hat{\theta}(k-1)\big) \\ K(k) = P(k-1)h(k)\big[\beta + h^T(k)P(k-1)h(k)\big]^{-1} \\ \qquad P(k) = \dfrac{1}{\beta}\big(I - K(k)h^T(k)\big)P(k-1) \end{cases}$$
$$P(0) = 10^6 I$$
$$\hat{e}(k) = 0, \ k < 0$$
$$\beta = 0.995$$

where I is an identity matrix with appropriate dimensions.

With the above-described algorithm and experiment data, we can readily obtain each driver's behavior model as shown in Eq. (1) with specified parameters.

3.2.4 Model Validation

As for model validation, hypothesis test (significance level $\alpha = 0.05$) was used to complete the task of model validation. In detail, if the model fitting residual signal is approximately a zero-mean white noise, then the model is validated (Zhou, 2006). To be specific, the model-fitting residual is given by

$$\varepsilon(k) = \delta(k) - \hat{\delta}(k)$$

(6)

There are different approaches to test if the model-fitting error is a white noise. With the help of MATLAB functions, the related hypothesis test can be readily done. Thus, the details are omitted here.

According to the above-mentioned methodology, all the five driver's experiment driving data and corresponding models have passed the model validation. That is to say that the modeling process (model structure identification and parameter estimation) is effective.

3.3 Abnormal Driving State Detection Based on SPC Control Charts

3.3.1 Statistical Process Control

The SPC is also known as statistical performance monitoring (SPM). It is used to monitor if the production process is satisfactorily under control. Its main measure is the control chart, which was originally proposed by Shewhart. Later, cumulative sum control chart (CUSUM) was put forward by Page (1961), exponentially weighted moving average control chart (EWMA) was also proposed by Roberts (1959), both are improvements of Shewhart control chart.

3.3.2 Shewhart Control Chart

Suppose process output data are mutually independent, and we have

$$Y_t = \mu + \varepsilon_t$$

where μ is a constant mean value, $\varepsilon_t \sim N(0,\sigma^2)$, then we have

$$Y_t \sim N(\mu, \ \sigma^2)$$

Randomly selecting N samples, denoted by Y_{ti}, from the population with sample size n, then the sample mean is

$$\overline{Y_t} = \sum_{i=1}^{n} Y_{ti}/n \tag{7}$$

For the control chart, we adopt the statistic variable

$$\overline{\overline{Y}} = \sum_{t=1}^{N} \overline{Y_t}/N \tag{8}$$

Then the upper control limit (UCL), the lower control limit (LCL), and the centerline of the control chart are as follows:

$$\begin{matrix} UCL \\ LCL \end{matrix} = \overline{\overline{Y}} \pm 3\frac{\delta}{\sqrt{n}}, \quad CL = \overline{\overline{Y}} \tag{9}$$

In this research, Y_t refers to the model fitting error e_t, with $N=20$, and $n=60$. Then the following situations are regarded as abnormal:

(1) e_t is out of the boundaries of $\pm 3\sigma$.
(2) There are at least two points, out of three continuous points, within the range between $\mu \pm 3\sigma$ and $\mu \pm 2\sigma$.

3.3.3 EWMA Control Chart

The statistical variable of traditional Shewhart control chart is composed of current observations without considering history observations. EWMA is an improvement of Shewhart control chart, in which the statistic variable is

$$Z_i = r\overline{y_i} + (1-r)Z_{i-1}, \quad i=1,2,\ldots$$

where $Z_0 = \overline{\overline{Y}}$, $0 \leq r \leq 1$ is a smooth factor. The choice of r has a significant influence on the sensitivity of EWMA control chart.

The UCL, the LCL, and the centerline are as follows:

$$\begin{matrix} UCL \\ LCL \end{matrix} = \overline{\overline{Y}} \pm 3 \frac{\overline{R}}{d_2 \sqrt{n}} \sqrt{\frac{r\left(1 - (1-r)^{2i}\right)}{2-r}} \tag{10}$$

where $\overline{R}$ is the sample range, d_2 is a constant related to n, and $\sigma = \overline{R}/(d_2\sqrt{n})$.

The following situations of Z_i will be regarded as abnormal:

(1) Z_i is out of the boundaries of UCL and LCL.
(2) There are at least two points, out of three continuous points, within the range between $\mu \pm 3\sigma$ and $\mu \pm 2\sigma$.
(3) There are more than 10 continuous points ascending or descending.

3.3.4 Experimental Results

According to the SPC-based abnormal state detection methods for five drivers' experiment driving data, we obtain the abnormal driving state detection results; see from Figs. 6–10, respectively. Each subfigure consists of two parts: the upper part is the alarm points detected by eye closure time period, which is a standard method for driver's fatigue detection; the lower part is the alarm points proposed by five rules of two control charts (Shewhart control chart and EWMA control chart) of model fitting error. Eye closure time period and frequency

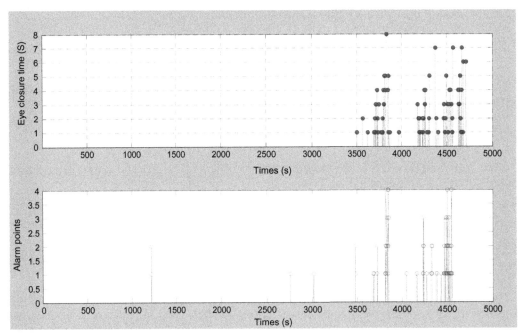

FIG. 6   Abnormal state detection for driver 1.

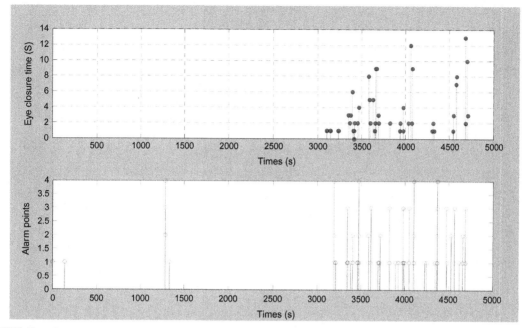

FIG. 7   Abnormal state detection for driver 2.

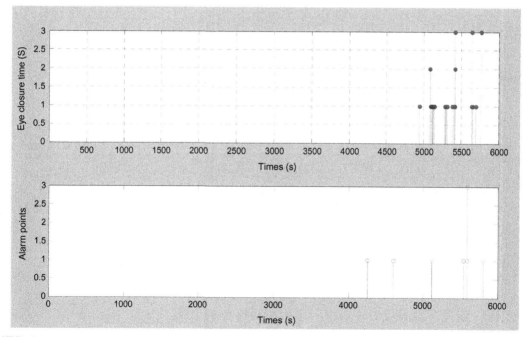

FIG. 8   Abnormal state detection for driver 3.

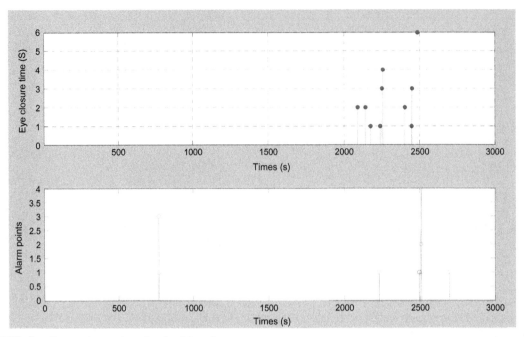

FIG. 9   Abnormal state detection for driver 4.

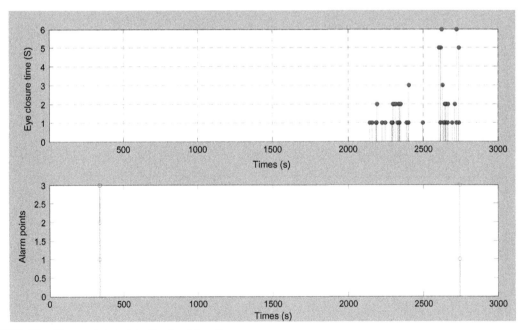

FIG. 10   Abnormal state detection for driver 5.

are the well-known standard for detecting abnormal driving state. The experiment results show that the abnormal driving state detection method proposed in this research is effective.

From Figs. 6–10, we may conclude the effectiveness of the proposed model/system performance evaluation in abnormal state (fatigue) detection of driver's behavior over experiment data.

4 CONCLUSIONS

A control system model validation approach is proposed based on system performance evaluation (with the help of SPC control chart). As an example, an abnormal driving state detection model is verified with five drivers' experiment data which were collected on the driving simulator VR-4. It should be pointed out that the sample size of driving experiments is limited. In the future research work, some more general driving experiments and other process control system monitoring experiments will be conducted to validate and improve the proposed system performance evaluation methodology.

References

Akaike, H., 1969. Fitting autoregressive models for prediction. Ann. Inst. Stat. Math. 21 (1), 243–347.

Chen, L.K., Ulsoy, A.G., 1997. In: Driver model uncertainty.Proceedings of American Control Conference, San Diego, CA, USA, pp. 714–718.

Chen, L.K., Ulsoy, A.G., 2001. Identification of a driver steering model, and model uncertainty, from driving simulator data. J. Dyn. Syst. Meas. Control 123, 623–629.

Harris, T.J., Boudreau, F., MacGregor, J.F., 1999. A review of performance monitoring and assessment technique for univariate and multivariate control systems. J. Process Control 9 (1), 1–17.

Huang, B., Shah Sirish, L., 1999. Performance Assessment of Control Loops: Theory and Application. Springer-Verlag, London.

Li, P., Hu, D., 2006. Basis for System Identification. China Water and Power Press, Beijing.

Liang, J., Qian, J., 2003. Multivariate statistical process monitoring and control recent developments and applications to chemical industry. Chin. J. Chem. Eng. 11 (2), 191–203.

Page, E.S., 1961. Cumulative sum charts. Technometrics 3 (1), 1–9.

Pilutti, T., Ulsoy, A.G., 1999. Identification of driver state for lane-keeping tasks. IEEE Trans. Syst. Man Cybern. A Syst. Hum. 29 (5), 486–502.

Roberts, S.W., 1959. Control chart tests based on geometric moving averages. Technometrics 1 (3), 239–250.

Yang, W., Gu, L., 1986. Time Series Analysis and Dynamical Data Modeling. Beijing Institute of Technology Press, Beijing.

Zhang, W., 2010. On the stability and convergence of self-tuning control—virtual equivalent system approach. Int. J. Control. 83 (5), 879–896.

Zhang, L., 2011. In: Model engineering for complex system simulation. Proceedings of 58th Forum on New Academic Views, October 15, 2011. China Science and Technology Press, Lijiang.

Zhang, J., Yang, X., 2000. Multi-Variable Stotistical Process Control. Cheimical Industry Press, Beijing.

Zhou, Y., 2006. Stochastic Process Theory, second ed. Publishing House of Electronics Industry, Beijing.

Further Reading

Liu, H., 1996. System Identification and Parameter Estimation. Metallurgical Industry Press, Beijing.

Index

Note: Page numbers followed by *f* indicate figures, *t* indicate tables, and *np* indicate footnotes.

Printed in the United States
By Bookmasters